AMERICA AND THE ISLAMIC BOMB

AMERICA AND THE ISLAMIC BOMB

THE DEADLY COMPROMISE

DAVID ARMSTRONG AND JOSEPH TRENTO

A project of the National Security News Service

STEERFORTH PRESS
HANOVER, NEW HAMPSHIRE

Library of Congress Cataloging-in-Publication Data
Armstrong, David, 1960–
America and the Islamic bomb : the deadly compromise / David Armstrong and Joseph Trento. — 1st ed.
p. cm.
"A project of the National Security News Service."
Includes bibliographical references and index.
ISBN-13: 978-1-58642-137-3 (alk. paper)
ISBN-10: 1-58642-137-9 (alk. paper)
1. United States — Foreign relations — Pakistan. 2. Pakistan — Foreign relations — United States. 3. United States — Foreign relations — 1945–1989. 4. United States — Foreign relations — 1989– 5. Khan, A. Q. (Abdul Qadeer), 1936– 6. Nuclear weapons — Pakistan — History — 20th century. 7. Pakistan — Military policy. 8. Technology transfer — Islamic countries. 9. Nuclear nonproliferation — Political aspects. 10. Nuclear terrorism. I. Trento, Joseph John, 1947- II. Title.

E183.8.P18A76 2007
327.7305491–dc22

2007031728

FIRST EDITION

For Noëlle, Guthrie, and Eliza
David Armstrong

For Susan
Joseph Trento

CONTENTS

Introduction

"Putting nuclear weapons in someone else's hands could be quite disastrous."

— A.Q. KHAN, 1999[1]

IN EARLY 2004, the world learned of a grave menace: A Pakistani black market network had provided nuclear-weapons know-how and technology to the rogue states of Libya, Iran, and North Korea. The implications were ominous. Headlines from Europe to Asia raised the nightmare specter of Third World despots — and possibly even terrorists — armed with atomic devices. Foreign capitals reacted with alarm.[2]

In Washington, however, the response was muted. Although the Bush administration had placed halting the spread of weapons of mass destruction (WMD) at the top of its priority list, US officials went out of their way to praise Pakistan's handling of the scandal, even refusing to take issue with President Pervez Musharraf's decision to pardon the leader of the nuclear smuggling ring, Dr. Abdul Qadeer Khan.[3] As a State Department spokesman explained, "I don't think it's a matter for the United States to sit in judgment on."[4]

Other American officials cast the Khan affair in a positive light. CIA Director George Tenet, eager to convey good news in the face of continuing criticism over the faulty intelligence about Saddam Hussein's alleged WMD programs that led to the invasion of Iraq, portrayed the exposure of Khan's dealings as a triumph attributable to the US and British intelligence services. "Our spies penetrated the network through a series of daring operations over several years," Tenet said. "Khan and his network have been dealt a crushing blow."[5] President Bush and other administration officials echoed that judgment. "The A.Q. Khan network is out of business," the president said. "We have ended one of the most dangerous sources of proliferation in the world, and the American people are safer."[6]

But those upbeat appraisals were misleading. True, Khan had been exposed and placed under house arrest, several of his collaborators were in custody, and elements of their nuclear smuggling operation were shut down. But all of that could have been accomplished far earlier, before much of the damage was done. As Tenet himself noted, US and British intelligence maintained close surveillance of Khan's activities for years. They knew exactly what he and his cohorts were doing. British and American spies had, in fact, monitored much of the transfer of nuclear technology. Yet they resisted taking action against Khan's gang, arguing that moving "too soon" would harm efforts to corral the whole network and would endanger intelligence sources. Rather than act to dismantle large portions of a dangerous nuclear smuggling ring and expose the budding weapons programs of its rogue customers, Washington and London chose to wait and continue collecting information. Khan's deadly trade was allowed to continue.

Even after the network was finally outed, the United States refused to push for an independent inquiry to ensure that all of the participants were accounted for and that their activities had ceased. Instead, Washington accepted President Musharraf's assurances that his government had the situation under control. The Bush administration has not even challenged Musharraf's refusal to grant American and international investigators direct access to Khan, nor Pakistan's declarations that the investigation is officially closed.[7] And the administration rolled over when Islamabad prevented an official US investigation into a recent nuclear smuggling case from going forward in Pakistan.

In what was surely a welcome outcome for both the Bush administration and Musharraf, A.Q. Khan became the sinister emblem of nuclear proliferation, an easily recognizable and convenient villain. The portrayal of Khan as a lone wolf operating outside of government supervision and control provided the media — and the public — with a major diversion from the truth. But the reality of the proliferation saga is far darker and more malign: Khan was just one part of a wide-ranging network. And although the public was not aware of it, the network operated for years with the full knowledge of Western authorities, who tolerated it to avoid losing Pakistan's support for short-term foreign policy objectives, including the backing of anti-Soviet Muslim militants who became the precursor to al Qaeda.

The unpleasant truth is that what is now known as the A.Q. Khan scandal is, fundamentally, a scandal of US foreign policy. US officials, and anyone else paying attention, knew for decades that Pakistan was trafficking in nuclear technology. Yet successive US administrations looked the other way as their sometime-ally first developed and then sold the building blocks for the ultimate weapons of mass destruction. The United States did so in pursuit of short-term, strategic policy goals, first in the Cold War and more recently in the war on terror. But in their efforts to counter the known threat posed by the Soviet Union, US presidents from Dwight Eisenhower forward followed a path that would lead to a post–Cold War environment far more dangerous than that which had existed *during* the Cold War — a world in which rogue states and borderless militant groups can obtain nuclear weapons.

A critical moment came during the Carter administration. In 1979, National Security Advisor Zbigniew Brzezinski convinced President Jimmy Carter to overlook Pakistani nuclear proliferation in exchange for Islamabad's cooperation in supporting the Islamic holy warriors struggling against the Soviet invasion of Afghanistan. Pakistan's assistance in the US- and Saudi-backed anti-Soviet campaign carried, in other words, a nuclear price tag.

Carter's decision to ignore Pakistan's nuclear weapons development soon snowballed. The Reagan and Bush I administrations embraced the strategy as a cornerstone of America's aggressive anti-Soviet efforts. Pakistan took full advantage of the situation, using America's willingness to look the other way as an opportunity to develop a full-fledged nuclear weapons capability. By the 1990s, with the Soviets routed from Afghanistan and official relations between Washington and Islamabad eroding, the nuclear technology Pakistan had developed with tacit US approval was being sold and traded to some of the world's most dangerous regimes. In the meantime, the Afghan jihadis had begun their transformation into al Qaeda and would soon set their sights on the world's lone remaining superpower, the United States.

Over the past three decades, official investigations in the United States and Europe came close on several occasions to pulling the lid off of Pakistan's nuclear smuggling network. But to protect strategic relationships with Islamabad, those inquiries were scuttled by intervention at the highest levels of government. The opportunities to roll up the network were lost and the spread of nuclear technology continued.

Suggestions by President George W. Bush and members of his administration that the Khan case is a success story would be laughable if the implications were not so dire. Many of the Khan network's operatives remain free and live openly in Europe, Asia, and even the United States. The underground trade in nuclear technology continues, and the opportunities for terrorists to get their hands on atomic weaponry are expanding. Many in Pakistan's military, intelligence, and scientific communities are closely allied with the Taliban and al Qaeda, groups that US policy in the years before 9/11 helped foster. The lure of profits, combined with ideological, religious, and ethnic loyalties creates conditions for potentially deadly cooperation between those with access to nuclear technology and our bitterest enemies.

Leaders from across the political spectrum agree that the greatest danger facing the United States is terrorists armed with nuclear weapons, a threat that they suggest could materialize on American soil in the near future.[8] But it is a threat the United States helped bring about. By turning a blind eye to Pakistan's nuclear trafficking, the United States unleashed a nuclear juggernaut that it could not control. And in pursuit of its Cold War objectives, Washington — with Islamabad's assistance — helped fortify the radical Islamic forces that would grow into the terrorist threat we face today. The gravest dangers now confronting the West — Islamic terrorism and nuclear proliferation — are, in short, the bastard children of foreign policy decisions made long ago.

1

Nuclear Politics

IT HAD BEEN cold and overcast in London all week, typical English autumn weather. Now dark clouds were rolling in. As British Customs agent Atif Amin headed home for the weekend, he gazed out the window of the train, keeping a close eye on the increasingly threatening sky. By the time he reached his station outside the city, a steady rain had begun to fall. [1]

Amin, a tall, sturdy native-born British citizen of Pakistani descent, climbed down from the train and made his way along the covered walkway and through the small station. He paused briefly at the exit, then stepped out into the rain and walked quickly toward his car, parked on a side street a few blocks away. As he made the short drive home, rain splattered on the windshield.

Amin pulled up to the curb in front of his house, got out of his car, then dashed across a narrow strip of pavement. He slipped through his front door just as the real downpour began.

Amin was glad to be home, away from the city and the harsh realities with which his work brought him in contact every day. Although just thirty-five years old, he had already spent years tracking some of the world's most sinister characters — opium-smuggling warlords in Afghanistan, sanctions-busting arms dealers supplying weapons to paramilitary groups in Montenegro and Africa,[2] and a shadowy international network of nuclear arms traffickers. While he could claim real and meaningful successes, he was also frustrated by the degree to which politics often interfered with his investigations.

As Amin settled in for the evening, he flipped on the television. The evening news was just coming on the BBC. The lead stories that night concerned the fighting in Iraq and Prime Minister Tony Blair's surgery for treatment of a heart ailment. But the program also devoted considerable time to coverage of the US presidential debate the night before.[3] Interest in the upcoming 2004 US election was remarkably high in Britain. President George W. Bush had squandered the goodwill that existed toward the United States in the aftermath of 9/11 and had grown increasingly unpopular in the UK and much of Europe due to his unilateral approach to world affairs and the US-led war in Iraq. The British people, in particular, had come to the conclusion that they had been drawn into the war under false pretenses, and while they held Prime Minster Blair responsible for going along with the scheme, they saw George Bush as the primary culprit.[4] The widespread belief throughout England and Europe was that Bush could not possibly win reelection. There was great curiosity, therefore, about the man they assumed would replace him, Senator John Kerry of Massachusetts. Although Kerry had served in the US Senate for nearly twenty years, he was an unknown commodity to most Europeans.

The debate featured on the news that night was the first between Bush and Kerry. The topic was US foreign policy and homeland security. Amin watched closely, anxious to learn what the candidates would say. Kerry began by offering his vision for protecting Americans in the post-9/11 world. He said he could "make America safer" than the president had by strengthening international alliances that Bush had left in "shatters," improving US military and intelligence capabilities, and cracking down more effectively on terrorist financing. He also said that he could do a better job in Iraq than the president had done and pledged to improve the training of Iraqi soldiers so they could defend their country.

Bush responded by emphasizing his "multipronged strategy" for keeping America safe and touted his administration's record on hunting down al Qaeda, ousting the Taliban from Afghanistan, and toppling Saddam Hussein's regime in Iraq. But then the president said something that caught Amin by surprise. In describing what he called his "policy of disrupting those who proliferate weapons of mass destruction," Bush declared that the black market nuclear arms network headed by Pakistani scientist A.Q. Khan had been "brought to justice."[5]

Amin couldn't believe what he had just heard. The president was taking credit for exposing Khan's illicit trade in nuclear weapons. Amin found Bush's comments outrageous. He knew from his own investigations that Khan's sales of nuclear technology had taken place while US and British intelligence had the Pakistani nuclear smuggling network under close surveillance. Yet there had been no serious attempt by either government to halt Khan's activities. In fact, Amin knew that the opposite was true. Several years earlier his superiors, with the encouragement of British intelligence, had pulled him off an investigation just as he began to uncover evidence of Khan's nuclear sales. Britain's Secret Intelligence Service, known as MI6, had even warned Amin not to go near a company that turned out to be involved in Khan's nuclear trafficking. The implication was that MI6 and the CIA had the situation under control and that Amin should steer clear to avoid exposing their monitoring operation. Now, however, years later, with the smuggling network belatedly exposed and the proliferation damage done, Bush was attempting to portray the Khan affair as an intelligence coup for which he could take credit. The picture he painted suggested that intelligence operatives had swooped in on the black market network at the first sign of trouble and headed off a potential disaster. In fact, Amin knew they had stood by and watched the disaster unfold, doing nothing to stop it.

Angry and disgusted, Amin turned off the television. Looking through the window, he could see that the storm outside had passed. Yet the sky remained cloudy. More rain was on the way.

Atif Amin's reaction to President Bush's remarks is entirely understandable. Amin knew that if his investigation of A.Q. Khan's smuggling network had been allowed to continue, it could have resulted in halting the transfer of nuclear technology to Iran, Libya, and North Korea years sooner. But that had not happened. The network was allowed to remain in place and the flow of nuclear equipment and know-how to rogue states continued.

Amin's experience was not unique. Since the mid-1970s, numerous official European and American investigations had come close to fully exposing Pakistan's nuclear smuggling operation. But time after time those inquires fell victim to politics. Washington intervened in the investigations. US foreign policy considerations trumped efforts to prevent the spread of nuclear weapons. A.Q. Khan and his cohorts were allowed to continue their deadly trade, and proliferation of nuclear technology went on unabated. The obstacles Amin and other investigators ran into were the upshot of government policies adopted decades earlier.

The Prospect of Cooperation

America's interest in halting the spread of nuclear weapons has always been tempered by political and foreign policy considerations. In the aftermath of World War II and the unleashing of the revolutionary destructive power of the atom against the Japanese cities of Hiroshima and Nagasaki, US officials began seriously debating America's nuclear policy. Although the United States had emerged from the war as the world's lone nuclear power, it was a position that could not last. Advisors to President Harry Truman understood that the technical knowledge required to produce an atomic bomb would not remain the sole possession of the United States indefinitely. America's wartime allies — the Soviet Union, Britain, France, and Canada — were well poised to develop nuclear arsenals.

Within Truman's administration, two distinct schools of thought emerged about the appropriate disposition of the bomb. Both were heavily influenced by concerns about America's postwar relationship with Moscow. Truman's secretary of war, Henry Stimson, and undersecretary of state, Dean Acheson, believed it was vital that the United States maintain cooperative relations with the Soviets. Failure to do so in the newly emergent age of nuclear weapons was simply too dangerous. The key to preserving trust between the two nations, Stimson and Acheson asserted, lay in Washington's handling of the nuclear question. They recommended that as a gesture of good faith, the US share information about peaceful applications of atomic energy with its World War II partners, including the Russians. In a memo to Truman, Acheson dismissed suggestions that the US develop the bomb in partnership with Britain and Canada — both of which had joined in the American-led wartime effort to build an atomic weapon — arguing that the Soviets would interpret such an arrangement as "unanswerable evidence of an Anglo–American combination against them" and would feel compelled to take action to restore the balance of power.[6]

That line of reasoning met fierce opposition from Navy Secretary James Forrestal. A virulent anticommunist, Forrestal viewed the bomb as "the property of the American people" and rejected any notion that the Soviets could be trusted.[7] He urged Truman to retain US nuclear monopoly. Many in Congress and in the press echoed Forrestal's call.

The president, while sensitive to the political mood of the country and, himself, increasingly distrustful of Soviet intentions, nevertheless adopted a stance that was broadly in line with the approach proposed by Stimson and Acheson. In an address to Congress on October 3, 1945, Truman declared that, "The hope of civilization lies in international arrangements looking, if possible, to the renunciation of the use and development of the atomic bomb, and directing and encouraging the use of atomic energy and all future scientific information toward peaceful and humanitarian ends." While making clear that there would be no sharing of information about the development of nuclear weapons, Truman proposed opening discussions with Britain and Canada, and then with "other nations," including the Soviet Union, aimed at reaching an agreement on "the conditions under which cooperation might replace rivalry in the field of atomic power." Although acknowledging that there would be great difficulties in following such a course, he insisted they would have to be overcome. The alternative, he said, would be "a desperate armament race which might well end in disaster."[8]

Trumans's speech was notable on several counts. Coming at a time when the United States maintained exclusive control over the power of the atom, it held out a vision of peaceful East–West collaboration. Had such a strategy been pursued, it might have not only warded off the nuclear arms race that Truman feared but also, in the process, eliminated the source of much of the mistrust that fueled the Cold War. The speech also marked the first formal proposal for the sharing of nuclear energy for peaceful purposes, a concept that — however well intentioned — would have disastrous long-term repercussions, including, predictably, the proliferation crisis we face today. Before any of that could transpire, however, the concepts imbedded in Truman's speech would be transformed by proto–Cold War politics.

Shifting the Debate

Truman's lead negotiator on the issue of international control of atomic energy was Secretary of State James Byrnes, a political conservative who took a dim view of the Soviets. Unlike the president, Byrnes opposed dealing directly with Moscow. As talks with Britain and Canada got underway in November 1945, Byrnes latched on to a proposal for debating the nuclear question in the newly formed United Nations. Although Truman had explicitly rejected such a concept in his address to Congress, Byrnes pressed ahead. For Byrnes, the plan — originally floated by Vannevar Bush, a top government scientific advisor who had been one of the primary organizers of the atomic bomb program — held the dual virtues of avoiding politically difficult discussions with the Soviets and of allowing the United States to maintain its hold on nuclear secrets until Moscow's intentions became clear. But Byrnes's proposal also eliminated the gesture of good faith inherent in Truman's speech and thereby risked inflaming tensions with the Soviets. As Dean Acheson noted, by moving the discussion of nuclear control to the

UN, Byrnes was following a path directly opposite the one he and Stimson had laid out.[9]

In other quarters, Byrnes's concept received a more favorable reception. The British, who by then were preparing to develop their own bomb, and the Canadians both embraced the UN strategy as the best means for dealing with the nuclear question. With them on board, Byrnes traveled to Moscow and, surprisingly, won the Russians' agreement to help establish a UN commission on atomic energy. The Soviets' only stipulation was that the commission be placed under the jurisdiction of the UN Security Council, where both Washington and Moscow held veto power. Although angered by Byrnes's backroom maneuvering, Truman reluctantly accepted the U.N. scheme.[10]

While Byrnes had succeeded in shifting the nuclear discussion to the UN, American objectives remained unclear. Neither Truman nor Byrnes nor anyone else had articulated a formal strategy. To solve the problem, Byrnes turned to Acheson, appointing him to head a committee charged with drafting a US policy. As its starting point, the committee was charged with conceiving an international arrangement aimed at preventing the use of atomic energy for destructive purposes while promoting its use "for the benefit of society."[11] Recognizing the need for technical expertise, Acheson convened a board of scientific consultants to develop a proposal. To serve as chairman he selected David Lilienthal, the former head of the Tennessee Valley Authority. The panel's most influential member was J. Robert Oppenheimer, the former scientific director of the Manhattan Project.

After more than seven weeks of intense effort, in March 1946 Lilienthal's panel presented its findings to Acheson's committee. The Acheson-Lilienthal report, as it became known, offered a stark and prescient appraisal of the dangers of nuclear weapons development and the risks of proliferation. The report concluded that because atomic energy programs could easily be diverted into nuclear weapons initiatives, any effort to ban the bomb exclusively through a system of international inspections would inevitably fail. "There is no prospect of security against atomic warfare in a system of international agreements to outlaw such weapons controlled only by a system which relies on inspection and similar police-like methods," the report said. "National rivalries in the development of atomic energy readily convertible to destructive purposes are the heart of the difficulty. . . . A system of inspection superimposed on an otherwise uncontrolled exploitation of atomic energy by national governments will not be an adequate safeguard."[12] The obvious solution of banning or controlling nuclear energy programs as a means of halting development of atomic weapons was precluded by the committee's mandate to promote the peaceful use of nuclear power. As a result, the committee instead recommended that the United States voluntarily relinquish its nuclear monopoly and invest its hopes for controlling the bomb in a system of international cooperation administered by a single authority—namely, the UN.[13] Whatever prospects for success such an arrangement may have had, they were quickly undone by Truman's choice of salesman for the plan.

"The Worst Mistake I Have Ever Made"

On the very day that Acheson submitted his committee's final report, Byrnes informed him that Truman had selected Wall Street mogul and political contributor Bernard Baruch to market the plan to the world. Acheson and Lilienthal were shocked. A former member of President Franklin Roosevelt's "Brain Trust," and later shunned by FDR, Baruch was renowned for his oversized ego. Acheson protested the appointment, but to no avail. Truman and Byrnes believed that Baruch's substantial influence with members of both parties in Congress — stemming largely from his extravagant campaign contributions — would help the administration push through legislation related to the nuclear initiative. Lilienthal wrote in his diary that he became "quite sick" upon learning of Baruch's involvement. He viewed Baruch as a vain self-promoter whose overt contempt for international cooperation threatened to scuttle the entire initiative by convincing Soviets that he was simply out to "put them in a hole."[14]

Lilienthal's concerns were well founded. Baruch made clear from the start that the plan submitted to the UN would be his own, not that of Acheson and Lilienthal. He insisted that he could divine an appropriate course of action purely by instinct and, as if to prove his point, refused to consult scientific advisors. Baruch ultimately set forth his vision at the inaugural session of the UN's Atomic Energy Commission in New York on June 14, 1946. It differed from the Acheson-Lilienthal report's recommendations in several significant respects. It scrapped, for example, the call for international control over the means of producing nuclear fuel as being inconsistent with American free-market values. Under the Baruch Plan, moreover, the United States would retain its nuclear arsenal, which at the time totaled nine warheads, until it had received assurances that no other nation could develop a bomb. In the meantime, the US would continue to expand its atomic stockpile. Baruch also tacked on two provisions that — deliberately or not — proved to be poison pills for the Soviets. One was the inclusion of "immediate, swift, and sure punishment" for those who violated any agreements that were eventually reached.[15] (When asked by Truman what that punishment would be, Baruch answered, "war."[16]) The other was a stipulation that such sanctions not be subject to Security Council veto — a direct contradiction of the agreement Byrnes had previously worked out with the Soviets.[17]

Despite the take-it-or-leave-it tone of Baruch's presentation, the Soviets offered a counterproposal that called for a prohibition on the use and development of atomic weapons and the destruction of all existing nuclear arsenals.[18] But Baruch was unwilling to negotiate and demanded a vote on his plan by the end of the year. The results were predictable. The UN commission approved the Baruch Plan, with only the Soviets and Poland abstaining. The Soviets then killed it with a veto in the Security Council. The best hope for meaningful cooperation on the control of nuclear weapons had been lost. Truman later described his selection of Baruch as the person to lead the effort as "the worst mistake I have ever made."[19]

Nuclear Rivalry

Whatever his concerns about Baruch, Truman himself did little to improve the prospects for nuclear cooperation. Just weeks after Baruch presented his plan, Truman signed legislation that enshrined secrecy and the retention of America's nuclear monopoly as official US policy. The Atomic Energy Act of 1946, signed by Truman on August 1 of that year, established the terms by which the United States would manage and control its nuclear weapons and technology. First and foremost, the act transferred control of all nuclear matters out of the hands of the military and placed it under civilian authority. For that purpose, it created the US Atomic Energy Commission (AEC), a five-member board of civilian nuclear governors, along with a Congressional panel known as the Joint Committee on Atomic Energy (JCAE) that would wield authority over nuclear legislation. The activities of both the AEC and the JCAE were shielded from public view, placing all aspects of America's nuclear program — from uranium mining to the production of nuclear fuel and weapons — behind a heavy veil of secrecy. The act also prescribed the death penalty for anyone found guilty of passing nuclear secrets to any foreign power, whether enemy or allied.[20]

The secrecy provisions of the Atomic Energy Act provided the clearest indication to date of America's determination to maintain its exclusive hold over nuclear weapons and technology. Rather than ensuring US security, however, this stance only heightened the concerns of foreign governments and spurred their efforts to develop their own atomic bombs. Realizing that the United States would neither give up its weapons nor place them under international control, the British resolved to press ahead with their nuclear weapons program. More ominously, the Soviet Union now redoubled its bomb-building efforts. The nuclear arms race, and with it the Cold War, were on.

Knockout Blow

The American nuclear monopoly ended on August 29, 1949, just four years after the United States leveled Hiroshima with an atomic bomb, when the now-reviled Soviet Union conducted a successful test of its own nuclear weapon. Coming years earlier than even the most prescient American analysts had predicted (thanks in part to information Russia's atomic spies had pilfered from the Manhattan Project) the Soviet test sent shock waves through America's military and policy-making communities. In response, the United States began a major expansion of its capacity to produce and stockpile nuclear weapons and launched a crash program to develop a thermonuclear hydrogen bomb.[21]

The American and Soviet nuclear arsenals grew quickly through the early 1950s. As they did, US military planners began raising alarms about the prospect of Russia delivering a nuclear "knockout blow" that could destroy America's industrial infrastructure and retaliatory ability. In the face of such concerns, Acheson, now the secretary of state, established a committee in April 1952 to consider new approaches to nuclear control and disarmament. To head the panel, Acheson selected Robert Oppenheimer. The Oppenheimer committee's report would have a major influence on the direction of US military and civilian nuclear policy.[22]

After more than a year and a half of deliberations, the Oppenheimer panel found that major increases in US and Soviet production of fissionable material, combined with lax accounting practices, made verification of any future nuclear disarmament treaty all but impossible. With no real prospect for arms control in sight, the committee concluded that the Soviets could develop a knockout-blow capability within a few years. That risk would remain until the United States developed adequate defensive and deterrent capabilities. But the committee also stressed the need for renewed arms control and disarmament efforts. Winning public support for such initiatives would be crucial.[23]

The Oppenheimer panel's final report offered four major recommendations: first, that there be a public discussion about the looming nuclear crisis; second, that the government release information about the size of the US nuclear arsenal and the scope of its weapons manufacturing capability, both to inform the public and to dispel any notion the Soviets might have that they already possessed a knockout capability; third, that the United States enter into arms control negotiations with the Soviets; and finally, that the United States decrease its dependence on nuclear weapons by developing a wider array of military capabilities.[24]

Operation Candor

The Oppenheimer committee submitted its report on January 15, 1953, just five days before Dwight Eisenhower was sworn in as president. Reading the report shortly after taking office, Eisenhower, who had taken a strong interest in nuclear issues since his days as commander of Allied forces during World War II, was impressed. Convinced that a new approach to arms control was required, Eisenhower instructed one of his senior advisors, psychological warfare specialist C.D. Jackson, to prepare a presidential address based on the Oppenheimer panel's proposals. Jackson, a former Time-Life executive who now headed Eisenhower's Psychological Strategy Board, was a master at developing ideas for scoring propaganda victories over the Soviets. Working with advisors from both inside and outside government, Jackson put those skills to good use as he set about crafting Eisenhower's speech, a project he dubbed Operation Candor.[25]

Jackson and his team had completed several drafts of the speech when, on August 12, 1953, the Soviets announced they that had successfully tested a thermonuclear bomb, just nine months after the first American test of a similar device.* News of the Soviet superweapon only reinforced the generally pessimistic view many of Eisenhower's advisors already held about the prospects for achieving a meaningful arms control agreement with the Russians. Eisenhower, however, maintained a more hopeful outlook and remained determined to engage the Soviets in an effort to curb the nuclear arms race.[26] He also sought a proposal that could calm world fears in the wake of the Soviet hydrogen bomb test.

A key moment came in September 1953, when discussions between Eisenhower and his advisers yielded a new concept. The idea would be for both the United States and the Soviet Union to set aside for peaceful purposes specified amounts of the uranium from their weapons programs. This so-called uranium bank would be administered by an international agency. Because it was believed at the time that uranium was extremely scarce, Eisenhower envisioned the proposal as an opportunity to reduce the risk of nuclear war by diverting bomb material to peaceful programs.[27]

As work on the bank concept went forward, it became increasingly clear that uranium was far more abundant than originally thought.[28] This, of course, significantly undercut the notion that the initiative could limit the production of nuclear arms, since uranium used for civilian purposes would not meaningfully affect the amount available for weapons. As a result, the entire proposal was left to stand on the thin hope that the feel-good proposition of using nuclear energy for peaceful purposes could engender enough goodwill to stave off an arms race with the Soviets. But even the concept of a peaceful atom was misguided. Since at least the time of the Acheson-Lilienthal report in 1946, it had been well understood that civilian nuclear programs could easily be channeled into weapons production. Thus, Eisenhower's initiative carried in it the seeds of a proliferation crisis.

*The Soviet test of August 12, 1953, was not of a true fusion bomb, but of a boosted fission weapon. With a yield of four hundred kilotons, it was considered a "low-yield" thermonuclear device. The American test of November 1, 1952, just days before the US presidential election, was of a true thermonuclear device. Its yield was 10.4 megatons, 750 times more powerful than the bomb dropped on Hiroshima. The Soviets first test of a true thermonuclear weapon came on November 22, 1955. Its yield was 1.6 megatons. (Robert Norris and Hans Kristensen, "Nuclear Pursuits," *Bulletin of the Atomic Scientists*, September/October 2003.)

Atoms for Peace

Eisenhower unveiled his Atoms for Peace concept in an address to the UN General Assembly on December 8, 1953. On the surface, it was an arms control plan that held out a hopeful vision to a world frightened by the prospect of nuclear war. But while Eisenhower was undoubtedly sincere in his desire to curb nuclear weapons, the proposal was far more effective as a propaganda tool. It dangled the appearance of an arms limitation plan that did not require inspections, while demonstrating American goodwill and the promise of peaceful nuclear power. If the Soviets balked at the idea, the United States could accuse them of them of being opposed to the control of nuclear weapons. As in the past, however, the Soviets would refuse to take the bait.

In his speech, Eisenhower mixed dark warnings with optimistic vision. Acknowledging the grave danger of nuclear war and the spread of nuclear technology, he went on to declare that the United States sought "more than the mere reduction or elimination" of nuclear bombs. "It is not enough to take this weapon out of the hands of the soldiers," he said. "It must be put in the hands of those who know how to strip it of its military casing and adapt it to the arts of peace." Specifically, Eisenhower proposed that the United States and the Soviet Union transfer uranium and other nuclear materials from their stockpiles to a newly created international atomic energy agency under the auspices of the UN. That new agency would be responsible for developing safeguarded methods for making the material available "to serve the peaceful pursuits of mankind" in fields such as agriculture and medicine. Eisenhower placed special emphasis on the goal of quickly providing "abundant electrical energy in the power-starved areas of the world," regions where the Soviets had, not coincidently, made significant inroads. In addition, although the original concept of Operation Candor had fallen by the wayside, Eisenhower's speech offered a none-too-subtle accounting of America's nuclear arsenal that sent a clear message to the Soviets about US military power.[29]

Eisenhower's speech was well received both at home and overseas, thanks in large part to a massive public relations campaign orchestrated by C.D. Jackson. The Soviets, however, viewed the uranium bank initiative as a propaganda vehicle. While agreeing to talks on the subject, they faulted the underlying concept. Four months after Eisenhower's address, the Soviets sent the United States a note pointing out that because plutonium is produced in the generation of nuclear power, using atomic energy for peaceful purposes would do nothing to reduce the amount of material available for weapons, the very point made by the Acheson-Lilienthal report years earlier. Despite this concern, the Eisenhower administration continued to promote Atoms for Peace.[30]

Atoms for Industry

C.D. Jackson's media blitz for Eisenhower's speech galvanized public interest in civilian uses of atomic energy. Companies already involved in America's nuclear weapons program now saw an opportunity to parlay their technical ability into potentially lucrative commercial ventures, particularly in the field of nuclear power generation. The fledgling industry objected, however, to the provisions of the 1946 Atomic Energy Act that would have required government ownership of nuclear power facilities. Both the business-friendly Eisenhower administration and the Republican-dominated Congress shared the industry's view. To facilitate commercial nuclear energy production, therefore, Congress rewrote the rules of the game. In doing so, it also opened the floodgates of nuclear proliferation.[31]

The Atomic Energy Act of 1954 eased many of the restrictions originally put in place to protect America's nuclear monopoly. Under the revised act, commercial utilities could own and operate nuclear power reactors using fuel supplied by the Atomic Energy Commission. In a further boost to industry, the act also prohibited the government from selling power generated by its military and research nuclear facilities.[32]

The 1954 act also allowed the transfer of nuclear technology to foreign countries. Many government and industry officials viewed the Atoms for Peace initiative as an opportunity for the United States to establish itself in the international nuclear power market. Industry executives saw this as a business opportunity. The Eisenhower administration considered it a Cold War priority. The administration hoped that demonstrating the commercial viability of nuclear power, particularly in the developing world, could offset Soviet efforts to establish civilian nuclear programs in Cold War battle zones. To accommodate those interests, the 1954 Atomic Energy Act allowed the transfer of nuclear material and technology to foreign countries for civilian purposes. The transfers would take place through agreements arranged by the AEC, and the United States would retain the right to ensure that materials involved were not used for military purposes.[33]

The Atoms for Peace program that emerged from this process was far different from the one Eisenhower had originally envisioned. Gone was the concept of a uranium bank overseen by an international agency. In its place stood a series of state-to-state agreements for direct transfers of nuclear materials and technology. Each participating country would have an autonomous nuclear program, reined in only by a system of international inspections and so-called safeguards. This, of course, was the very arrangement that the Acheson-Lilienthal report had warned would be unable to prevent the spread of nuclear weapons.[34]

In the years that followed, the United States would enter into nuclear technology sharing agreements with dozens of countries. American companies sold nuclear technology for use in US-sanctioned programs that allowed foreign scientists and engineers to learn the secrets of nuclear energy. Many of those scientists and engineers received training in the United States or with US funding. And several of the countries that participated in the program used it as a springboard for developing nuclear weapons and became the source of the proliferation crisis we face today.[35]

Turkey was the first country to sign an agreement with the United States to participate in the Atoms for Peace program. The country would eventually become a central hub in A.Q. Khan's nuclear black market network. Israel became the second country to join the Atoms for Peace initiative. It is now known that Israel's chief weapons scientist at that time viewed Atoms for Peace as an opportunity to obtain a reactor from the United States that could be modified to generate plutonium for use in a nuclear weapons development effort.[36]

India and Pakistan also entered into nuclear sharing agreements with the United States as part of Atoms for Peace. Legions of Indian and Pakistani scientists flocked to the US for training in nuclear technology. The US also supplied both countries with equipment, financing, and technical assistance, helping them establish nuclear programs that quickly became the basis for efforts to build nuclear weapons. As A.Q. Khan himself later noted, this was the inherent risk in the Atoms for Peace strategy. "Knowledgeable and sane people never overlooked the fact that there was a grave danger of nuclear weapons proliferation with the spread of peaceful uses of nuclear energy," Khan wrote in 1986. "Once you know how to make reactors, how to produce plutonium and reprocess it, it becomes a rather easy task to produce nuclear weapons."[37] And yet the United States blithely ignored this danger.

By the late 1950s, France, China, and Israel had each initiated nuclear weapons development programs. A National Intelligence Estimate prepared in the final days of the Eisenhower administration found that other countries might soon follow suit in an attempt to keep pace. As a result, the estimate warned, nuclear proliferation would be one of the major problems facing the world in the coming decade.[38] It was a prophetic warning, one that would be echoed repeatedly in the years ahead. But it would be heeded, unfortunately, only after it was too late.

Today, the world must confront the consequences of America's failure early on to adequately address the risks associated with the widespread diffusion of nuclear technology. In its eagerness to promote the benefits of civilian nuclear energy out of a combination of idealistic, commercial, and political motives, the United States became the Johnny Appleseed of a technology that could easily be channeled into the development of atomic weapons. The results were predictable. Atoms for Peace helped establish nuclear programs in some of the world's least stable, most trouble-prone regions. Those ostensibly civilian programs allowed ambitious leaders seeking power and prestige to develop nuclear weapons that would otherwise have remained far beyond their reach. Smaller, threatened states soon sought the same capability. Thus, a program intended to promote the "peaceful" application of nuclear energy became a major spur for the rapid spread of nuclear weapons. Those weapons have now become so ubiquitous that there is a significant risk that they will fall into the hands of a terrorist organization. The greatest threat we face today is, in short, the upshot of a policy devised over a half century ago.

2
Nuclear Relations

ON AUGUST 11, 1955, the United States entered into a nuclear cooperation agreement with Pakistan as part of the Atoms for Peace program. The agreement, signed in Washington by Assistant Secretary of State George V. Allen and Pakistan's ambassador to the US, S. Amjad Ali, called for the US to supply Pakistan with a research reactor and the enriched-uranium fuel needed to operate it. The United States would also assist Pakistan in the design, construction, and operation of such research reactors.[1] Washington would contribute $350,000 to the research effort and Pakistan would contribute a roughly equal amount.[2] It was this modest civilian nuclear program that laid the foundation for Pakistan's development of nuclear weapons.

The 1955 nuclear deal was the culmination of a long seduction of the United States by Pakistan. The mating dance had distinctly political overtones. Since the end of World War II, US foreign policy had focused almost exclusively on countering the expansion of communist influence in Europe, East and Southeast Asia, and the Middle East. The Asian subcontinent received far less attention. To the extent Washington concerned itself at all with the region, India was generally seen as the key to furthering US interests. But New Delhi had adopted a stridently independent posture and expressed open disdain for the United States, undermining whatever slim prospect existed for a meaningful alliance between the two nations. Pakistan sensed an opportunity and exploited it adroitly.[3]

In the early 1950s, Pakistani leaders began openly vying for Washington's attention. They lavished praise on the United States, proclaimed their opposition to communism, and contrasted their own pro-Western orientation with India's tilt toward socialism and loose alignment with the Moscow. They also stressed the strategic value of their nation's proximity to the Soviet Union, China, and the Persian Gulf.[4]

Pakistan's approach had the desired effect. Many in Washington came to see the Islamic republic as an ideal Cold War partner, providing a natural roadblock against Soviet access to the Arabian Sea and a potential ideological and military bulwark against communist expansion in Central Asia and the Middle East. But it was only after President

Eisenhower took office that a formal US–Pakistani alliance emerged. The new administration set out to contain communism through a global web of strategic partnerships. In an effort to capitalize on that strategy, Islamabad sent Army Chief of Staff Gen. Mohammad Ayub Khan to Washington in the fall of 1953 to lobby for American military assistance. The White House embraced the idea, and in 1954 the United States and Pakistan entered into a mutual defense agreement — a move that set off bitter protests from India and soured US–Indian relations for years.[5] In 1955, Pakistan became a member of the US-backed South East Asian Treaty Organization and joined with Iran, Iraq, and Turkey in the Baghdad Pact, later known as the Central Treaty Organization, for which the United States provided security assurances. Mohammad Ayub Khan, who ruled Pakistan from 1958 to 1969 after seizing power in a military coup, later quipped that during the 1950s Islamabad became America's "most allied ally in Asia."[6]

As US–Pakistani relations strengthened during the mid-1950s, American military and development aid flowed freely, reaching five hundred million dollars a year by 1957. In exchange for this largesse, the United States hoped that it had enlisted a reliable Cold War partner. But whatever its advantages in the short term, this marriage of convenience would eventually yield ominous results.

Strategic Maneuvering

It was in this environment of political-military alliance that the United States helped Pakistan launch its nuclear program. After signing the nuclear cooperation agreement with Washington in 1955, Islamabad established the Pakistan Atomic Energy Commission (PAEC) the following year to administer the effort. In the early stages the PAEC focused on the development of human talent. It sent hundreds of aspiring nuclear scientists and engineers to the United States and other Western countries for training. The stated goal of the Pakistani program was to develop nuclear technology for application in the fields of agriculture, medicine, and industry, as well as to meet the country's growing energy needs.[7] But attaining a nuclear weapons capability remained a constant subtext.

Hints of Pakistan's interest in using its nascent civilian nuclear program as a springboard for developing atomic weapons appeared early on. The

1955 agreement with the United States provided only enough funding for Pakistan to obtain a small research reactor. The agreement made clear, moreover, that US assistance could not be used to purchase reactors that could be used for anything other than civilian purposes. But the governing board of the PAEC chaffed at those restrictions. According to the book *Long Road to Chagai,* a history of Pakistan's nuclear weapons program, the PAEC board complained repeatedly to officials in Islamabad that the proposed research reactor had only "limited capability" and urged Islamabad to allocate funds for the purchase of a much larger power reactor from either the United States or Canada.[8] Such a reactor would produce more spent fuel, which could be reprocessed to extract plutonium that could be used for nuclear weapons. The PAEC's interest in obtaining a power reactor may have been spurred by developments in India. The same month that Pakistan signed the cooperation agreement with Washington, India finalized a deal to obtain a large research reactor from Canada that would produce significant quantities of plutonium that could be used for nuclear weapons. As part of the agreement, the United States would supply the heavy water used to moderate the reactor, known as the Canada-India Reactor, US (CIRUS). The plutonium reprocessed from the reactor would eventually be used in India's first nuclear test in 1974.[9]

Despite the PAEC's objections to the US deal, Islamabad balked at the proposal of laying out additional funds for a power reactor. The two sides went around and around for three years before the PAEC finally relented. Even then, the PAEC expressed its "disappointment" at the government's position.[10]

By the time the Pakistani reactor deal was complete, the United States had begun to question the wisdom of its alliance with Islamabad. American development aid had done little to improve Pakistan's fundamental economic and political weaknesses, while US military assistance had helped create a Pakistani defense behemoth that Islamabad could not sustain without continued American support. There was also a growing realization in Washington that improved relations with India were vital to US efforts to counter communist influence in the region and that American military aid to Pakistan hurt Washington's ability to develop closer ties with New Delhi. President Eisenhower increasingly shared that view. By 1957, he had concluded that the US alliance with Pakistan had been a "terrible error." Yet he also knew that, having started down that path, the US could not simply walk away from its commitments to Islamabad.

And at a practical level, Eisenhower understood that the United States could not bear the cost of taking on India as a full-fledged ally. The policy the administration ultimately adopted was to accept India's nonaligned status and offer New Delhi significantly greater economic and military aid while simultaneously halting further increases in military assistance to Pakistan. The administration also chose to "actively encourage" India to accept American assistance in developing atomic power reactors, stressing US willingness to share large volumes of previously classified nuclear information and to provide training to Indian nuclear scientists.[11] And in a further bow to Indian interests, the United States accepted New Delhi's demands for a weaker system of safeguards to be imposed by the newly created International Atomic Energy Commission (IAEA), the UN agency responsible for oversight and inspection of nuclear programs that had emerged from the Atoms for Peace initiative.[12] While this policy shift succeeded in drawing India toward the US orbit, it also had the predictable effect of galvanizing Pakistan's determination to develop nuclear weapons.

As US–Indian relations improved during the late 1950s, Washington's doubts about its dealings with Pakistan grew deeper. American concerns became particularly acute in the face of Ayub Khan's military takeover in 1958, an event that underscored Pakistan's profound instability. Yet Pakistan managed to recoup its standing with Washington by once again positioning itself as a vital Cold War ally. In 1959, Islamabad entered into an agreement with the United States that allowed the CIA and the National Security Agency (NSA) — the group responsible for US communications intelligence — to use Pakistan's northern provinces as a jumping-off point for anti-Soviet operations, a base for U-2 spy flights, and a listening post for eavesdropping on Russian nuclear facilities.[13] The US–Pakistani relationship was salvaged and American support continued. The long-stalled research reactor deal now went forward, and in August 1960 the United States released the funds for the project.[14] As a result, by the early 1960s Washington had maneuvered itself into the awkward position of subsidizing the military and nuclear programs of the Asian subcontinent's two archrivals.

While the United States continued to dangle nuclear technology as an enticement to potential Cold War allies during the late 1950s, American security officials began raising concerns about the possible spread of nuclear weapons. In 1957, growing unease among State Department officials about the effects of proliferation on the United States' position in the world prompted a call for an official assessment of the issue.[15] The result was a National Intelligence Estimate (NIE) produced by the CIA in June of that year. The estimate suggested that in the coming decade, as many as ten countries would be able to produce at least a few "nominal" atomic weapons "by exploiting the potential of their nuclear research and power programs," the ostensibly peaceful applications of atomic power the US actively promoted. It also noted that it would become increasingly difficult to achieve effective international controls over those weapons.[16] Just over a year later, amid mounting evidence that France was on the verge of becoming the fourth member of the nuclear club, the CIA issued a new NIE that offered an even bleaker assessment. It concluded that the acquisition of nuclear weapons by any fourth country was "certain to produce difficulties" and would "tend to increase the chances of general war by an expansion of local conflicts."[17] American national security officials would issue increasingly sharp warnings about the dangers of proliferation in the years ahead. Yet Washington remained stubbornly committed to its policy of assisting foreign nations in their development of nuclear technology.

"We Will Get One of Our Own"

For American officials, promoting Pakistan's nuclear program was a cold political calculation. They viewed it as part of a larger effort to win partners and ward off communist influence in South Asia. While the danger of a nuclear energy initiative being channeled into a weapons program was well understood by Washington, that concern took a backseat to Cold War objectives. For Pakistan, on the other hand, developing a nuclear capability was a matter of national pride — and survival.

Pakistan's quest for nuclear weapons is in many ways a legacy of the circumstances of its birth. The country came into being in 1947 as the result of a hastily conceived partition of British India. As Great Britain prepared to withdraw from the colony after World War II, India's ancient religious and ethnic factions fell into squabbling over a proposed power-sharing arrangement. When the situation threatened to devolve into civil war, the British chose to divide the colony into two independent states, predominantly Muslim Pakistan and predominantly Hindu India. Pakistan itself was split into two distinct regions separated by the vast new nation of India. One sat on India's western flank, along the Indus River, and the other, thousands of miles to the east, in the Ganges River delta in Bengal. Worse yet, the British partition plan failed to resolve the status of Kashmir, a predominantly Muslim northern region claimed by both India and Pakistan. The disputed territory has remained a source of constant tension between the two nations ever since and has been the root cause of three full-scale wars, a limited war, and many intense clashes, some of which have come perilously close to erupting into nuclear conflagrations.

Partition itself was a bloody affair. Tearing apart regions that had been bound together for millennia, it set off widespread rioting, massacres, and ethnic cleansing. One of those caught in the mayhem was a ten-year-old schoolboy named Abdul Qadeer Khan. The young A.Q. Khan grew up in the predominantly Muslim state of Bhopal, India. Although considered a relative safe haven for Muslims following partition, the region was not immune to ethnic violence. Khan claims to have witnessed trainloads of dead Muslims massacred by Hindus coming into the Bhopal station. When he migrated to Pakistan several years later, he reportedly saw fellow Muslims being mistreated and attacked by Indian soldiers and he himself suffered indignities at the hands of Hindus. "It was something I'll never forget," he told an interviewer decades later.[18]

Khan's sense of vulnerability was representative of his new country's predicament. Pakistan was born in a position of inferiority. It emerged from partition smaller in both size and population than the new state of India. The territory comprising Pakistan, moreover, was less developed than the land India inherited, and it lacked basic infrastructure. Living in the shadow of its larger, richer, and more powerful rival, Pakistan's primary concern, from the start, was security. Many of Pakistan's leaders quickly identified the acquisition of nuclear weapons as a vital element of the nation's effort to offset India's conventional military superiority. The man who would lead Pakistan's early drive to obtain those weapons was himself a product of the West.

Zulfikar Ali Bhutto, the future Pakistani prime minister, saw himself from an early age as a man of destiny. The heir to a powerful, aristocratic family, he was born in 1928 in the Sindh province of what is now southeastern Pakistan. His father was an influential figure in the British colonial government, known as the Raj, and was knighted for his service to the Crown. After completing his early education at an elite British school in Bombay, the young Bhutto traveled to the West in 1947 to continue his studies. He earned a bachelor's degree in political science in 1950 from the University of California, Berkeley, then went on to study law at Oxford University, where he graduated with honors. After being called to the bar in London in 1953, Bhutto returned to Pakistan to teach and practice law. His first appearance on the world stage came in 1957 as a member of the Pakistani delegation to the UN. Bhutto's impassioned speeches before the world body earned him a reputation as a brilliant and charismatic orator and a champion of anticolonial causes. Following the Pakistani coup in October 1958, Bhutto joined the military government of Mohammad Ayub Khan, becoming the youngest cabinet minister in Pakistan's history.[19]

Bhutto served in a series of key posts in Ayub Khan's government, eventually becoming Minister of Fuel, Power and Natural Resources in 1960. In that position, he quickly emerged as the leading advocate for his country's nuclear program. During his studies in the West, Bhutto had developed a firm belief in the virtues of political independence and economic self-reliance, themes that would become mainstays of his government career.[20] Mastering the secrets of the atom was, in his view, essential to bringing those blessings to Pakistan.

As Minister of Fuel, Power and Natural Resources, Bhutto held effective control over the Pakistan Atomic Energy Commission. He later wrote that when he took charge of the commission, it was "no more than a signboard of an office," chronically underfunded and lacking direction. Bhutto was determined to change that situation. "Assiduously and with granite determination," he wrote, "I put my entire vitality behind the task of acquiring nuclear capability for my country."[21] Bhutto lobbied tirelessly on behalf of the PAEC and won it a substantially bigger budget.[22] He also closed the deal for the long-delayed American research reactor, allowing Pakistan's nuclear program to begin in earnest.[23] And with Bhutto's active encouragement, the PAEC began sending hundreds of scientists and engineers, many of whom would be instrumental in Pakistan's development of the bomb, to the West for training in nuclear technology.[24]

Bhutto's focus on nuclear issues gained new significance in 1963 when Ayub Khan tapped him to become Pakistan's foreign minister. In his new post, Bhutto grew increasingly concerned with China's steady movement toward testing a nuclear weapon and suggestions that India, having lost a war to China the year before, would attempt to follow suit. Bhutto and his government colleagues began seriously discussing whether Pakistan should initiate a nuclear weapons program of its own. Pakistani concerns intensified in 1964, when India's leading nuclear scientist, having learned from the United States that a Chinese test was imminent, held a press conference to announce that India could produce an atomic bomb within eighteen months.[25] Less than two weeks later, on October 16, 1964, China detonated its first nuclear device. The dual events sent shock waves through the Pakistani government. In a meeting with US State Department and CIA officials the following month, Ayub Khan expressed "deep concern" at the prospect of India diverting material from its civilian nuclear program to produce a bomb.[26]

By early 1965, Pakistani leaders had become obsessed by the specter of a nuclear-armed India. To help offset the threat, they turned to India's nemesis, China. In March, Bhutto and Ayub Khan traveled to Beijing to meet with Chinese premier Chou En-lai and enlist support for Pakistan. The Manchester *Guardian* reported afterward that the discussions had been "dominated" by the issue of nuclear weapons. Bhutto later hinted that during the visit he sought China's help in developing a bomb.[27] Upon his return from the China venture, Bhutto uttered a prophetic oath, vowing to match India's nuclear capability. "If India builds the bomb, we will eat grass or leaves, even go hungry, but we will get one of our own," he declared. "We have no alternative."[28] China would play a key role in Pakistan's eventual acquisition of the bomb.

The obsession with India's nuclear capability may, paradoxically, have inspired Pakistan's ill-fated effort to wrest Kashmir from India in 1965. In August of that year, Pakistan, acting on a plan advocated by Bhutto, sent thousands of guerrilla fighters into Indian-controlled Kashmir in an attempt to ignite a Muslim uprising in the region.[29] Author and South Asia expert George Perkovich has suggested that Pakistan may have launched the attack in hopes of ousting India from Kashmir before New Delhi acquired a nuclear weapon. "Given how focused Bhutto and others were on the feared Indian rush to build the bomb, and the deterrent effects an Indian bomb would have on Pakistan, it is inconceivable that this factor did not enter into the Pakistani decision to launch the 1965 war," Perkovich writes.[30]

The Kashmir adventure did not work out as Pakistan had planned. The hoped-for uprising failed to materialize and India fought back with unexpected determination. After seven weeks of conflict, the two sides retreated to their original positions. In the meantime, the United States had cut off economic and military assistance to both countries. The negotiated settlement, mediated by the Soviets in early 1966, called for India and Pakistan to refrain from future uses of force and the repatriation of prisoners from the conflict. It did nothing, however, to resolve the underlying dispute over Kashmir.[31]

The 1965 India-Pakistan war marked a major shift in America's South Asia policy. After more than ten years of engagement in the region and repeated pledges by the Kennedy and Johnson administrations to protect Pakistan from both "communist aggression" and attack by India, Washington began to back away.[32] During the conflict, the United States took a relatively minor role in efforts to end the hostilities, leaving it largely to the UN and the Soviet Union to carry out negotiations. Increasingly focused on the war in Vietnam and frustrated by a lack of progress on Kashmir, the US turned its attention to other Cold War battlegrounds.

While America's desire to distance itself from the problems of South Asia is understandable in political terms, it is also clear that Washington's own policies helped create the very difficulties it sought to escape. The United States, after all, had supplied the weapons Pakistan used to prosecute the war. And the United States had also helped India develop the nuclear capability that fueled Pakistan's anxiety. Although John F. Kennedy had warned of the dangers of nuclear proliferation while campaigning for the presidency in 1960,[33] his own administration helped move India toward acquisition of an atomic bomb. In 1961, shortly after Kennedy took office, the United States entered into negotiations to build two nuclear power reactors for India despite the clear understanding on both sides that the facilities could be used to produce material for atomic weapons.[34] More remarkably, that same year State Department officials concerned by the prospect of China testing a nuclear device in the near future seriously considered helping India produce a bomb. According to a declassified State Department memorandum, the officials contended that it would be "desirable if a friendly Asian power beat Communist China to the punch" by conducting a nuclear test first and that the most likely candidate for doing so was India. The top-secret memo went on to suggest that the United States consider providing technical assistance that could allow India to detonate an atomic device ahead of China.[35] While the administration ultimately chose not to follow through with this proposal, the very fact that the United States entertained such an idea indicates a willingness to consider setting aside nonproliferation goals to win short-term Cold War advantages. That attitude had helped bring about the problems the United States confronted in South Asia.

In early 1963, a Pentagon estimate predicted that over the next decade eight countries, including India, would be able to develop nuclear weapons along with crude delivery systems. The military analysts noted that many countries had "reduced the lead time and cost of acquiring nuclear weapons by getting research reactors and starting nuclear power programs."[36] Despite that well-understood finding, the United States later that year finalized the agreement to construct the two atomic power reactors in India.[37] It was, of course, concern about the prospect of India diverting its rapidly expanding civilian nuclear program into military applications that stoked Pakistan's fears during this period,[38] propelled Islamabad into the arms of the Chinese, and may have contributed to the decision to initiate the 1965 war. Not long after the fighting in that war stopped, the State Department issued an intelligence report indicating it was likely Pakistan would launch a nuclear weapons program if it became clear India was pursuing a bomb. The report also suggested that Islamabad might seek "tangible assistance" from China in the form of fissionable material, technical aid, or even a completed weapon.[39] It was a prescient assessment. The fact that it played out as predicted was in large measure due to US policy, past and future.

The Man with the Plan

For Pakistan, the loss of the 1965 war was a national trauma. Not only had the country failed to win control of Kashmir, it had suffered a demoralizing defeat at the hands of archrival India, a sign of martial impotence. For Ali Bhutto the war was a turning point. He viewed the negotiated settlement as an effective capitulation to India and a calamity for Pakistan.[40] He also saw the US failure to back Pakistan, despite the long-standing security guarantees, and the decision to cut off aid as evidence that Washington could not be relied upon as an ally, and he determined that Islamabad needed to steer a more independent course, less beholden to the vagaries of Western interests.[41] China, India's rival, looked increasingly attractive as a partner. The bitter loss to India also galvanized Bhutto's belief in the need for Pakistan to obtain a nuclear capability. In a National Assembly debate just weeks after the fighting ended he repeated his oath: If India armed herself with nuclear weapons, Pakistan would get an atomic bomb "no matter if she has to starve."[42] He also vigorously backed a proposal to purchase a French nuclear fuel reprocessing plant as a building block for an atomic weapons program. Ayub Khan, however, proved unenthusiastic and chose not to pursue the deal.[43]

Increasingly at odds with Ayub's government, Bhutto resigned as foreign minister in June 1966 to form an opposition movement. The following year, he launched the Pakistan People's Party (PPP), the nation's first mass political organization. Skillfully positioning himself as a populist leader, the flamboyant and eloquent Bhutto toured Pakistan promising "Food, Shelter, and Clothing" for the masses. He called for the introduction of "Islamic socialism" and eternal vigilance toward India, all the while railing against Ayub's military regime.[44]

Bhutto laid out his vision for Pakistan's future in a book titled *The Myth of Independence*. Finished in 1967 and published in 1969, the book presented Bhutto's prescription for improving Pakistan's condition while preserving its independence and identity. A key thesis of the book was the need for Pakistan to obtain nuclear weapons in order to stand up against the industrialized world and protect itself from a nuclear-armed India.[45]

As Bhutto's political movement gained strength, it drew the wrath of Ayub's government. In November 1968, Bhutto was arrested for political agitation and spent three months in jail. But the imprisonment only added to Bhutto's popularity, and upon his release he took up his campaign with renewed vigor. His activities, combined with growing calls for autonomy in East Pakistan, lead to Ayub's resignation in March 1969.[46]

Ayub handed over power to a new military junta headed by the army chief of staff, Gen. Agha Mohammad Yahya Khan. But the change in leadership did little to quell the unrest in East Pakistan. Fearing a rebellion, Yahya called for nationwide elections in December 1970. Bhutto's PPP won a majority of votes in West Pakistan, but the pro-autonomy movement dominated the more populous East Pakistan, giving them enough parliamentary seats to control the new government. When efforts to reach a power-sharing arrangement broke down over the question of autonomy — which Bhutto opposed as a threat to Pakistani unity — rioting erupted in the East, along with calls for a fully independent state of Bangladesh. In March 1971, Yahya ordered a military crackdown in the East that quickly devolved into a massacre. The army moved in with bloody efficiency, mowing down insurgents, murdering pro-independence leaders, raping women, and butchering unarmed men. Bhutto remained loyal to Yahya, declaring after the frenzy of violence began, "Pakistan is saved." He later accepted an appointment as deputy prime minister in Yahya's government and defended the slaughter in a defiant speech at the UN.[47]

After nine months of bloody civil war between Yahya's army and East Pakistani rebels, India intervened on the side of the pro-independence forces in December 1971. The Indian army quickly overwhelmed Yahya's troops, and East Pakistan emerged as the new nation of Bangladesh. Pakistan had suffered another humiliating defeat at the hands of its bitter enemy. Its military went into open rebellion. On December 20, 1971, four days after Pakistan's surrender, Yahya resigned. Before leaving office he designated Bhutto as the country's new president and martial law administrator.[48]

The Islamic Bomb

Bhutto inherited a nation on the verge of collapse. With its military vanquished, its eastern wing shorn, and more than half its population gone, Pakistan's morale had been shattered. Bhutto's first challenge, therefore, was to restore the country's sense of pride and dignity. As daunting as that task may have appeared, Bhutto was uniquely suited to tackle it. Having won the previous year's elections in the West, he could legitimately claim a mandate to lead the country. More important, perhaps, was Bhutto's sense of personal destiny. "I was born to make a nation, to serve a people, to overcome an impending doom," he later wrote.[49] Remarkably, he would do just that.

Supremely self-confident, Bhutto set about implementing his long-standing vision for Pakistan. In a flurry of domestic activity, he introduced sweeping economic and land reforms, nationalized key industries, and overhauled the education system. He then lifted martial law and initiated work on a new constitution that would provide a foundation for Pakistan's transition to civilian, democratic rule. In doing all this, writes former US State Department South Asia specialist Dennis Kux, "Bhutto enjoyed his finest hour."[50]

On the international front Bhutto set Pakistan on a course of nonaligned neutrality. Even before taking office, he began mending fences with the United States, informing President Nixon and other US officials that his days of Yankee bashing were behind him, and that he now sought good relations with Washington. Yet Bhutto made clear that he would not be a US lapdog. He maintained and even strengthened ties with China, and established diplomatic relations with North Vietnam and North Korea. He also withdrew Pakistan from the British Commonwealth and from the US-backed Southeast Asia Treaty Organization (SEATO).[51] "Pakistan was once called the most allied ally of the United States," Bhutto later quipped, referring to Ayub Khan's well-known comment. "We are now the most nonallied."[52]

Beyond his ambitious reforms of domestic and foreign policy, Bhutto's most dramatic effort to restore his country's honor came in the nuclear realm. Long the leading proponent of Pakistan acquiring nuclear weapons to offset India's superior strength, Bhutto now viewed development of the bomb as an opportunity to bring glory to his country and his fellow Muslims. The goal, he later wrote, was for Pakistan to have a "full nuclear capability," thereby setting "Islamic civilization" on an equal footing with "the Christian, Jewish and Hindu civilizations," as well as the "communist powers."[53] Just one month after taking power, Bhutto embarked on a mission to build an Islamic bomb.

On January 20, 1972, Bhutto convened a secret meeting of Pakistan's top scientists and government officials in the city of Multan, a commercial and industrial hub near the center of the country in Punjab province. It was a beautiful, clear day, and Bhutto assembled the group of about fifty under a canopy on the lawn of an old British colonial mansion. The scene was reminiscent of an old-time revival meeting, with the Reverend Bhutto preaching fire and brimstone of a decidedly man-made variety. Summoning all of his great oratorical skill, he began by recounting Pakistan's humiliation at the hands of hated India and pledged to remove the stain of shame from his country. Pakistan would build an atomic bomb, he declared, and they, the assembled scientists, would be the ones to build it.[54]

It was an audacious vision. Pakistan had almost none of the resources needed for such a venture. It had no plutonium, no highly enriched uranium, little in the way of infrastructure or know-how, and, most significantly, no money. Yet Bhutto had laid down the challenge.

The scientists were dumbfounded. The president, the man they looked upon as a savior following the calamitous defeat by India, was asking them to build a nuclear bomb. It was a moment of which many of them had long dreamed. Some of the younger scientists, in particular, had complained for years about the government's failure to match India's efforts to develop a bomb. Now they would have an opportunity to engage in a patriotic mission to bring security and dignity to their country.[55]

"Can you give it to me?" Bhutto asked.

"Oh, yes, yes, yes," the scientists shouted. "You can have it."

"How long will it take?" Bhutto demanded.

The scientists hesitated. A debate broke out. The difficulties would be great, they said. A few expressed doubts it could be done at all. Predicting how long it would take would be nearly impossible. Finally, one of the scientists suggested that it might be done in five years.

"I want it in three years," Bhutto responded, thrusting three fingers in the air.[56] "I know you can do it. If the Americans can do it in three years, we also can."[57]

One of the scientists jumped to his feet. "It can be done in three years," he shouted.

Another scientist broke in. "It isn't like making firecrackers," he said. "We don't know how long it will take."

Another debate ensued. Bhutto listened. "So, can you do it?" he asked at last.

"Yes, we can do it," the scientists answered, "given the resources and given the facilities."

"I shall find you the resources and I shall find you the facilities," Bhutto replied.[58]

The scientists broke out in cheers. They danced and shouted, "Allahu Akbar! Allahu Akbar!" (God is great!), the ancient Muslim war cry.[59]

Bhutto would have his bomb.

Brotherhood of the Bomb

Just hours after rallying his scientific troops at Multan, Bhutto set off on a dizzying tour of world capitals. The twenty-nation junket was designed to bolster Pakistan's relations with its diplomatic and political partners. It would also set the stage for obtaining the promised financing necessary for the bomb program.

Bhutto's agenda focused heavily on Muslim and Arab states, several of which had provided assistance to Pakistan during the recent war with India. Stops included Libya, Iran, Saudi Arabia, Egypt, Syria, Jordan, and the United Arab Emirates, as well as Morocco, Algeria, Tunisia, and Turkey. In speeches along the way, Bhutto railed against what he described as the West's betrayal of his country and called for Islamic solidarity.[60] But Bhutto's trip did more than simply foster Muslim goodwill. It also cemented relationships that would prove vital not only in Pakistan's development of the bomb, but also in the creation of A.Q. Khan's nuclear supermarket.

Among the most portentous stops on Bhutto's tour was Libya. There, the new Pakistani president received a warm welcome from Libya's new revolutionary leader, Colonel Muammar el-Qaddafi. Despite their very different backgrounds, the two men quickly became close friends. Qaddafi, the son of nomadic Bedouin tribesmen, had come to power three years earlier, at age twenty-seven, in a military coup that had toppled the rickety government of Libya's King Idris. Qaddafi's ambition, like Bhutto's, was unbounded, and he shared the Pakistani leader's desire to dramatically enhance his country's position on the world stage. He also shared Bhutto's interest in acquiring nuclear weapons.[61] In early 1970, just months after taking power, Qaddafi naively sent an envoy to China to inquire about buying an atomic bomb. The Chinese politely explained that nuclear weapons were not generally available for sale. They did, however, express a willingness to assist Libya in the area of nuclear research.[62]

But Qaddafi was determined to buy a bomb, and in Bhutto's Pakistan he saw the perfect opportunity. By the time of Bhutto's visit, Libya was on the verge of becoming enormously wealthy, as Qaddafi led the effort of the oil-producing states that would soon break the hold Western oil companies had on the mineral resources of the Middle East. But while Libya had money, it lacked the technical and scientific infrastructure necessary for a nuclear weapons program. Pakistan, on the other hand, while poor, had — thanks to the Atoms for Peace program — a large number of highly skilled nuclear scientists and technicians.[63] For Bhutto and Qaddafi the way forward was clear. They would enter into their own version of a nuclear cooperation agreement.

Bhutto's visit to Tripoli led to a series of secret meetings in Paris in which Libyan and Pakistani officials worked out details of the nuclear pact. According to the 1981 book *The Islamic Bomb*, which first reported the meetings, representatives of several Gulf states that Bhutto had visited during his grand tour, including Saudi Arabia and possibly Iran, also participated in the sessions. The outlines of the accord were straightforward: Pakistan would build a nuclear weapon; Libya would provide the funding. But Qaddafi's representative made clear that Libya expected "full access" to the nuclear technology Pakistan developed. Qaddafi also wanted Pakistan to train Libyan scientists in the bomb-making arts. Saudi Arabia and the other Gulf states involved in the meetings also pledged money to the bomb-building effort, although they gave no indication they sought to take possession of an actual weapon and apparently preferred to leave the technology in the hands of the Pakistanis.[64]

Bhutto and Qaddafi personally finalized the bomb deal in February 1974. Qaddafi had traveled to Pakistan to attend the Organization of the Islamic Conference summit in Lahore. Bhutto, who had been elected prime minister the previous year, rolled out the red carpet for Qaddafi, taking the Libyan leader on tours of Pakistani military bases and armaments factories, as well as a cricket stadium that had been renamed in Qaddafi's honor. Bhutto also arranged for the head of the PAEC to take Qaddafi on a private tour of Pakistan's new Canadian-built nuclear power reactor in Karachi.[65] Qaddafi reciprocated the hospitality by declaring that Libya was prepared to sacrifice its peoples' blood if Pakistan was ever threatened. "Our resources are your resources," he told a huge crowd assembled at Qaddafi Stadium.[66]

But beyond these public displays of camaraderie and support, the real business of the trip was nuclear trading. In a series of private discussions, Bhutto and Qaddafi hammered out the final details of the bomb agreement. Qaddafi reportedly promised Bhutto as much money as was needed for the project. In return, he asked that Libya receive the first weapon. The two men sealed the deal with a handshake.[67] The bomb program took on added urgency for Pakistan just three months later, when India tested its first nuclear device.

By early the following year, Qaddafi was openly boasting of his plans to purchase nuclear weapons. "Tomorrow, we shall be able to buy an atom bomb and all its component parts," the French newspaper *Le Point* reported him saying in January 1975. Later that year, special couriers sent by Qaddafi reportedly began arriving in Karachi with suitcases filled with as much as $100 million, earmarked for Pakistan's nuclear program. But for all of Qaddafi's largesse, the question of access to the bomb and bomb-making technology would become a source of contention between Libya and Pakistan in the years ahead, as the Libyan leader became increasingly convinced that he was not getting everything he had bargained for.[68]

Libya was not the only country that contributed to Pakistan's bomb program. Bhutto's 1972 tour helped shore up relations with other nations that would play important roles in the project. Saudi Arabia was particularly supportive. The Saudis, like the Libyans, were oil rich but technologically poor. For them, backing Pakistan's development of the bomb was an opportunity to have access to the technology and, potentially, security guarantees from Islamabad, without having to run the risks associated with housing a nuclear weapons program on their own soil. They also shared Bhutto's vision of Muslim uplift through an Islamic bomb. King Faisal reportedly took a personal interest in the project and became a major backer of Pakistan's effort to build the weapon.[69] The Saudis would also help finance Pakistan's purchases of nuclear and missile technology from China.[70] While there is no public evidence Saudi Arabia expected access to any weapons Pakistan produced, French and US intelligence officials who monitored contacts between the Saudis and Bhutto's regime became convinced that Riyadh viewed Islamabad's nuclear program as an opportunity to have a proxy atomic bomb. "The Saudi attitude was [that] they could not afford to publicly undertake a program," says a top French intelligence official, "but the most extreme elements in the Arab world would be reminded that it was the Saudis who were footing a large portion of the bill."

Bhutto had undertaken his Islamic diplomacy mission at an opportune moment. By the early 1970s, Israel was widely known to be on the verge of attaining a nuclear capability, a prospect that created consternation in the Middle East. Those concerns intensified in the wake of Israel's decisive victory in the 1973 war against a coalition of Arab nations. The defeat was also a stinging blow to Muslim and Arab pride. For the Gulf states, many of them newly rich as a result of the 1973 oil crisis, support for Pakistan represented an opportunity both to restore Islamic dignity and to place the Muslim and Arab world on an even footing with the Jewish state.[71]

Whether explicitly to support development of the bomb or as a show of Muslim solidarity, Arab and Gulf-state money began pouring into Pakistan in 1973. The shah of Iran was the largest known contributor, donating hundreds of millions of dollars in just a two-year period.[72]

According to a high-level US intelligence official assigned to Tehran in the 1970s, the shah viewed the Pakistani program much as the Saudis did. "A pan-Islamic bomb program was something even Persian Shiites and Arab Sunnis could agree on," he said.

Overall, Pakistan received several billion dollars in donations from Muslim and Arab nations during the mid-1970s, just as Bhutto's bomb program was gaining momentum.[73] Much of the Gulf-state money destined for Pakistan coursed though the Bank of Credit and Commerce International (BCCI), a shadowy, Pakistani-held institution with close ties to Islamabad and Persian Gulf sheiks that — in addition to serving as a hub for terrorist financing, arms trafficking, money laundering, and funding of US and Saudi covert operations — helped finance Pakistan's bomb project.[74]

For all of Bhutto's talk of an Islamic bomb, it remains an open question whether he would actually have shared nuclear weapons with his Muslim and Arab backers. While the concept may have made a good fundraising pitch, Bhutto was surely clear-eyed enough to realize that such a move would have been not only highly provocative, but also detrimental to Pakistan's own security. Still, it is not unreasonable to imagine that once he had a working weapon in hand, the temptation to earn extra cash for his impoverished nation by placing the bomb on the auction block would have been considerable. But Bhutto's former press secretary, Khalid Hasan, a man privy to Bhutto's thinking on the issue of the bomb, insists the prime minister would not have taken such a step. "He would never have done that," Hasan told the authors of *The Islamic Bomb*. "He would have used it [the bomb] for diplomatic and political advantage with the Arabs, and even with the Western countries. But he would never have put it up for auction."[75] Yet despite Hasan's certainty, what Bhutto would actually have done can never be known, since he was overthrown and hanged before the bomb project came to fruition. What *is* known is that once the bomb was completed, other Pakistanis would have had less compunction about selling it than Hasan ascribes to Bhutto.

Whatever Bhutto might have done in the future, his efforts to attain and fund the bomb did not escape the attention of the United States. In January 1975, a secret State Department background paper reported that Pakistan had begun a nuclear program and was negotiating for the purchase of facilities that would allow it to produce enough plutonium to build an atomic weapon. It also noted that "Arab financing" was available for the project and that the Pakistanis appeared "quite prepared to proceed to a weapons capability."[76] Despite this, the very next month the US chose to lift the embargo on military aid that had been imposed on Pakistan and India during the 1965 war.[77] In early 1976, at the time that George H.W. Bush became director of the CIA, US intelligence estimates concluded that Pakistan was involved in "a crash program to develop nuclear weapons."[78] By May, an intelligence report found that Libya had "agreed to finance the Pakistani [plutonium] reprocessing project in return for some unspecified future nuclear cooperation."[79] Still, US aid to Pakistan continued.

3

The Age of Proliferation

AT 8:05 IN THE morning on May 18, 1974, a huge underground explosion rocked the desolate northwestern Indian village of Pokhran, ninety miles from the Pakistani border. A giant dome of sand and earth rose above the desert floor. From a distance, Indian Prime Minister Indira Gandhi looked on in approval. Within minutes, a prearranged message went out to the Foreign Ministry in New Delhi: "The Buddha is Smiling." India had successfully conducted its first nuclear test.[1]

For India, the Pokhran test was cause for celebration. Headlines around the country lauded the feat as proof of Indian ingenuity and resolve. The *Indian Express* declared that the blast had catapulted the country to the "front rank" of nations. India could no longer be "dismissed as a 'pitiful giant,'" it added.[2] Elsewhere, the test was viewed with alarm. The fact that a desperately poor, developing nation had built such a weapon by piggybacking off its civilian nuclear power program suggested others might soon follow suit. The long-ignored danger inherent in the Atoms for Peace concept was now all too apparent. Austrian journalist and atomic weapons expert Robert Jungk deemed the Indian test the opening of "the Second Atomic Age, the Age of Proliferation," an era in which nuclear weapons would become widespread.[3]

In an attempt to allay such fears India portrayed the test as a "peaceful nuclear explosion." The country's Atomic Energy Commission insisted New Delhi had "no intention" of producing nuclear weapons. And Indira Gandhi told reporters that her government was "firmly committed" to using atomic energy only for peaceful purposes. "There is nothing to get excited about," she said.[4]

But for Pakistan, there was a great deal to get excited about. Its much larger neighbor and longtime enemy — a country with which it had already fought three wars in a little more than twenty-five years — had just acquired the most powerful destructive force on earth. Bhutto decried the Indian test as a "fateful" development. "A more grave and serious event has not taken place in the history of Pakistan," he said. "Pakistan will not succumb to nuclear blackmail."[5] Bhutto also challenged India's claim of peaceful intent. "It is an incontrovertible fact, stressed by the nuclear powers themselves, that there is no difference between tests for so-called peaceful purposes and military purposes," Bhutto said. "The technology is the same."[6] It was a point Bhutto understood only too well.

The Indian test added new urgency to Bhutto's longstanding quest for nuclear weapons. The previously hypothetical threat of a nuclear-armed India had become a reality. As a result, the bomb program begun two years earlier at Multan was now the top national priority. Publicly, Bhutto insisted that while Pakistan would continue to develop its "peaceful" nuclear program, it would not seek to explode a bomb, despite India "brandishing the nuclear sword."[7] But behind the scenes, he pursued a very different course. "Now, nobody is going to stop us from taking a similar step," he reportedly said.[8]

Bhutto issued the official go-ahead for the bomb project just weeks after the Pokhran test. He announced the decision in mid-June at a meeting of his top military, political, and scientific advisors in Rawalpindi, the military town adjacent to Islamabad. The session took place at Prime Minister House, Bhutto's official residence, which had once served as the headquarters of the British military commander for the region. In typically flamboyant style, Bhutto recounted how a hostile and powerful India, a country that had played a major role in Pakistan's loss of its eastern wing, had now added nuclear capability to its overwhelming conventional superiority. The situation, he said, represented a dire security threat for Pakistan. Bhutto went on to remind the participants of his own calls to develop nuclear weapons years earlier and of Ayub Khan's failure to heed his advice.[9]

The group also learned that diplomatic efforts to win security assurances for Pakistan in the wake of the Indian test had come to naught. China, France, Great Britain, and the United States had all refused to guarantee Pakistan's security in the event of an Indian attack. And in a further insult, each nation — itself a nuclear power — insisted that while it deplored India's decision to test an atomic device, Pakistan had a moral obligation not to follow suit.[10] But for Bhutto and his advisors, the message was just the reverse: If the declared nuclear-weapons states would not provide protection, Pakistan had no choice but to protect itself. Pakistan would proceed with the bomb.

The Plutonium Route

Although the official decision to build the bomb came on the heels of India's nuclear test, efforts to develop the weapon had been ongoing since the Multan meeting more than two years earlier. While Bhutto had focused on raising funds for the project, the Pakistan Atomic Energy Commission (PAEC) had been hard at work doing the research and development necessary to make the bomb a reality. Hundreds of newly minted nuclear scientists and engineers had returned from training in the West to join in the great effort.

One of the first issues the PAEC bomb makers confronted once Bhutto set them on their mission at Multan was determining the type of weapon they would build. They had two options. One was to create a bomb based on highly enriched uranium (HEU). The other was to build a plutonium device. Producing the nuclear explosive material for either type of weapon would involve mastering a number of highly sophisticated and complicated technologies.

Both HEU and plutonium are derived from uranium, a naturally occurring, mildly radioactive element found in many parts of the world. Natural uranium is made up of two forms, or isotopes, of the element. One is U-238, the relatively benign form that accounts for 99.3 percent of the raw element. The other is U-235, which makes up the remaining 0.7 percent of raw uranium and is the only isotope found in nature in any appreciable quantity that can be readily split, or fissioned, to create a nuclear chain reaction.

Producing highly enriched uranium involves separating the two isotopes and creating high concentrations of U-235. This is done through one of two processes. In the gaseous diffusion method, raw uranium is converted to gas and forced though a semiporous membrane, creating a slight separation of the U-238 and U-235 isotopes. By repeating the process through thousands of stages, successively higher concentrations of U-235 are obtained. The other method of uranium enrichment involves the use of large numbers of interconnected ultra-high-speed centrifuges. In this process, gaseous uranium, known as uranium hexafluoride or UF_6, is channeled into specially designed cylinders capable of spinning faster than the speed of sound. The tremendous centrifugal force created by the rotation pushes the gas molecules containing the heavier U-238 isotopes toward the outside of the cylinder, while the molecules containing

U-235 collect closer to the center. The stream containing the slightly higher concentration of U-235 is withdrawn and passed on to the next centrifuge in a string of thousands, and the process is repeated until the desired concentration is achieved. Uranium enriched to 3 percent U-235 is sufficient to run a nuclear power reactor. Uranium enriched to 93 percent U-235 can be used to make a bomb. The first of only two nuclear weapons ever used in warfare, the bomb dropped on the Japanese city of Hiroshima by the United States in August 1945, was a highly enriched uranium device. That bomb, code-named Little Boy, used forty-four pounds of HEU and created a blast equivalent to about fifteen thousand tons of TNT. Subsequent improvements in design have allowed for much higher explosive yields from equivalent amounts of fissionable material.

The second option for the PAEC scientists was to build a plutonium bomb. The plutonium isotope required for a nuclear weapon, Pu-239, does not exist in nature. It is a man-made element that is produced as a byproduct of burning uranium fuel in a nuclear reactor. The newly created Pu-239 can be harvested through a technique known as reprocessing. This involves removing the spent and now highly radioactive fuel from a reactor, chopping it up, and dissolving it in nitric acid. The Pu-239 is then separated from the spent uranium and other ingredients of this toxic soup through a chemical process developed by the United States and promoted around the world as part of the Atoms for Peace program.[11] The resulting plutonium can be used either as fuel for a nuclear reactor or as the explosive material for a nuclear weapon. Plutonium was used as the core for the second (and, so far, last) atomic device used in war, the bomb dropped by the United States on the city of Nagasaki, three days after Hiroshima. The Nagasaki bomb, code-named Fat Man, used about thirteen pounds of plutonium and yielded an explosion equivalent to twenty-one thousand tons of TNT. India based its bomb program on plutonium.

After weighing the pros and cons of both routes to the bomb the PAEC chose to pursue a plutonium weapon. The scientists believed it would prove cheaper and easier than developing enrichment technology. Many of the elements needed for plutonium production were within relatively easy reach.[12] The primary challenge would be acquiring a reprocessing plant. And for that, as with much of the rest of its nuclear program, Pakistan would turn to the West.

Responsibility for obtaining the all-important reprocessing facility fell to the man Bhutto had tapped to head the PAEC at the time of the Multan meeting, Dr. Munir Ahmad Khan. Khan, a US-trained nuclear engineer who had just returned from a post at the UN's nuclear watchdog agency in Vienna, the International Atomic Energy Agency (IAEA), made the task his first order of business. For several years prior to Khan's arrival, the PAEC had been engaged in fruitless negotiations with both Britain and France for supply of a reprocessing plant. The discussions had stalled as the PAEC debated the merits of each country's offer. The British proposed selling Pakistan a small-scale plant incapable of producing sufficient plutonium for a bomb. Some within the PAEC argued that such a facility could be upgraded after it was built to meet Pakistan's weapons needs. Others argued that such a solution was at best impractical and at worst unrealistic. They favored accepting the French offer to supply a much larger plant capable of producing weapons-grade plutonium without alteration.[13]

Upon taking over the PAEC, Munir Khan quickly broke the logjam. Despite the objections of some PAEC engineers, he ended further consideration of the British proposal and stepped up negotiations with the French.[14] In March 1973, Pakistan signed a contract with the French firm Saint-Gobain Techniques Nouvelles (SGN), the world's chief exporter of reprocessing technology, for preliminary designs of a reprocessing facility.[15] But from his years at the IAEA, Khan must clearly have understood this move would set off concerns in the West about Pakistan's nuclear ambitions. Even as the deal with the French went forward he authorized secret negotiations with a Belgian firm for a small-scale reprocessing facility. Although Pakistan finalized the agreement for the French plant in October 1974, just six months after India's nuclear test, Khan kept the Belgian option open. After the West's predictably negative response to the French deal the Belgian backup plan became a crucial factor in Pakistan's eventual acquisition of a reprocessing plant.[16]

While efforts to obtain a reprocessing facility were underway, PAEC scientists were busy working out the technical details of building an actual bomb. Much of their activity focused on weapons design. Despite the PAEC's aggressive training campaign during the 1960s, at the time Munir Khan took over the commission it still did not have a single scientist schooled in the fine art of bomb making. To address that shortcoming, Khan established a Theoretical Group within the PAEC to carry out research and development on the bomb.[17]

The man Khan chose to lead the Theoretical Group was Dr. Riazuddin, a Western-trained mathematician and physicist who headed the physics department at Quaid-i-Azam University in Islamabad. Riazuddin had earned his doctorate from Cambridge University in 1959 and had also studied at Imperial College in London under the renowned Pakistani physicist and future Nobel Prize winner, Dr. Abdus Salam.[18] Salam, who had played a central role in establishing the PAEC, later served as scientific advisor to both Ayub Khan and Ali Bhutto. It was Salam who, following the Multan meeting, in which he had been a participant, informed Riazuddin of Bhutto's decision to build the bomb and that he, Riazuddin, would lead the PAEC team charged with designing the weapon.[19]

Just months after taking up his new post, Riazuddin paid a visit to the United States. It was familiar turf for the physicist, who had served as a research associate at several American universities during the 1960s and early 1970s. But on this trip he was engaged in research of a different nature. During his stay, Riazuddin obtained copies of all the declassified reports on the Manhattan Project available from the Library of Congress and other US repositories. He then returned to Pakistan, reports in hand, to plunge back into the work of building a bomb.[20]

Throughout 1973 and early 1974, Riazuddin and his team made steady progress. After determining early on that plutonium was the preferred route to the bomb, the scientists began working out the optimal method of detonation and the type of explosive to be used to initiate a chain reaction. Early designs showed considerable promise. During the June 1974 meeting at which Bhutto announced the decision to build the bomb in the wake of India's nuclear test, Riazuddin briefed the participants on the work of his group and gave an upbeat assessment of the prospects for quickly delivering a working weapon. He also brought them up to date on the status of the negotiations with France for the plutonium reprocessing facility.[21]

The first outward signs of Pakistan's stepped-up nuclear efforts came in October 1974 when Islamabad announced plans to build as many as twenty-four new power reactors by the year 2000 and to master all aspects of the production and use of nuclear fuel. That same month, Pakistan signed the agreement with France for the long-sought reprocessing plant.[22]

The United States kept a close eye on these developments. In January 1975, in preparation for a visit by Bhutto to Washington the following month to discuss lifting the ten-year-old ban on military aid, the State

Department drafted a secret background paper on Pakistan's nuclear program. It noted Islamabad's plans to build more power reactors, the deal for the French reprocessing plant, and Pakistan's efforts to obtain additional nuclear facilities from the West. Once in place, the paper said, that infrastructure would give Pakistan "a virtually independent nuclear fuel cycle and the opportunity to separate sufficient plutonium to build a nuclear weapon." The paper added that because of India's nuclear test the Pakistanis had "a solid incentive" to develop a bomb. "They may well have already decided to produce a weapon," the paper said, "and they have clearly decided to have the capability to build one."[23] Yet despite all of this the US chose to resume military aid to Pakistan.[24]

But even as these events played out, developments far removed from Washington and Islamabad had already begun to shift Pakistan's nuclear program in a new and ultimately far more dangerous direction.

Motive and Opportunity

The same forces that propelled Pakistan's quest for nuclear weapons also set in motion a crosscurrent of activity that would fundamentally alter the nature of Islamabad's bomb program and give rise to its nuclear smuggling network. Just as Pakistan's bitter loss to India in the 1971 Bangladesh war and New Delhi's 1974 nuclear test had sealed Bhutto's determination to build a bomb, they also convinced many ordinary Pakistanis, including those living abroad, of their country's need for a nuclear deterrent. Among them was A.Q. Khan.

At the time of the Bangladesh conflict, Khan was a thirty-five-year-old doctoral student at the historic Catholic University of Leuven in Belgium. Since earning his bachelor's of science degree from Karachi University in 1960, Khan, like hundreds of other bright, young, scientifically inclined Pakistanis, had been living and studying in Europe. He had first gone to Germany in 1962, where he studied metallurgical engineering at Technical University in West Berlin, then to Holland, where he earned a master's degree in engineering from the prestigious Delft Technical University in 1967, and finally to Belgium to work toward his doctorate in physical metallurgy.[25] Along the way the affable Pakistani had developed a wide range of contacts, including individuals who would later emerge as part of his smuggling network. Although living abroad and, by all accounts, blending in comfortably to Western society — even marrying a Dutch–South African woman — Khan kept close tabs on events in his homeland. Following Pakistan's defeat in the 1965 war with India, he wrote letters to several Dutch newspapers disputing what he regarded as misrepresentations of Islamabad's role in initiating the conflict.[26] Six years later, while living in Belgium, Khan intently followed news of the war in East Pakistan and was devastated by the loss to India. He later recalled watching television coverage of the humiliating surrender ceremony in which the commander of the Pakistani forces was stripped of his medals and insignia by an Indian general. Scenes followed of Pakistani prisoners being kicked and beaten by Indian troops. The images reportedly left Khan badly shaken. "He could not concentrate on his work for many days," according to Zahid Malik, Khan's authorized biographer. "All he thought of was to make Pakistan so strong that it would never have to face such a trauma again."[27] He would soon have an opportunity to help do just that.

In early 1972, just months after the loss to India and Bhutto's subsequent declaration at Multan that Pakistan would build a bomb, Khan received a job offer from a Dutch engineering firm known as the Physical Dynamics Research Laboratory (FDO). Although Khan had completed his doctorate and was looking for work, the position with FDO was not one for which he had applied. Rather, the offer came about through the scientific old-boy network. One of Kahn's former classmates from Delft headed the metallurgical section at FDO and was casting about for a good, young engineer to work on an important new project. He contacted Khan's mentor at Leuven, Professor Martin Brabers, seeking recommendations. Brabers — who later became the first head of a Pakistani institute closely associated with Khan and Islamabad's nuclear program — suggested his protégé, and Khan's old schoolmate extended the offer.[28] It was an early indicator of the networking skills for which Kahn would become infamous.

For both Khan and his wife, Henny, the offer from FDO was appealing. It would allow them to return to Holland, the country where they'd met and been married, and to be near Mrs. Kahn's parents, who lived in The Hague.[29] But for Khan, the job held added benefits.

FDO was a subcontractor to one of the partners in a major European nuclear project. Two years earlier, the governments of Britain, West Germany, and the Netherlands had joined forces in an effort to end

European dependence on the United States for the enriched uranium fuel used in power reactors. To do so, the three nations formed a consortium known as Urenco and built a state-of-the-art enrichment facility in the Dutch town of Almelo, near the German border. There, Urenco scientists developed highly classified ultra-centrifuges capable of meeting Europe's enriched-uranium needs.[30] But those same centrifuges could also be used to produce the highly enriched uranium used for atomic bombs, a fact that was not lost on the Europeans, who sought to keep their nuclear weapons options open, or on A.Q. Khan, who would soon gain access to Urenco's most sensitive technology.[31]

Easy Access

Khan, his wife, and their two young daughters, Dina and Ayesha, arrived in Holland in April 1972. They moved into a small, two-story house in the village of Zwanenburg on the outskirts of Amsterdam, not far from FDO's headquarters. The girls enrolled in school, Khan and the Dutch-speaking Henny made friends with the neighbors, and the family settled into a quiet, seemingly unexceptional existence. On weekends and holidays they might drive to the shore or to visit Khan's friends in Leuven. Khan also joined in neighborhood volleyball games. The Khans were, by all appearances, a perfectly ordinary family.[32]

Khan's work life also appeared unexceptional. He started at FDO on May 1, 1972.[33] He had been hired under FDO's contract with Ultra-Centrifuge Nederland NV (UCN), the Dutch partner in Urenco, to evaluate the high-strength metals used for centrifuge components.[34] It was an ideal opportunity for the young metallurgist, allowing him to expand his understanding of the properties of specialized metals while learning about the workings of the new top-secret centrifuges.[35] He would prove an unusually attentive student.

Due to the extremely sensitive nature of the Urenco project, the partner states had agreed to adopt strict security measures. Complete background checks were required for all employees of Urenco and for employees of all its subcontractors who were not British, Dutch, or West German citizens. For Khan, this proved no obstacle. The Dutch internal security service and the Ministry of Economic Affairs ran checks on him but found nothing out of the ordinary. He was granted a limited security clearance known as "Secret Inclusive." A Dutch government investigation later determined that because Khan was assigned to an area classified merely as "restricted" and had indicated he hoped to become a citizen of the Netherlands, only a cursory background review had been conducted.[36]

Despite Khan's relatively low-level security clearance, in practice his movements at FDO were almost entirely unrestricted. The atmosphere was one of easygoing collegiality, and the charming young Pakistani quickly made friends with his fellow workers. Few paid any attention to his comings and goings. In that respect, he was like any other FDO employee. As the authors of the *Islamic Bomb* later explained, "No one at FDO seems to have taken security as a serious matter." That lax attitude would soon result in Khan gaining access to the Urenco centrifuge plant itself.[37]

After only a week on the job, FDO sent Khan on a visit to the Urenco facility in Almelo. Khan's managers wanted him to have a firsthand look at Urenco's operation and develop a better understanding of the new ultra-centrifuges and the enormous stress the machines placed on the metals he would be evaluating. Although Khan was not cleared to enter the ostensibly high-security plant, he was nonetheless admitted. He spent two days touring Urenco, learning its procedures, and meeting with the scientists and engineers who worked on the centrifuges. It was the first of many visits.[38]

Khan also had access to Urenco centrifuge data at FDO. As part of its contract with Urenco, FDO translated highly classified technical documents. Khan, who as a result of his long study in Europe had become literate in German and English and had a solid command of French and Dutch, was often asked to assist. With FDO's permission, moreover, he frequently took the documents home, although this was a clear violation of security protocols. Over the next several years, Khan would enjoy easy access to highly sensitive, detailed information about some of the world's most advanced uranium enrichment technology and, equally important, to the lists of manufactures supplying the centrifuge components.[39]

A Spy Is Born

The lax security at FDO and Urenco might have been dismissed by history as a potentially dangerous but ultimately insignificant matter had it not been for the influence of events taking place far from Holland.

Just as India's nuclear test propelled Pakistan's fledgling bomb-building effort from an important strategic objective to an urgent national imperative, it also transformed Khan from an obscure expatriate scientist into a prospective nuclear spy. Under those circumstances, the failure to adequately protect Urenco's centrifuge secrets took on new significance.

Khan had been working at FDO for two years when India detonated its nuclear device in May 1974. Upon hearing the news, Khan, according to his authorized biography, resolved to make "every possible effort" to help his country.[40] "The young Pakistani scientist, working his life away in an alien country, ardently believed that if he could place his knowledge and expertise at the feet of his country, Pakistan would be put on the road to a nuclear status," writes Zahid Malik.[41] In fact, Khan would provide Pakistan with much more than just knowledge and expertise.

Khan's first known attempts to offer his services to Pakistan were awkward and amateurish. Malik reports that following the Indian test, Khan began meeting with Pakistanis who were visiting Holland, and expressing his desire to return to his homeland. These efforts, however, proved fruitless. In one instance, he approached a pair of Pakistani military scientists who were in Holland to purchase a wind tunnel and told them of his wish to help Pakistan become "competent in nuclear technology." The visitors offered Khan no encouragement, saying his talents would be unappreciated in Pakistan and he would have difficulty finding employment.[42]

Undeterred, Khan took a bolder approach. In the summer of 1974, he sent a letter to Prime Minister Bhutto through the Pakistani ambassador in Belgium. In it, Khan presented his credentials, highlighting his knowledge of centrifuge enrichment technology, and urged Bhutto to consider enriched uranium as the preferred path to the bomb. Khan argued that a uranium weapon could be produced more cheaply than a plutonium bomb. He also suggested that he had the expertise to handle the job.[43]

Khan's argument apparently resonated with Bhutto. Already concerned that the PAEC's plutonium program was in trouble, due to Canada's decision in the wake of India's nuclear test to begin withholding support for the reactor it had built for Pakistan, Khan's proposal for an enriched uranium weapon undoubtedly sounded like a welcome alternative.[44] "He seems to be talking sense," the prime minister scrawled in the margin of Khan's letter. Bhutto then sent the missive on to Munir Khan, the leading advocate for a plutonium bomb, asking him to arrange a meeting with the brash young scientist. The PAEC chief dutifully agreed to do so, but before proceeding asked his liaison at the Pakistani embassy in Belgium, Siddique A. Butt, a British-trained physicist who had been assigned to the diplomatic corps to help arrange the purchase of a plutonium reprocessing plant for Pakistan's bomb program, to look into A.Q. Khan's claims. After investigating the matter, Butt reported that A.Q. Khan was, in fact, engaged in "pioneering work on centrifuges" and could be "helpful in Pakistan's nuclear program." Butt himself would go on to develop an elaborate nuclear procurement network that played a central roll in Pakistan's development of the bomb.[45]

Although still lukewarm to the idea of an enrichment program, upon receipt of Butt's findings Munir Khan asked one of the PAEC's top scientists, Sultan Bashiruddin Mahmood, a British-trained nuclear engineer who had participated in the Multan meeting, to prepare a report on the feasibility of developing a uranium bomb. Mahmood's analysis, presented to Munir Khan in November 1974, found that with a centrifuge-enrichment plant Pakistan could produce enough weapons-grade uranium to build a bomb by 1979.[46] Mahmood would soon become the first director of Pakistan's uranium enrichment program. Decades later, he and another Pakistani nuclear scientist would be detained by Pakistani authorities after meeting with Osama bin Laden and other senior al Qaeda figures and engaging in discussions about the terrorist organization's efforts to obtain nuclear, chemical, and biological weapons. Today, Mahmood is reportedly under permanent house arrest.[47]

Munir Khan hand-delivered Mahmood's report to Bhutto in a private meeting at the prime minister's personal residence in his hometown of Larkarna. Based on Mahmood's findings, Bhutto gave preliminary approval to proceed with the development of a centrifuge enrichment program.[48]

Having received his marching orders from Bhutto, Munir Khan began making arrangements to meet with A.Q. Khan upon the latter's return to Pakistan. In the interim, the aspiring spy had gained access to information that would prove invaluable to Pakistan's uranium enrichment effort.

The Brainbox

In the period just after A.Q. Khan vowed to help Pakistan acquire a nuclear capability, a unique opportunity came his way. In October 1974, Urenco enlisted Khan's help in translating a top-secret report on a major advance in uranium enrichment technology. It concerned a technical breakthrough in the design of a sophisticated German centrifuge known as the G-2, a machine capable of enriching more uranium faster than any other then in existence.[49] UCN and Urenco were preparing to switch their centrifuge program from an earlier Dutch model to the new German machine and were anxious to have the report translated quickly.[50]

To work on the report, Urenco assigned Khan to a high-security section of the Almelo facility called the brainbox, where the consortium's most sensitive work was carried out. Although strict security procedures existed for the brainbox, they were almost universally ignored. As a Dutch government investigation later determined, the brainbox was essentially "an open environment."[51] Scientists and engineers talked freely about their work and highly classified information was left unprotected. For sixteen days, over a period of several weeks, Khan had virtually unlimited access to both the brainbox and Urenco's main centrifuge plant. He apparently took full advantage of the situation. Colleagues often saw him wandering the premises, notebook at the ready. In one instance, a member of the staff spotted Khan jotting notes in a foreign language, presumably Urdu. Asked what he was writing, Khan smiled and said it was a letter to his family in Pakistan.[52]

Strange things were happening around Khan's home as well. In the latter part of 1974, around the time S.A. Butt was conducting his investigation of Khan, neighbors began noticing cars with diplomatic license plates from France and Belgium parked at the house. The visitors often stayed until the early hours of the morning. It was out of the ordinary, but no one thought too much of it at the time. Only years later did the significance become apparent.[53]

In December 1974, Khan and his family traveled to Pakistan for their annual holiday visit. But this trip was hardly routine. During the stay, Khan reportedly met with both Bhutto and Munir Khan. According to Malik, in the course of the discussions with the prime minister, Khan argued that the deal for the French plutonium reprocessing plant had been ill advised and would not give Pakistan a nuclear weapons capability. As he had in his letter to the prime minister earlier in the year, Khan again urged Bhutto to consider developing a uranium bomb, which he said could be done relatively cheaply within five years. Despite some initial skepticism, Bhutto apparently found merit in Khan's arguments. In the end, Malik writes, the prime minister encouraged Khan to help Pakistan establish a uranium enrichment facility.[54]

After the first of the year, Khan met several times with Munir Khan in Karachi. The two men huddled in secret sessions for hours at a time, discussing the fine points of plutonium versus uranium weapons.[55] The PAEC chief allowed the eager young scientist to understand that his services could indeed be valuable to Pakistan but that he should continue his work in the Netherlands for the time being and learn as much as possible about the uranium enrichment process.

Toward the end of his stay in Pakistan, A.Q. Khan toured several major PAEC installations, including the Pakistan Institute of Nuclear Science and Technology (PINSTECH) outside Islamabad, where the American-made research reactor was housed. He reportedly came away unimpressed. According to Malik, Khan concluded that while Pakistan possessed the basic infrastructure of a nuclear program, it was nowhere near building a bomb. Nevertheless, because the government was committed to the bomb project and was prepared to provide the necessary facilities, Khan believed that with hard work the "objective could be achieved." He was not shy about sharing his views with Munir Khan and offered extensive suggestions about which facilities were required and how they might be obtained. A.Q. Khan returned to Holland convinced, according to Malik, that the "moment he had been waiting for" was at hand and that he would soon be called to serve his nation. In the interim, he reportedly spent all his spare time "reading articles" about enrichment technology.[56] In fact, he was doing much more than just reading about the subject.

Project 706

On February 15, 1975, just weeks after his meetings with A.Q. Khan, Munir Khan presented Bhutto with a formal proposal for the establishment of a uranium enrichment program. The $450-million plan called for construction of a centrifuge plant, development of a uranium mine in Pakistan's Punjab province, and the creation of a facility for the production of uranium gas.[57] Together, these elements could provide Pakistan with an indigenous source of uranium fuel for a bomb. The timing of the proposal was significant. Just ten days earlier, during a visit to Washington, Bhutto had informed President Gerald Ford that he would place all of Pakistan's nuclear reactors under international safeguards as a means of ensuring that Pakistan would not secretly develop a plutonium bomb if the United States lifted its ten-year-old ban on military aid to Islamabad and provided sufficient conventional arms to meet Pakistan's security needs. Ford told Bhutto that he would give "active consideration" to the idea and, in fact, lifted the ban later that month.[58] But now, even before the embargo was lifted, Bhutto was already plotting a new route to the bomb, independent of plutonium.

Bhutto approved the enrichment plan proposed by Munir Khan. The PAEC quickly established a new unit, given the innocuous name of Directorate of Industrial Liaison (DIL), to carry out research and development on centrifuge technology. Sultan Bashiruddin Mahmood was placed in charge of the effort. But in an apparent indication of Munir Khan's lack of enthusiasm for the project, the PAEC assigned Mahmood's team to a leaky, World War II–era barracks full of bats and snakes at the Chaklala Airbase outside Islamabad.[59]

DIL's laboratory facilities were primitive. The unit, code-named Project 706, had almost none of the equipment needed for a meaningful research and development program. Early on, therefore, Mahmood traveled to Europe to shop for the necessary technology. While in Belgium he stayed at the home of S.A. Butt. A.Q. Khan was summoned from Holland to meet him, and the two men shared an attic room in Butt's house. Despite the cramped quarters, for Khan it was the opportunity he had long awaited.[60] He would soon begin supplying the information on which Pakistan's uranium enrichment program and his own career as an international nuclear smuggler were built.

After returning to the Netherlands, Khan began flooding Butt's office with information and material pilfered from Urenco. Working through intermediaries, Khan sent everything from centrifuge designs and technical literature to parts and lists of suppliers. He even sent blueprints of an entire uranium enrichment facility. In at least one instance, Khan sent Butt a discarded component from a Urenco centrifuge. That shipment was deemed so important that it was reportedly flown by military aircraft to Pakistan, where Munir Khan personally delivered it to Prime Minister Bhutto. As S.A. Butt later recalled, "A.Q. Khan did some daring things, risking his job and imprisonment."[61]

In fact, Khan's actions went beyond daring and bordered on reckless. Throughout 1975, his efforts to steal Urenco's nuclear secrets became increasingly blatant. At the office, he began speaking openly on the phone about centrifuge technology with people in Pakistan and at Pakistani embassies in Europe.[62] He also repeatedly asked his officemate, an FDO photographer and technician named Frits Veerman, to take pictures of Urenco centrifuges and components. Veerman, who was Khan's subordinate within the lab hierarchy, obliged.[63]

Khan became friendly with Veerman and sometimes invited him home for dinner or a cup of tea. Veerman recalls more than once spotting thick stacks of top-secret, blue-bound documents and highly classified centrifuge designs lying around Khan's house. Khan said that he brought them home to have his wife help with translation.[64] On one occasion, Khan asked Veerman out to the house to photograph a technical drawing he had brought home.[65] Veerman sometimes encountered other Pakistanis at Khan's house. He later reportedly learned from Dutch authorities that Khan's guests were Pakistani agents operating under diplomatic cover.[66]

Khan's behavior eventually aroused Veerman's suspicions. The conscientious Dutchman tried repeatedly to warn Urenco and FDO officials that Khan was a spy. The response was reportedly to deny that a problem existed and to scold Veerman for making such allegations against a superior.[67]

It was a pair of incidents in the fall of 1975 that finally forced FDO to pull its head out of the sand. The first was an attempt by the Pakistani embassy in Belgium to use a report that appeared to have come from UCN to order specialized wrapping foil used for ultra-centrifuges from a company in France. The French firm, Metalimphy, contacted UCN officials, who confirmed that the confidential report was indeed theirs and should not have been in the hands of the Pakistanis. Suspicion quickly fell on Khan as the source of the leak. In October, FDO alerted the Dutch intelligence service, then known as the BVD, to its concerns.[68]

Another red flag went up that same month when the BVD learned that Khan had been observed at a nuclear trade show in Switzerland asking suspicious questions about atomic weapons. Dutch authorities decided it was time to act. The BVD wanted Khan arrested. First, however, they checked with their counterparts at the CIA.[69] Word came back that the United States wanted the Dutch to back off. "The Americans wished to follow and watch Khan to get more information," said former Dutch prime minister Rudd Lubbers, who at the time served as the Netherlands' Minister of Economic Affairs.[70] Lubbers took the matter up with American officials who, he says, told him that blocking Khan's access to Urenco would be sufficient.[71] Lubbers has since suggested that Washington was motivated by a Cold War interest in helping Pakistan offset the nuclear threat from Soviet-leaning India.[72] It would not be the last time that Cold War considerations prompted the United States to interfere with an official investigation of Khan's activities.

Bowing to US pressure, Dutch authorities refrained from taking legal action against Khan. The best that Lubbers and his Economic Affairs Ministry could manage was to convince FDO to "promote" Khan to a new position where he would no longer have access to Urenco centrifuge data.[73] But the move seems to have had the opposite effect of that presumably intended by the United States. Cut off from the centrifuge technology, Khan was no longer engaged in the activity the CIA ostensibly hoped to monitor. Moreover, the shift in jobs may have alerted Khan to the fact that he was under surveillance. With his usefulness to Pakistan apparently at an end in Holland, Kahn decided to return home. Munir Khan would later say that the Pakistani nuclear spy chose to "come out of the cold."[74]

On December 15, 1975, just two months after first coming under scrutiny from Dutch authorities, Khan and his family left suddenly on a trip. Friends and neighbors received letters saying the family was in Pakistan for a vacation. Khan was said to have come down with yellow fever, delaying the family's return. In January, Khan submitted his resignation to FDO, effective March 1, 1976. He would not be returning to Holland.[75] What no one yet realized was that Khan had already absconded with the plans for almost every centrifuge on Urenco's drawing board, including the all-important G-2.[76] It would prove to be one of the greatest nuclear heists of all time.

4
Gearing Up

BACK IN PAKISTAN, Kahn quickly began helping to make use of the plans he had stolen from Urenco. Any ailment he may have suffered did not prevent him from diving headlong into the uranium enrichment effort. Shortly after his return in late 1975, Khan reportedly visited the facilities of Project 706. What he found, according to Malik, was that Sultan Mahmood's team had made almost no progress. "His disappointment was great at seeing how much time and money had been wasted," Malik writes. "Work was at a standstill."[1]

It is not clear whether the project was as moribund as Khan claimed or this statement was a self-serving depiction intended to enhance the appearance of his importance to the effort. But in either case, Khan was apparently successful in convincing Bhutto that the program had stalled. Malik reports that following the tour of Project 706, Khan received an invitation to meet with Bhutto in Islamabad. There, Khan informed the prime minister of the "sorry state of affairs" at Chaklala. For Bhutto, who had received far more upbeat appraisals from Mahmood and Munir Khan, this assessment was undoubtedly disturbing. According to Malik, Bhutto asked Khan to delay his scheduled return to Holland. Several days later, after conferring with his advisors, the prime minister formally requested that Khan remain in Pakistan and serve the nation by joining the uranium enrichment project. Khan, of course, agreed. Upon learning of Khan's decision, Malik writes, Bhutto pounded his fist on a table and declared, "I will see the Hindu bastards now."[2]

The "Political Scientist"

Before the ink was dry on his resignation from FDO, Khan was busy sharing Urenco's secrets with his new colleagues in the Pakistani enrichment program. Within days of his meeting with Bhutto, Kahn was huddling with Mahmood and other Project 706 scientists to study the centrifuge designs smuggled out of Holland. Of the entire group, Khan was the only one who had ever actually seen or worked with centrifuges. The others had only a limited understanding of the machines, based on old Manhattan Project literature and the blueprints Khan had stolen.[3] As a result, Khan quickly established himself as a kind of guru within the centrifuge program, despite Mahmood's more senior position as project director.

Unlike Urenco, security for Project 706 was extremely tight. Munir Khan had a well-known obsession with secrecy, and he imposed strict rules to ensure that the program remained under wraps. Information was rigidly compartmentalized, and the scientists and engineers involved in the effort were forbidden to discuss their work, even with colleagues. Many, including Mahmood, chaffed at the restrictions.[4]

A.Q. Khan's frustrations ran in a somewhat different direction. He believed that the enrichment effort was simply moving too slowly, due to excessive bureaucracy and incompetent leadership. He also resented the superior status enjoyed by Mahmood and Munir Khan, whose talents he considered far inferior to his own.[5] By early spring, he had begun actively plotting to push them aside.

In April 1976, Khan sent a letter to Munir Khan complaining about the enrichment program's lack of progress and threatening to quit. "I have tried to contribute as much as I could under the circumstances but, frankly, I am not at all satisfied with it and could have contributed at least ten times of what I have been able to do," Khan wrote. He maintained that the project was already two years behind schedule and, if things did not improve, would fail to meet the 1980 target date for producing enough enriched uranium to build a bomb. While suggesting a desire to bow out of the project, Khan sarcastically noted that new engineers joining the team would likely be able to see the effort through "under their 'able' and 'intelligent' director [Mahmood]," along with Munir Khan's guidance.[6]

Just days later, A.Q. Khan sent a similar letter to Bhutto, repeating his assertion that the enrichment effort was far behind schedule. "Each week passing is putting the project behind by at least 2 to 3 months," Khan wrote. He also reiterated his ostensive wish to leave the program.[7] In response to the letter, Bhutto reportedly called Khan to learn more about the source of the scientist's dissatisfaction. According to Malik, Khan told Bhutto exactly what was on his mind, unleashing a torrent of criticism aimed at Munir Khan and the PAEC leadership. "Munir Ahmad Khan and his people are liars and cheats," Malik reports Khan telling the prime minister. "They have no love for the country. They are not even faithful to you. They have told you a pack of lies. No work is being carried out and Munir Ahmad Khan is cheating you."[8]

Khan's outburst led to a meeting with Bhutto and Foreign Secretary Agha Shahi. At the session Khan claims to have bluntly stated that the prime minister was being duped by Munir Khan. Bhutto informed the upstart scientist that he had given the PAEC chief until the end of 1976 to carry out a nuclear test using a plutonium device. Khan scoffed at the idea and, by his own account, quizzed the prime minister about the feasibility of meeting such a deadline. Did Pakistan have access to plutonium, an unsafeguarded reactor from which to get spent fuel, or a reprocessing plant with which to produce plutonium? Bhutto acknowledged it did not. "Then how on earth do you plan to carry out the test by December?" Khan asked. "Mr. Prime Minister you are being taken for a ride."[9]

Khan had a point. The international safeguards on Pakistan's nuclear facilities would almost certainly have prevented testing of a plutonium device on a short timetable, assuming it could be done at all. But the truly remarkable aspect of Khan's performance was the sheer audacity of his willingness to take his concerns — and his complaints about his superiors — directly to the prime minister. Khan's ability to cozy up to Pakistan's political leaders and enhance his own position at the expense of his rivals would soon earn him a reputation among his colleagues as the "political scientist."[10]

But however unseemly, Khan's aggressive tactics paid off. He had succeeded in convincing Bhutto of the problems inherent in Munir Khan's plan for developing a plutonium bomb while also making clear his belief that the enrichment effort could move ahead much faster under new leadership. Bhutto was reportedly furious after hearing Khan's assessment of the nuclear program and briefly considered firing Munir Khan. In the end, however, the prime minister settled on a more pragmatic, and in one respect ingenious, course. On the advice of top aides, Bhutto opted to retain Munir Khan as head of the PAEC and leader of the plutonium program while placing A.Q. Khan in charge of the enrichment project.[11] This had the obvious effect of keeping open both paths to the bomb. But this dual approach yielded another, less obvious benefit: While the world remained focused on Pakistan's declared plutonium program as the probable route to a nuclear weapon, the undeclared and unsafeguarded enrichment effort would move forward relatively unhindered, at least in the initial stages. This provided valuable time, of which A.Q. Khan took full advantage.

Kahuta

On July 31, 1976, A.Q. Khan was formally installed as the director of Project 706, replacing Sultan Bashiruddin Mahmood. The enrichment program, now officially known as Engineering Research Laboratories (ERL), became an independent entity, outside the control of Munir Khan and the PAEC. Although a three-member board was established to oversee ERL's activities, its primary purpose was to secure funding and provide the project with protection against foreign political or military interference. Khan would answer directly to the prime minister, and to the prime minister alone.[12] As one colleague later put it, Khan enjoyed absolute, "no-questions-asked authority" over ERL.[13] It was from this base of autonomous power that Khan would launch his nuclear smuggling network.

Under Khan's leadership, ERL began a crash program to enrich uranium. It involved simultaneously procuring essential equipment and components from abroad, developing prototype centrifuges based on the stolen Urenco designs, setting up a pilot-scale enrichment facility, and developing plans for a full-scale centrifuge plant. The pilot facility would be based in the town of Sihala, a few miles southeast of Islamabad. While operations got underway there, Khan and his team began searching for a site for the permanent, large-scale operation.[14] After touring dozens of locations, they eventually selected a one-hundred-acre parcel of land near the village of Kahuta, a remote outpost in the low mountains about twenty miles southeast of Islamabad. The site, in Khan's view, offered the perfect combination of stealth and security. The hilly terrain provided considerable protection against enemy air strikes, while the isolated position ensured that there would be few foreigners lurking about. Khan also considered Kahuta's relative proximity to the capital an asset. Scientists working at the facility could be assured their families had access to good education and health care, and most important, Khan himself would remain close to the seat of power.[15]

Bhutto's government began buying up the land for the enrichment plant in September 1976. By offering generous compensation packages to the landowners, many of them members of the military, the process was completed in just a few weeks. With the site selection and land purchase settled, Khan next turned his attention to the planning and construction of the new facility and to rounding up the scientists and equipment needed to operate it.[16] Even as he did so, outside forces were once again pushing his enrichment program to the top of Pakistan's list of strategic priorities.

The Nonproliferation Regime

In March 1976, Pakistan had signed a contract with France to begin construction of the plutonium reprocessing plant called for in their 1974 agreement.[17] The facility was a crucial element in Munir Khan's strategy for building a plutonium bomb. But the construction deal contained a significant caveat. Under pressure from the United States, the French had agreed to proceed with the deal only on the condition that the facility be placed under international safeguards aimed at preventing the diversion of fissionable material for military purposes.[18] As A.Q. Khan had pointed out to Bhutto not long afterward, this effectively quashed the PAEC's plans for building a plutonium weapon, making the enriched uranium route Pakistan's best hope for attaining a bomb.

The US intervention in the reprocessing plant deal marked a dramatic shift in American policy. For more than two decades, Washington had vigorously promoted the "peaceful" use of atomic energy while blithely ignoring the risks of that same technology being used to develop nuclear technology. Now, belatedly, the United States was taking the lead in trying to prevent more countries from joining the nuclear-weapons club. Two factors lay behind the turnaround. One was a legitimate concern about the spread of nuclear technology. The other was rank politics. The first of these grew out of India's nuclear test of 1974. New Delhi's "peaceful nuclear explosion" in the Pokhran desert had shattered complacent notions about the wisdom of sharing nuclear know-how. India had, of course, drawn on its US-assisted nuclear power program to produce the plutonium used for the test explosion. The latent dangers of the Atoms for Peace scheme had been laid bare.

US anxiety about the implications of India's nuclear test only increased in the wake of Pakistan's reprocessing-plant deal with France and a West German agreement to sell enrichment and reprocessing technology to Brazil.[19] Taiwan, South Korea, and Argentina were also believed to be engaged in nuclear programs that could lead to the production of plutonium weapons. It was a disturbing trend that did not lend itself to easy solutions. In Washington, the Ford administration had come to recognize that, thanks to Atoms for Peace, the knowledge required to produce a nuclear bomb was no longer the exclusive province of a few Great Powers. It also understood that a growing number of developing nations sought the technology both as a means of alleviating concerns about energy dependence stemming from the 1973 oil crisis and as a possible path to the bomb. As they pursued that interest, neighboring countries, feeling threatened, would likely be compelled to head down the same path. The situation could easily snowball out of control.

The Ford administration approached the problem as an issue of supply and demand. However great the demand, the reasoning went, the supply could still be controlled, at least for the time being. The trick would be to convince supplier nations to resist the temptation of the huge profits that could be made by selling nuclear technology to shady customers. Toward that end, in April 1975 the administration convened a meeting in London of the major nuclear technology suppliers. In addition to the United States, the session included representatives from France, West Germany, Britain, Canada, Japan, and the Soviet Union. The US sought agreement on two points: first, that supplier states would not lower safeguard requirements as an inducement to potential customers; and second, that the export of reprocessing and enrichment technology be banned. The US had unilaterally placed tough new restrictions on its own sale of these technologies following the Indian test.[20]

Achieving an accord on the first issue proved relatively easy. Meetings continued throughout the summer and fall, with the group gradually expanding to include eight additional supplier states: Belgium, Sweden, Czechoslovakia, Italy, the Netherlands, Poland, East Germany, and Switzerland. By November, the Nuclear Suppliers Group, or London Club, as it was also known, had adopted a set of explicit export guidelines. The member states agreed that they would insist on International Atomic Energy Commission (IAEA) safeguards and additional physical protection measures before selling any of a long list of sensitive technologies to countries that did not already have nuclear weapons. They also agreed to obtain guarantees from their customers that any new facility using the exported technologies would be placed under safeguards and that the technologies would not be used to produce a nuclear explosive device.[21]

But the United States did not get everything for which it had asked. The other supplier states proved unwilling to require that importers place all of their nuclear facilities under safeguards. Most important, the US was unable win a ban on the sale of enrichment and reprocessing equipment.[22] That failure would provide A.Q. Khan with precisely the opening he needed to supply his uranium enrichment facility.

"A Horrible Example"

As a result of US pressure, the London Club guidelines were eventually incorporated into both the West German deal with Brazil and the French contract with Pakistan. But despite the beefed-up safeguard requirements, the Ford administration continued to express reservations about the French–Pakistani arrangement. Its concerns stemmed in part from the inadequacies of the London Club provisions, which could allow Islamabad to continue its pursuit of nuclear weapons. By early 1976, US intelligence had determined that Pakistan was involved in a "crash program" to build a bomb and by spring had discovered that Libya had agreed to finance the reprocessing project.[23] But in addition to this legitimate concern about the dangers of Pakistan's headlong rush for the bomb, the administration also had a less high-minded motive for continuing to press its objections to the deal: political survival.

Throughout 1976, Democratic presidential hopeful Jimmy Carter campaigned on the issue of halting the spread of nuclear weapons. In his election-year autobiography he declared "the unnecessary proliferation of atomic weapons" to be the greatest danger facing the world.[24] Carter, who had done post-graduate work in nuclear science, was particularly critical of both President Ford's failure to publicly condemn India's nuclear test and the administration's slow response to Pakistan's reprocessing plant deal.[25] Shortly after becoming president, Carter threatened to cut off US supplies of nuclear fuel and technology to countries that did not accept international safeguards on their use.[26]

Congressional Democrats were also pressing the proliferation issue. In May 1976, Senator Stuart Symington of Missouri introduced legislation barring US assistance to any country found trafficking in enrichment equipment or reprocessing technology outside of international safeguards. The Democratically controlled Congress passed the Symington Amendment the following month, effectively putting both Pakistan and the Ford administration on notice that nonproliferation would now be taken seriously.[27]

Anxious to demonstrate his administration's commitment to the issue, President Ford sent Secretary of State Henry Kissinger to Pakistan in August in hopes of persuading Bhutto to cancel the contract for the reprocessing plant. It was the secretary's third trip to the country, and he and Bhutto knew each other well. Kissinger had worked closely with Bhutto in trying to contain the 1971 Bangladesh crisis, and Bhutto had opened doors to China for Kissinger as part of the Nixon administration's strategy for ending the Vietnam War. This time, however, Kissinger had come to seek a major concession.[28]

During a private meeting with Bhutto in Lahore on August 8, Kissinger made clear that the United States knew Pakistan sought the reprocessing plant for its nuclear weapons program and urged the prime minister to cancel the deal. As an incentive, the secretary offered Bhutto 110 fully loaded Corsair A-7 attack jets, which Pakistan's air force coveted to bolster its ability to strike India. But Kissinger also let Bhutto know that there would be consequences for failing to comply. He warned the prime minister that Pakistan would face far harsher treatment at the hands of Jimmy Carter and the Democrats, who were favored to win the upcoming elections, including a possible loss of American aid under the Symington Amendment. As Bhutto later told the story, Kissinger threatened that the United States would "make a horrible example" of him if he continued to pursue acquisition of the reprocessing facility. Yet neither carrots nor sticks were enough to dissuade Bhutto. He defiantly told Kissinger that the US should not interfere in Pakistan's affairs.[29]

Unable to sway Bhutto, Kissinger tried a different approach to halting the reprocessing plant deal. Flying off to Paris, he pressed the government of Prime Minister Jacques Chirac to back out of the contract with Pakistan. The French reacted angrily to the suggestion and, like Bhutto, denounced the United States for its meddling. But Ford and Kissinger kept at it and eventually won a shift in French policy. In mid-December, France announced that it would no longer export reprocessing technology. It refused, however, to make the ban retroactive, allowing the Pakistani deal to proceed, at least for the time being.[30] Pakistan's plutonium program suffered a blow the following week, when Canada announced it would no longer provide assistance for the nuclear power reactor it had furnished to Islamabad and halted the supply of uranium fuel rods.[31]

For the Ford administration, these limited victories came too late. Jimmy Carter and the Democrats dominated the November elections.

Carter entered the White House in January 1977, pledging to make nuclear nonproliferation a top priority of his administration. But Carter's good intentions would soon give way to political expediency. Efforts to prevent the spread of nuclear weapons would once again take a backseat to Cold War considerations, allowing Pakistan free rein to pursue the Islamic bomb.

End Run

The Ford administration's fixation on Pakistan's plutonium program throughout 1976 proved a boon for A.Q. Khan. While Henry Kissinger scrambled to kill the reprocessing-plant deal, Khan was left alone to assemble the elements of his uranium enrichment facility. And as US-backed safeguards transformed the French plant into a nuclear-weapons white elephant, funds earmarked for the project were diverted to Khan's program.[32]

As planning and construction got underway at Kahuta, Khan and his Project 706 colleagues turned their attention to acquiring the equipment needed to outfit the centrifuge plant. It was a vast undertaking. Hundreds of highly sophisticated parts and pieces of machinery were required, most of which lay far beyond the capacity of Pakistan's still-developing industrial base to produce. "A country which could not make sewing needles, good and durable bicycles or even ordinary durable metalled roads was embarking on one of the latest and most difficult technologies," Khan later explained in a newspaper article. "It was not possible for us to make each and every piece of equipment or component within the country. Attempts to do so would have killed the project in the initial state."[33]

The solution was straightforward. Khan and his team would purchase what they needed on the open market from the more advanced, industrialized states. At the same time, they would begin a program to reverse engineer the components, allowing them to switch to "indigenous production" when, and if, the rest of the world caught on to their scheme. A key to the strategy's success was the list of suppliers Khan had stolen from Urenco. "My long stay in Europe and intimate knowledge of various countries and their manufacturing firms was an asset," Khan explained with masterful understatement.[34]

The Kahuta procurement effort also benefited from a major loophole in the London Club guidelines. In drafting the protocols, the supplier nations had banned the export of finished products, such as centrifuges, that could be used in the production of nuclear weapons, but not their component parts. As a result, Pakistan was able to purchase virtually everything it needed for the enrichment program piece by piece.[35]

Khan also shrewdly exploited other weaknesses in the export control system. One was the willingness of greedy suppliers to look the other way on suspect deals in order to make a profit. Another was the West's low opinion of Pakistan's scientific and engineering capabilities. Western companies, Khan later said, willingly sold Pakistan equipment that could be used in the enrichment of uranium "in order to mint money," while supplier states "ignored" the purchases in the naive belief that the technology would "go [to] waste."[36] Khan also claimed that once news of the enrichment effort got out, suppliers began lining up to sell Pakistan their wares. It was behavior that even Khan found unconscionable. "They begged us to purchase their goods," he said. "For the first time the truth of the saying 'They will sell their mother for money' dawned on me."[37]

Human Resources

Beyond developing strategies for rounding up the materials and equipment, Khan also focused on recruiting human talent for the enrichment program. With virtually unlimited resources and the full backing of the government, Khan was able to lure top people to the project both from within Pakistan and from overseas. Khan raided the ranks of the PAEC for its best scientists and convinced many highly trained Pakistani engineers to give up lucrative jobs in the United States, Britain, Canada, and elsewhere in the West to return home and join in the great national effort.[38] "He could choose the people he really wanted," said Khan's former mentor in Leuven, Professor Martin Brabers. "He knew who the good people were. He gave them good salaries so they would not want to leave the job."[39] Brabers himself was well acquainted with the enticements Khan could offer. He visited Khan in Pakistan several times as a guest of the government just as the enrichment project was getting underway.[40] Brabers was later quoted as saying that he had a "moral duty" to help his former students when they asked. He denied, however, that he helped Pakistan build an atomic bomb. In the early 1990s, Brabers served as the first rector of the Ghulam Ishaq Khan Institute of Engineering, Science and Technology, an organization of which A.Q. Khan was project director for many years. The institute became the main training ground for Pakistan's aspiring nuclear weapons and missile engineers. The United States later placed the institute on a list of entities suspected of involvement in Pakistan's nuclear weapons and missile programs.[41]

Khan also reached out to other friends and colleagues in Europe for help. Following his abrupt departure from the Netherlands in late 1975, he wrote several times to his friend and FDO officemate, Frits Veerman. At first, the letters were relatively innocuous. Khan informed Veerman of his health, asked for help rounding up items left behind in the office, chatted about the pleasant weather in Islamabad, and invited Veerman to visit him in Pakistan. The tone changed dramatically, however, shortly after Khan took charge of the enrichment project. In July 1976, two of Khan's top ERL lieutenants visited the Almelo centrifuge plant, accompanied by a Urenco official who had himself met with Khan in Pakistan. The two Pakistani scientists hand-delivered a letter to Veerman from Khan.[42] "Very confidentially," Khan wrote, "I request you to help us." Khan explained that there were several things he "urgently" needed for his "research program." These included the technical specifications for etching the tiny steel balls used in the bottom bearings of Urenco centrifuges, photo negatives of the etching pattern, and three or four samples of the etched ball bearings. Khan also asked Veerman to get him samples of the metal membranes and steel springs used in the centrifuge damping mechanism. "Frits, these are very urgently required," Khan wrote. Without them, Khan explained, his research would come to a halt. "I am sure that you can provide me with these," Khan added. "I hope you will not disappoint me."[43]

But Veerman did disappoint. The Dutchman angrily told the two Pakistani envoys that the information Khan wanted was secret and that he would not provide it.[44] Veerman then turned the letter over to FDO officials.* Seven years later, the Dutch government finally used the letter

*Although Khan's Pakistani colleagues returned home without further incident, Veerman was not so lucky. He showed Khan's letter to his FDO supervisor. The supervisor reportedly told him to destroy it or he would be arrested, a prediction that proved prescient. FDO eventually let Veerman go. He was later picked up by Dutch security and held for two days. Government agents reportedly accused him of spying, but Veerman turned the tables, accusing them of allowing dangerous technology to leave the country. The agents finally sent Veerman home, telling him to keep his mouth shut. "You may not talk about this anymore," Veerman later recalled the agents saying. "It is dangerous for Holland." (William Langewiesche, "The Wrath of Khan," *Atlantic Monthly*, November 2005. Also see Shahid-Ur-Rehman, *Long Road to Chagai*, p. 58.)

to bring charges of "attempted espionage" against Khan. He was convicted in absentia in November 1983 and sentenced to four years in prison. An appeals court overturned the verdict two years later, finding that prosecutors had not properly served Khan with a summons. Khan went on national television in Pakistan to declare that he had been "vindicated."[45] The Dutch considered reopening the case in 1986 but backed off at the request of the CIA, according to then-Prime Minister Ruud Lubbers.[46] Lubbers has suggested that the United States wanted Khan left alone in part because Pakistan had by then become a key US ally in the battle against the Soviet occupation of Afghanistan.[47]

While the approach to Veerman did not work out as anticipated, Khan nevertheless found willing accomplices within the Dutch centrifuge program. During the second half of 1976, reports began circulating within the Dutch government of FDO personnel visiting Pakistan. The government consulted Urenco officials about the damage the company might suffer as a result of FDO employees aiding Pakistan.[48] Yet even as those concerns were being weighed, Khan continued to get access to information about the Urenco operation through two Pakistani nationals employed by FDO as quality inspectors. The Dutch government investigation later found that the two Pakistanis were "active helpers of Khan."[49]

Khan also got help from Pakistanis living in the United States and Canada. In one instance, he wrote to an associate asking him to send the names and addresses of Pakistanis attending a North American Muslim engineers' conference. Such tactics paid off, and Khan later reported that "slowly but surely" people were joining his enterprise.[50]

One of those who joined was Abdul Aziz Khan, a Pakistani-born electrical engineer working in Canada. In 1977, A.Q. Khan wrote urging him to return to Pakistan to work on a "project of national importance" for which Pakistani scientists and engineers in Britain and the United States were being recruited. Despite considerable financial inducements and other perks, A.A. Khan declined. He agreed, however, to collect technical information in the US and Canada and to travel to Pakistan during vacations to help train A.Q. Khan's young engineers. The two men maintained an active correspondence, and A.A. Khan later became the central figure in a scheme to send American-made equipment used in centrifuges to Pakistan via Canada.[51]

The Network

While continuing to enlist scientists and engineers for the uranium enrichment project, Khan also assembled a vast network of middlemen and front companies to supply the Kahuta operation. Armed with the designs and supplier lists Khan had stolen from Urenco, dozens of purchasing teams fanned out across Europe, Asia, the United States, and Canada to buy materials and equipment for the program on the open and gray markets. Existing Pakistani companies lent their services to the cause, often posing as recipients of the goods. Shell companies were established in Europe, the Middle East, and Asia to further obscure the trail. And when Western embassies in Islamabad began denying visas to Pakistanis suspected of involvement in the nuclear program, Khan's agents invited executives of the American and European supplier companies to less discriminating locales, such as Turkey, Dubai, Singapore, and Hong Kong, to negotiate lucrative business deals.[52] Independent South Asian and European engineers and businessmen, many of them old friends and associates of Khan, also signed on as consultants and purchasing agents for the rapidly expanding network.

Khan enjoyed the full backing of the Pakistani government and its diplomatic corps for his purchasing campaign. Military aircraft were requisitioned for shipments of particularly large or sensitive equipment, and PAEC officials were provided diplomatic credentials and installed in Pakistani embassies in the West to facilitate purchases for the bomb program. Diplomatic pouches were sometimes used to smuggle proscribed items into Pakistan.

Within months of taking over the uranium-enrichment effort, Khan had established a highly efficient, covert supply network that would quickly provide all of the material needed for Pakistan's centrifuge program. It would also serve as the basis for what the world eventually came to know as the A.Q. Khan nuclear supermarket.

Pakistan's nuclear shopping spree got underway in Europe. In 1976, just as the centrifuge program was ramping up, a three-man delegation headed by Dr. G.D. Alam, a centrifuge expert at ERL, embarked on a mission to locate a supplier of high-precision vacuum valves used in the enrichment process. The team began its quest in Switzerland's so-called Vacuum Valley, a region of small industrial towns along the Rhine that specialized in the technology. Their first stop was in the village of Haag, near the border with Liechtenstein, at a company called Vakuum Apparat Technik (VAT), a well-known supplier of pumps and valves to nuclear and other industries around the world. Dr. Alam, a soft-spoken man in his late thirties with a beard reminiscent of an Arab sheikh, was perfectly direct about what he was after. He reportedly told VAT executives he wished to purchase several thousand vacuum valves for a Pakistani centrifuge enrichment plant. VAT expressed no objections. Before proceeding, however, company officials checked with the Swiss government to make sure the sale would not violate export regulations. The bureaucrats in Berne informed VAT that while the London Club guidelines prohibited the sale of complete centrifuges to unsafeguarded facilities such as Pakistan's, there were no restrictions on the export of vacuum valves or other centrifuge components, even if they were intended for a uranium enrichment program. With the tacit approval of the government, VAT officials informed Alam that they would be glad to fill his order.[53]

One VAT employee developed an ongoing association with Pakistan. During a trip to Switzerland in the 1970s, A.Q. Khan met VAT's export manager, an engineer named Friedrich Tinner. The two men struck up a relationship and continued meeting over the years. Tinner later formed his own engineering firm through which he is alleged to have done business with Khan during the 1980s. In 1996, Tinner reportedly came under scrutiny by Swiss authorities for attempting to ship valves for uranium enrichment centrifuges to Iraq. He was never found to have violated any law. He and his two sons, also engineers, would later emerge as central figures in the investigation of Khan's nuclear smuggling network.[54]

Pakistan's 1976 purchasing campaign included a stop at the firm of CORA Engineering in Switzerland's Chur Valley. There the Pakistanis sought a major component for the enrichment program: a large facility used to feed uranium gas into centrifuges and then transform the enriched gas into a solid. Once again, the Pakistanis were reportedly perfectly clear about the equipment's intended use. CORA executives didn't blink. Like VAT, they checked with Berne to determine what export regulations would allow. As with the vacuum valves, the London Club protocols did not cover the conversion facility, and the Swiss government found no problem with the sale going ahead. As it turned out, an export license was not even required.[55] Anxious to promote Swiss industry, Berne adopted the attitude that if an export was not explicitly prohibited, it should be allowed. Some within the Berne government also believed that Switzerland should maintain the ability to develop nuclear technology for a weapons program, an option it did not entirely abandon until the end of the Cold War.[56]

CORA completed the uranium handling plant in the summer of 1978. It was so large that three specially chartered C-130 cargo planes were required to transport it to Pakistan. CORA itself expressed no regrets about the sale. "We are not producing revolvers or cannons, and we are not producing bombs," a CORA official told BBC reporters in the early 1980s. "We are not involved in nuclear weapons in any respect, because we wouldn't know how to make a nuclear weapon."[57]

From Switzerland, G.D. Alam and his purchasing team traveled to West Germany in search of additional supplies for the enrichment program. One firm they visited was Hanau-based Leybold Heraeus, another well-known manufacturer of vacuum technology and a subcontractor to Urenco. From Leybold the Pakistanis purchased vacuum pumps and uranium gas purification equipment worth about six million deutsche marks ($2.3 million).[58] According to the book *Long Road to Chagai,* the Pakistanis also ordered a second uranium gas conversion facility. Leybold had supplied a similar plant to Brazil and gladly tweaked the plans they had on hand to Alam's specifications. The company eventually shipped the facility through an Austrian subsidiary to Pakistan via Dubai. It arrived in 1978, several months ahead of the Swiss plant.[59] As with VAT and CORA, no export licenses were required for any of the equipment purchased from Leybold. The United States would later accuse Leybold of providing nuclear technology to Pakistan, Iraq, and South Africa.[60]

During the dealings with Pakistan, a young Leybold engineer and salesman by the name of Gotthard Lerch traveled to Islamabad and became acquainted with A.Q. Khan. It marked the beginning of a decades-long relationship. Lerch left Leybold in 1985 to set up his own company in Switzerland's Vacuum Valley. In the late 1980s, he came under investigation by German authorities on suspicion of assisting Pakistan's nuclear program, but he was never charged with a crime. Iranian officials have since reportedly identified Lerch to IAEA inspectors investigating Tehran's nuclear program as one of their business contacts. Lerch currently faces charges in Germany of joining with A.Q. Khan and others in supplying Libya's secret uranium enrichment program starting in the late 1990s. He denies the charges.[61]

Pakistan also did business with other West German companies. One, Aluminium Walzwerke, based in Singen, filled large orders for specialized items, including rolled metal rods and tiny, welded aluminum devices based on highly detailed designs supplied by the Pakistanis. None of the sales, reportedly totaling forty million deutsche marks ($15.5 million), were prohibited under the existing export guidelines.[62]

Some items the Pakistanis sought, however, were restricted. In 1977, the chief purchasing agent for Pakistan's nuclear program, the physicist-cum-diplomat S.A. Butt, contacted the West German firm Rohstoff-Einfuhr seeking fifteen tons of concentrated uranium ore, known as yellow cake, or an equivalent amount of uranium gas. Gaseous uranium would be useful only for a centrifuge enrichment facility. Rohstoff-Einfuhr's president, Alfred Hempel, a reputed ex-Nazi who was alleged to have supplied the nuclear programs of India and South Africa, contacted sources in South Africa, one of the world's leading suppliers of yellow cake, to try to arrange the deal. But he also checked with the West German government to make sure the sale would be allowed. Bonn told him that, unlike the sale of individual centrifuge components, the transaction was prohibited under international agreements. South Africa also refused to participate in the deal after learning from Hempel that the customer for the uranium was Pakistan. Some things, it seemed, were not for sale. But the Pakistanis would find other ways to fill their uranium needs, and Western companies, along with lax export controls, would help them do it.[63]

A.Q. Khan also took part in the European shopping campaign. In 1976 he traveled to Belgium for discussions with Belgonucleaire, one of the world's leading suppliers of nuclear fuel and equipment. There, Khan reportedly negotiated the purchase of glove boxes used to protect lab workers from radiation.[64] The following year, Khan returned to Holland to work out details of a deal for specially hardened steel tubes used in ultracentrifuges. The year before, S.A. Butt had placed an order for the tubes with the Dutch firm of Van Doorne Transmissie (VDT) the year before. When Khan arrived at VDT he increased the order to sixty-five hundred tubes, a number so large employees of the company referred to the deal as "the Pakistani Pipeline." Holland's Ministry of Economic Affairs learned of the order and instructed VDT not to make the delivery. Although VDT was aware that the tubes were for use in a Pakistani enrichment program, it chose to ship them anyway.[65]

The Dutch government eventually took VDT to court, but was unable to win a conviction. Because the tubes were not "specifically developed" for use in centrifuges, they were not covered by existing export laws and no license was required to ship them.[66]

Khan's former Dutch employer, FDO, also cashed in on the Pakistani buying spree. In 1977, the company agreed to sell Khan and his associates specially designed computerized measuring devices that were originally intended for Urenco. When the consortium backed out of the deal, Pakistan's willingness to step in and take the equipment was like "a gift from heaven" for FDO. An FDO sales manager traveled to Pakistan to arrange the deal and stayed at Khan's house. In 1983 the sales manager and an FDO director were indicted on charges of failing to obtain proper export licenses for the equipment. Both were acquitted when a Dutch court rejected the government's argument that because some components of the devices required a license, the complete units did also, an odd interpretation of the export laws for which FDO was no doubt grateful.[67]

Guilty Knowledge

A.Q. Khan has long boasted that Pakistan's uranium enrichment program remained hidden from the West for nearly three years.[68] Western officials have largely substantiated that claim.[69] It is, of course, false; but it has been a useful fiction for all concerned. It has allowed Khan to portray himself as a wily scientist who outwitted Western intelligence agencies and cleverly evaded stringent export restrictions, and it has permitted the West to plausibly explain why it failed to prevent Pakistan from developing an enrichment capability.

But the fact is that early on the West knew what Pakistan was up to and failed to take appropriate action. The Dutch became suspicious of Khan's activities while he was still in Holland and clearly understood the implications. Yet they went along with an American scheme that allowed Khan to remain free so that his actions could be "monitored." As a result, Khan was able to return to Pakistan, taking the secrets of enrichment technology with him. The CIA obtained a full set of plans for the Kahuta enrichment facility while it was still under construction.[70] Yet operations there continued unabated.

Even after Pakistan's development of uranium enrichment technology had become obvious, the Western response was myopic. Security at Urenco remained lax, allowing Pakistan continued access to sensitive information. And when international export guidelines proved inadequate for stopping the flow of nuclear technology to Pakistan, Western governments failed to respond in a timely fashion. Western businesses, meanwhile, used the existing guidelines to justify their dealings with Pakistan's nuclear program.

The United States, Britain, and other Western governments would later accuse Holland of not doing enough in response to Pakistan's efforts to obtain enrichment technology. But those same governments would soon exhibit a similar inability to confront Pakistan's procurement efforts. In March 1978, the British government warned the United States about suspicious Pakistani purchases in England of equipment that could be used in a uranium enrichment program. Yet London failed to impose effective controls on such exports for another seven months, and Washington would not do so for an entire year, largely due to foreign policy considerations. In the meantime, critical technology made its way to Kahuta from companies in Britain, Europe, and the United States, and Pakistan passed a major milestone in its quest to develop an atomic bomb.[71]

5
The Compromise

ON JANUARY 11, 1977, an accredited Pakistani diplomat by the name of Ikram ul-Haq Khan took up residence in Apartment C at No. 8 Hauptstrasse in Wachtberg-Pech, West Germany, an elite enclave outside the capitol city of Bonn. The apartment, twenty miles from the Pakistani embassy, would serve as Khan's home as well as his place of business. His business was purchasing material for Pakistan's uranium enrichment program.[1]

I.H. Khan was an agent of Pakistan's Special Works Organization (SWO), a military engineering unit established by Prime Minister Bhutto — reportedly at the request of A.Q. Khan — to build the Kahuta enrichment facility.[2] But SWO also became a conduit for the parts and equipment used to outfit the plant. I.H. Khan headed what amounted to a special purchasing office for the effort. From his small apartment on Hauptstrasse, he helped arrange much of Pakistan's acquisition of material and components for the centrifuge program from companies in West Germany, Holland, Switzerland, and beyond.[3]

Of the many deals in which I.H. Khan was involved, one would have particularly far-reaching implications. Shortly after setting up shop in Germany, Khan reportedly helped arrange the purchase of thirty-one high-frequency electrical inverters from Emerson Electric Industrial Controls in Swindon, England. Inverters are sophisticated voltage regulation mechanisms that are essential to controlling the speed of ultra-centrifuges. Emerson, a subsidiary of the giant American corporation Emerson Electric, was a major manufacturer of the devices and supplied them to British Nuclear Fuels Ltd., the UK partner in the Urenco consortium.[4]

To arrange the inverter order, the Pakistanis worked through a firm of commission agents, Team Industries of Leonberg, West Germany. One of Team's principals, Ernst Piffl, reportedly made the initial approach to Emerson. He then returned, an Emerson official later recalled, with "a Mr. Khan and three other Pakistanis." A deal was struck and the Pakistanis agreed to pay Emerson £133,566 ($258,000) for the inverters.[5]

Emerson could hardly have missed what the Pakistanis were up to. A former Emerson employee later told the authors of *The Islamic Bomb* that higher-ups in the firm would automatically have assumed the inverters were intended for a uranium enrichment program. The company's sister operation in the United States, Emerson Electric Industrial Controls Division of Santa Ana, California, was also approached by the Pakistanis, but reportedly rejected the order. Yet the British firm accepted. It did so, according to the former employee, in the belief that the Pakistanis did not have the competence to utilize the complex technology and that the inverters would simply rust away in their crates. Emerson officials would soon be disabused of that notion but, nevertheless, continued working with the Pakistanis.[6]

The Pakistanis reportedly originally informed Emerson that the inverters were to be delivered to the firm of Scimitar Engineering Company Ltd., a small industrial equipment supplier in the Welsh seaport of Swansea. But before the deal was finalized, Scimitar was replaced by another Swansea company, Weargate Ltd.[7] Weargate was one of a string of newly formed companies headed by a pair of seemingly unlikely business partners, both of whom would have ongoing dealings with Pakistan. One was a Welsh engineer named Peter Griffin who had worked for Scimitar. The other was a naturalized British citizen of Indian Muslim origin named Abdus Salam (not to be confused with Dr. Abdus Salam, the Nobel Prize–winning physicist who helped found Pakistan's nuclear program).[8] Salam, a north London businessman and self-described engineer, was reportedly an old friend of A. Q. Khan.[9] The consignee for the order, according to Emerson's records, was the director general of the SWO in Rawalpindi, the garrison town just down the road from Kahuta.[10]

While working on the inverters, Emerson accepted two more orders from the Pakistanis. The first came from Team Industries. It was for thirty-one sets of input transformers, devices used to control electrical current in industrial facilities. These were air-freighted from London to Rawalpindi by a British subsidiary of the West German national railway. The next order was processed through Weargate in the spring of 1978. It was for one hundred additional inverters, along with a set of spare parts. The units, valued at £1.25 million ($2.75 million), were to be sent out in five shipments over a period of four months starting in December 1978.[11] But those shipments would never be sent.

Too Little, Too Late

In late March 1978, just as Pakistan was arranging its second order of inverters from Emerson, an official from the British embassy in

Washington visited the US State Department to discuss a new Carter administration initiative to put greater emphasis on halting development of enriched uranium weapons. In the course of the meeting, the British official shared "some disturbing information" with his American hosts: Pakistan had placed a suspicious order for inverters with a company in England. The Americans were puzzled. "We didn't even know what an inverter was," a State Department official later told the *Washington Post.* The British would educate their American counterparts through a series of diplomatic and intelligence exchanges in the months that followed, and other governments began to take notice. In the meantime, Emerson continued processing the inverter orders for Pakistan.[12]

Emerson's dealings with the Pakistanis became public in July when a British politician raised the matter in Parliament. Frank Allaun, a leading member of the Labour Party, had learned that Emerson had inverters marked "Pakistan Special Project" sitting on its production line.[13] Allaun, who had a long-standing interest in nuclear disarmament, immediately understood the significance of this information and began asking questions in the House of Commons. Was the British government aware, he asked, that Emerson was supplying Pakistan with inverters that could be used for uranium enrichment? Had British Customs approved their export?[14]

While waiting in vain for answers, Allaun took up the issue with Energy Minister Tony Benn, another longtime opponent of nuclear weapons. Benn quickly launched a full-scale investigation. But even as Benn's inquiry was underway, Emerson completed the first order of inverters for Pakistan and continued working on the second.[15]

Despite the public attention brought by Allaun and Benn's ongoing investigation, the first thirty-one inverters were shipped to Rawalpindi in August.[16] Shortly after they arrived, Emerson received a telex from the Pakistanis in which they sought a long list of highly technical modifications. Any illusions Emerson officials still harbored about the Pakistanis' inability to make use of the equipment were shattered. Yet the company continued working on the second order.[17]

Tony Benn's investigation eventually determined that inverters were not covered by existing export controls. Their sale to Pakistan was entirely legal. In October, the British government belatedly added inverters to the list of controlled items, seven months after first raising the issue with Washington. While that effectively blocked further inverter shipments, the damage was already done. The delay had allowed the first order to go through, and the Pakistanis were soon hard at work learning how to build inverters on their own. Moreover, the new restrictions quickly proved insufficient. The Pakistanis discovered additional loopholes in the export laws and continued purchasing unrestricted items for use in their centrifuge program. As a result, in early 1979, Britain was forced to expand its list of banned exports twice more, once to include parts that could be used to build inverters, and then to cover equipment specifically designed for use in a centrifuge enrichment program.[18]

Cold War Calculus

As delinquent as the British had been in taking action against Pakistan's efforts to acquire enrichment technology, the United States was even slower. Months after London added inverters to its list of prohibited exports, Washington still had not acted. The delay was not the result of inattention or a lack of information. Rather, the cause lay in the realm of foreign policy. By the summer of 1978, Pakistan had begun to take on new importance in Washington's Cold War calculations. A coup in neighboring Afghanistan in April 1978 had brought a communist regime to power and, with it, growing Soviet influence. Meanwhile, the shah of Iran, Mohammad Reza Pahlavi, America's foremost ally in the region, was showing signs of weakness in the face of a political upheaval that would eventually drive him from power. Pakistan, as a result, looked increasingly like America's last best hope in South Asia. The Carter administration was therefore reluctant to challenge Islamabad too aggressively about its nuclear activities and chose instead to try to curb Pakistan's pursuit of atomic weapons through offers of military and economic aid. Although the policy proved short-lived, it foreshadowed much of what was to come.

The strategy took shape in the second half of 1978. Despite mounting evidence of Pakistan's efforts to develop an enrichment capability, the Carter administration elected not to publicly criticize Pakistan over its increasingly obvious pursuit of a uranium bomb. It did, however, continue the behind-the-scenes diplomatic campaign to block Pakistan's acquisition of the plutonium reprocessing plant from France. In response to US pressure, France ultimately canceled the contract for the facility in August 1978. Although Pakistan had already received almost all of the designs for the plant by the time France pulled out of the deal and had everything it needed to proceed with the facility on its own, US officials portrayed the move as a major nonproliferation victory.[19] The administration also used the moment as an opportunity to try to win future Pakistani cooperation. The day after news of the French decision broke, the State Department announced that the United States was considering resuming aid to Pakistan, which had been cut off the year before as part of the effort to pressure Islamabad into backing out of the reprocessing-plant agreement.[20]

Yet even as the administration pondered that move, there were growing indications — including the revelations from London — of Pakistan's determination to develop a uranium weapon. By early fall, the CIA had begun warning members of Congress that a "preponderance of evidence" indicated Pakistan was secretly attempting to build a uranium enrichment plant.[21] Nevertheless, the administration announced in October that it would restore $69 million in aid to Pakistan.[22] The following month, the State Department formally extended an offer to supply Pakistan with fifty American-made F-5 fighter jets.[23] Although little noticed at the time, news reports hinted that the decision to resume aid had been prompted by Washington's growing concerns about Afghanistan's continuing drift toward the Soviet Union and the shah's increasingly untenable position in Iran.[24] But whatever assistance the United States may have hoped for on the Cold War front, it soon became clear that the resumption of American aid had done nothing to deter Pakistan from its nuclear ambitions.

By early 1979, evidence of Pakistan's continued pursuit of uranium enrichment technology had become undeniable. In January, the United States quietly opened high-level diplomatic talks with Islamabad on the matter.[25] The following month, the State Department sent "urgent" notices to France, Switzerland, and West Germany, warning that Pakistan's nuclear procurement activities extended "far beyond" efforts to purchase inverters, and followed up by sending high-level briefing missions to those countries.[26] Yet the administration still did not add inverters to the export control list or publicly express its concerns about Pakistan's enrichment program.

Finally, in early March, Deputy Secretary of State Warren Christopher traveled to Pakistan in an effort to persuade Islamabad to place its enrichment facilities under international safeguards.[27] The mission came at a dramatic moment. As Christopher sat down to meet with the Pakistani leadership, Zulfikar Ali Bhutto, long the driving force behind his nation's nuclear weapons program, sat in a jail cell awaiting execution after being ousted in a military coup in 1977. Bhutto had asserted — without any evidence — that the United States had orchestrated his overthrow as punishment for his refusal to give up the French reprocessing-plant deal.[28] In a remarkable political testament smuggled out of his prison chamber the previous fall, the deposed leader had acknowledged that Pakistan was on the verge of "full nuclear capability" at the time of the coup and suggested that the military government of Gen. Mohammed Zia ul-Haq would abandon the bomb program. "What difference does my life make now when I can imagine 80 million of my countrymen standing under the nuclear cloud of a defenseless sky?" Bhutto wrote.[29]

While the United States had no apparent role in Bhutto's ouster, American officials may privately have hoped early on that the change in leadership would, in fact, bring about a shift in Islamabad's nuclear policy. But any such hopes were quickly dashed as Zia, whose slicked-down hair, dark eyes, and prominent mustache gave him the appearance of a villain straight out of central casting, not only continued pursuit of the reprocessing plant but also ramped up A.Q. Khan's uranium enrichment effort. More ominously, Zia had issued public statements that sounded for all the world like Bhutto's calls for an Islamic bomb. "China, India, the USSR, and Israel in the Middle East possess the atomic arm," Zia declared in 1978. "No Muslim country has any. If Pakistanis had such a weapon, it would reinforce the power of the Muslim World."[30]

From the time of the coup, the United States had employed a wide range of tactics — from inducements and diplomatic pressure to embargos and threats — to dissuade Zia from continuing his quest for the bomb. But Zia, perhaps realizing that Washington would inevitably back down when its Cold War interests were threatened, remained unswayed. Nevertheless, Christopher would try once more. He informed Zia that Pakistan faced yet another suspension of aid unless President Carter received "reliable assurances" that Islamabad was not seeking nuclear weapons. Zia readily "assured" his guest that Pakistan's nuclear program was "entirely peaceful." But he would not rule out a possible "peaceful" atomic test and refused to allow international inspections of his country's nuclear sites, including the Kahuta enrichment facility.[31]

In the wake of Christopher's failed mission, the Carter administration finally added inverters and other key uranium enrichment components to the list of controlled exports in late March, a full year after the British had first alerted the United States to Pakistan's procurement efforts.[32] But even then the administration, still hoping to preserve US–Pakistani relations, did not formally cut off aid to Zia's government. That move came only in April, two days after Bhutto was hanged, despite appeals for clemency from President Carter and other world leaders. In response to reporters' questions, the State Department acknowledged that under terms of the Symington Amendment, the 1976 law prohibiting US aid to any country that acquired unsafeguarded enrichment or reprocessing equipment, the United States would withhold roughly $85 million in assistance slated for Pakistan.[33]

The Carter administration's decision not to take action against Pakistan's enrichment program earlier came at a heavy price. The previous fall, while the administration was busy trying to woo Islamabad with offers of military and economic assistance, the Pakistanis had purchased about a half dozen inverters from a California company that US officials would not identify.[34] Moreover, by the time the administration finally got around to expanding the export control list, the CIA had already determined that A.Q. Khan's far-flung purchasing network had succeeded over the previous year in acquiring almost all the equipment required to put the Kahuta plant into operation.[35] (Left unsaid, of course, was that the agency's supposed monitoring of Khan, begun in 1975, had presumably failed to detect those purchases until alerted to them by European intelligence services.[36]) In early May, the *Washington Post* reported that Assistant Secretary of State Thomas Pickering had told a Congressional panel that the United States caught Pakistan's end runs around international export controls "too late" to prevent Islamabad from attaining the means to build nuclear weapons within two to five years.[37] In fact, the US had not caught the matter too late; it had simply acted too late.

But what the United States did not yet know, and would not learn for some time, was that the previous June, more than two months after the British had alerted Washington to Pakistan's inverter purchases, A.Q. Khan and his cohorts had succeeded in separating tiny amounts of the fissionable U-235 isotope from U-238. They had, in fact, enriched uranium, albeit to a very low level.[38]

Pakistan did not yet have the bomb, but it had taken significant strides toward developing one. And it would be the Carter administration, despite all its concerns about nuclear proliferation, that would give Islamabad the green light to proceed.

Ace in the Hole

By all outward appearances, US–Pakistani relations had reached an all-time low by the spring of 1979. Diplomatic relations were severely strained, and the Carter administration was contemplating new measures to counter Islamabad's enrichment program. The situation grew more tense in June, when, after driving past the Kahuta enrichment facility, the French ambassador to Pakistan and his top aide were dragged from their car and severely beaten, apparently by Pakistani security forces.[39] Several weeks later, a British journalist was beaten up by security guards when he attempted to interview A.Q. Khan outside the scientist's home in Islamabad.[40] Similar incidents would follow.

Over the summer, the Carter administration launched what it called a last-ditch effort to prevent Pakistan from developing a nuclear weapon. As part of that process, the White House established an interagency task force to generate ideas for derailing Islamabad's bomb program. The panel, headed by State Department arms-control expert Gerard Smith, considered a variety of carrot-and-stick approaches, ranging from the conventional to the provocative. One idea the group debated was to try to reduce Pakistan's desire for nuclear arms as a hedge against India by offering Islamabad advanced American conventional weapon systems, such as F-16 fighter jets. Another was to tighten the economic screws on Pakistan by prohibiting American investment and blocking international loans. But by far the most controversial concept the group floated was a proposal to mount a covert operation to destroy Kahuta using paramilitary forces.[41] Although there is no evidence the administration ever considered pursuing that option, an uproar nevertheless ensued when word of the group's work leaked out in the press. The Pakistanis took the matter seriously enough to lodge a formal complaint with the US ambassador in Islamabad.[42] They also installed a battery of French-built Crotale antiaircraft missiles near Kahuta and increased air defenses in the area to protect against an enemy air strike.[43]

But even as these contentious incidents played out in public, Zia knew he had an ace in the hole. For all the Carter administration's tough talk about halting Pakistan's nuclear program, the general understood that Washington had another, more pressing agenda. While nonproliferation remained a high priority, much of the administration's attention was dominated by two mounting Cold War concerns. One was the situation in Iran. There, the Islamic revolution that drove the shah from power at the beginning of the year resulted in the loss of key American listening posts that had been trained on the Soviet Union. For Zia, this presented an opportunity. When American officials approached him about installing new intelligence collection facilities in Pakistan, he readily agreed.[44] It was a familiar role for Pakistan, harkening back to the U-2 spy missions flown out of Pakistan in the 1960s and Henry Kissinger's approaches to China through Pakistani intermediaries in the early 1970s. Zia's cooperation with American intelligence agencies allowed him to maintain back-channel influence in Washington, despite the chill in official relations.[45]

Zia's strategy quickly paid off. The second Cold War battlefront preoccupying the Carter administration was Afghanistan, a country long coveted by the Soviets as a stepping-stone on a path toward a warm-water port on the Indian Ocean which would lead directly through Pakistan. The United States, of course, bitterly opposed any such Soviet expansion. Throughout the first half of 1979, therefore, the White House watched nervously as Moscow ramped up support for the revolutionary communist government in Kabul. A ray of hope emerged in early March, when the CIA learned that an anti-Soviet Afghan revolt was gaining strength. Soon after, the agency drafted a proposal for secret US backing of the anticommunist Afghan Muslim guerrillas soon to become widely known as the mujahedin, a group already supported at a low level by Pakistan's intelligence service. Such an effort, the agency reasoned, might not only slow down Soviet progress in Afghanistan but also help deflect some of the energy of Middle Eastern Muslims, inspired by the Iranian revolution, away from the United States and toward the Russians. But there were also risks. Any evidence of US interference in Afghanistan would surely lead to Soviet retaliation. The United States would have to operate through a "cutout," a proxy through which money and weapons could be secretly supplied to the mujahedin. And the CIA had just the candidate: Zia's much-feared secret service, Inter-Services Intelligence (ISI).[46] The Pakistanis, in fact, had already approached the agency about providing assistance to the Afghan rebels, but indicated that without a solid commitment from the US they "could not risk Soviet wrath."[47]

The CIA's proposal found favor with President Carter's National Security Advisor, Zbigniew Brzezinski, a Polish émigré and fierce anticommunist hawk. Brzezinski believed the Afghan situation offered the United States a rare opportunity to frustrate the Soviet's expansionist goals in the Third

World and potentially draw them into a quagmire. But he was also aware of concerns within the administration about the risk of escalation. He suggested, therefore, that the United States provide the fledgling Afghan resistance with "nonlethal" support, at least in the beginning. There was nothing to say, however, that guns and other weaponry could not come later. Throughout the late spring, as the CIA reported an increasing flow of Soviet armaments, equipment, and advisors into Afghanistan, Brzezinski encouraged the president to approve this modified version of the plan for covert assistance to the rebels.[48]

But Carter did not bite immediately. Along with his commitment to human rights and stopping nuclear proliferation, Carter had taken office pledging to reform American intelligence operations. High-profile Congressional hearings in the mid-1970s had exposed a wide range of intelligence scandals, including assassination plots against foreign leaders, rogue covert operations aimed at subverting foreign governments, and domestic spying on American citizens. Lawmakers, responding to a public already angered by abuses of government power in Vietnam and Watergate, enacted new legislation mandating greater congressional and presidential control over CIA operations. Yet Carter hoped to go further and overhaul an agency that, as he wrote in his memoir, had been involved in "plotting murders and other crimes."[49] As a first step, he had selected as CIA director an outsider, career Navy officer and former NATO commander Admiral Stansfield Turner. (Carter's initial choice for the job, former Kennedy hand Theodore Sorenson, had been deemed unacceptable by Congress.) Acting on Carter's mandate, Turner quickly cleaned house, firing hundreds of seasoned analysts and spies — along with many rogue elements — and placing new emphasis on electronic surveillance and intelligence gathering.* Yet even as Carter struggled with

*Carter's mass firing at the CIA became known as the Halloween Massacre. Many of the covert officers had been George H.W. Bush's subordinates when he was CIA director. When Bush ran for the Republican presidental nomination, Carter found himself with scores of former high-level intelligence officers doing everything they could to remove him from office. Led by Theodore C. Shackley, the CIA's Associate Deputy Director of Operations, many of these officers formed what amounted to a renegade CIA, accountable not to the president but to his political opponents. (Shackley had been deeply involved in a joint CIA–Saudi effort to conduct covert operations while Bush was CIA director.) The gutted Directorate of Operations provided the Carter administration with almost no useful intelligence out of Iran as the shah's regime collapsed. For a complete account of these events, see Joseph Trento, *Prelude to Terror: The Rogue CIA and the Legacy of America's Private Intelligence Networks* (New York: Carroll & Graf, 2005).

the ghosts of the CIA's past, he found himself embroiled in a variety of Cold War confrontations in Africa and the Middle East that demanded American action. But how to proceed without repeating the mistakes of the past? The CIA plan for Afghanistan, as modified by Brzezinski, seemed to offer a solution. It would allow the United States to conduct operations in such a way that, if they went wrong, someone else would take the blame — a model first perfected by Henry Kisssinger in the mid-1970s in the wake of the exposure of American intelligence scandals. It was in that context that Brzezinski's appeals to approve the Afghan plan resonated.[50] A president who had come to office determined to get the United States out of the dirty-tricks business would now set in motion what would eventually become the biggest covert operation in American history.

On July 3, 1979, President Carter signed a directive, known as a "finding," authorizing nonlethal, covert support for the anti-Soviet Afghan insurgents. The program called for the CIA to spend a little more than a half million dollars on psychological operations and propaganda activities in Afghanistan and to provide the rebels with cash, medical supplies, and radio equipment.[51] To disguise the US role, the equipment and cash would be channeled through intermediaries in foreign countries to Zia's ISI for delivery to the mujahedin.[52] It was a modest beginning, but it laid the foundation for much bigger things to come.

On the very day Carter signed the finding, Brzezinski claims to have sent the president a note in which he stated his belief that the aid would "result in military intervention by the Soviets." It was a view not universally shared within the administration. If Brzezinski, in fact, made such a prediction it was prescient. But whatever Brzezinski's powers of prognostication, no one in the administration had yet begun to consider seriously the potential repercussions of America's Cold War alliance with militant Islamists or the costs to the goal of nonproliferation at which that alliance would be formed.[53]

Hollow Threats

During the summer of 1979, US aid to the Afghan rebels began coursing through Pakistan's intelligence service. Insurgent activity in Afghanistan picked up, causing trouble for the Soviet-backed regime in Kabul. In addition, the CIA had gone beyond its authorized brief and, in conjunction with its ISI and Afghani cohorts, conducted reconnaissance and intelligence missions in and around the Hindu Kush mountains of western Afghanistan and beyond.[54] By the end of August, Zia and the ISI were pressuring the United States to provide arms and additional equipment to help expand the insurgency.[55] But despite growing cooperation between the CIA and ISI in Afghanistan, and the clear willingness of Washington and Islamabad to work together on issues of mutual interest, diplomatic relations between the two nations remained strained, particularly over the question of nuclear weapons.

Throughout the summer and fall, as the April cutoff of US aid began to take a toll on Pakistan's economy, Zia and other high-ranking officials vowed again and again that they would not succumb to outside pressure over their "peaceful" nuclear program. In July, shortly after the US and Pakistani intelligence services began supplying the mujahedin, Zia declared in an address to the nation that Pakistan had no option but to continue its development of nuclear technology. "We shall eat crumbs," he said in another eerie echo of Ali Bhutto, "but we will not allow our national interest to be compromised."[56]

In September, just weeks after the uproar over the leak of the proposed US plan for destroying Kahuta, President Carter received a secret State Department intelligence report detailing the weakness of Zia's domestic position. The study found that Pakistan faced "severe" economic difficulties and that Zia's government had become so unpopular that it might be forced from power by the end of the year. In the meantime, Pakistan's urge to assist in the Afghan campaign was tempered by the threat of retaliation from the Soviet Union. Zia wanted to use the United States as a "counterweight" to Moscow, according to the study, but US–Pakistani relations were being complicated by Zia's "determination" to develop nuclear weapons. The study noted that the Pakistanis seemed bent on "stringing out negotiations" with the United States and other countries in order to "gain time" for their nuclear scientists. There was conflicting intelligence as to when Islamabad might conduct a nuclear test, but the study concluded that it was unlikely that Pakistan would have technical capability before the end of the year, as had previously been suggested.[57]

It was against this background that the Carter administration — in a classic demonstration of Albert Einstein's famous observation that the definition of insanity is doing the same thing over and over and expecting different results — launched yet another diplomatic effort to convince

Pakistan to abandon its nuclear weapons program. In mid-October, Secretary of State Cyrus Vance hosted several days of talks in Washington with Zia's top foreign affairs advisor, Agha Shahi. With the situation in Iran sliding away from the United States and with Afghanistan heating up, the Pakistanis hoped the White House might be more sympathetic to Pakistan's security concerns and relent somewhat on the nuclear issue. But the administration, presumably hoping to capitalize on Zia's weakened political and economic circumstances, showed no outward sign of backing off.[58] Its resolve was undoubtedly further stiffened by news reports just days earlier that work at Kahuta had ground to a halt due to what was believed to be an "irregular supply" of parts from Europe resulting from the more stringent export controls.[59] Zia further undermined Pakistan's negotiating position by announcing during Shahi's talks with Vance that he was indefinitely postponing long-promised elections and that he planned to establish a "genuine Islamic order" in Pakistan. This latter move, in particular, only added to existing US concerns stemming from Zia's recent efforts to bolster domestic support through a program of "Islamization," in which he aligned himself closely with radical Muslim factions and causes — a tactic that would soon spin out of control. But the main topic of the summit was Pakistan's continuing drive for the bomb. Secretary Vance told Shahi that the United States sought three commitments: that Pakistan would not provide nuclear technology to other countries, that it would allow international inspections of its nuclear sites, and that it would not conduct a nuclear test. In what now seems an ironic gesture, Shahi assured Vance that Pakistan would not share its nuclear technology. On the other two points, however, Vance received no satisfaction. Shahi said that Pakistan would submit to inspections only if India did the same. Regarding testing, Shahi said Pakistan "had not yet reached that stage" and would make a decision only after it had the ability to explode a nuclear device.[60] Vance reportedly told Shahi and his colleagues that there would be no more American aid until Pakistan backed off its nuclear program.[61]

But the Pakistanis would not back off. They clearly understood that the mouthings of Vance and other administration officials were hollow, and they acted accordingly. In late October, just days after Shahi's combative summit in Washington, Zia pointedly refused to say that Pakistan would refrain from exploding an atomic device. He also backed away from previous claims that his country's nuclear program was entirely peaceful. Asked by reporters if he would make a no-test guarantee, Zia replied, "I am not closing my options." While insisting that Pakistan was not making a bomb and that its nuclear program was devoted to developing "energy sources," he added that if in the process of that development "steps have to be taken, we will take them." He then went on to say that in his opinion, "there is no such thing as a peaceful explosion," a point that while true, was not exactly what Washington wanted to hear.[62] Yet even as Zia figuratively thumbed his nose at the United States, events were already unfolding that would soon demonstrate the Carter administration's willingness to overlook virtually any sin to maintain Pakistan's cooperation.

Lofty Traditions

On October 22, 1979, Mohammed Reza Pahlavi, the deposed shah of Iran, secretly entered the United States for treatment of what would prove to be terminal cancer. After the shah fled Iran in January under pressure from the Islamic revolutionary movement, the Carter administration had attempted to appease the new Iranian government of Ayatollah Ruhollah Khomeini by insisting Pahlavi would not be welcomed into the United States.[63] When President Carter reneged on that pledge, admitting the former shah on humanitarian grounds, Khomeini railed against what he described as American plotting and urged his followers to demonstrate against the "Great Satan." On November 5, a group of about three hundred Iranian students stormed the American embassy in Tehran, ransacked its offices, and took dozens of hostages. It was the start of the Iranian hostage crisis, a human and political drama that would dominate American attention for the next 444 days. It was also the start of a chain of incendiary events that would soon bring US–Pakistani relations to their lowest point ever.

On November 20, 1979, as the Carter administration struggled in vain to help the more than four dozen American citizens held captive in the US embassy in Tehran, hundreds of armed Muslim fundamentalists seized the Grand Mosque in Mecca, Saudi Arabia, the holiest site in all of Islam. The militants, intent on overthrowing the Saudi royal family, engaged in a bloody, two-week battle with security forces. The fighting left hundreds dead, many hundreds injured, and the American-allied House of Saud grappling to maintain its hold on power.[64]

As the Mecca crisis began, a rumor flew from continent to continent that the United States and Israel were behind the Grand Mosque seizure.

In Washington, the State Department, still reeling from the takeover of the American embassy in Tehran only a little more than two weeks before, sent out urgent cables to US diplomatic outposts worldwide, warning of possible attacks. Anti-American protests soon erupted across the Muslim world. The most violent by far was in Islamabad. There, the CIA station had learned weeks earlier that radical students at the city's Quaid-i-Azam University might stage demonstrations at the American embassy in support of the hostage takers in Iran. The intelligence was correct. The Pakistani students were members of the youth wing of the fundamentalist Islamic political organization Jamaat-e-Islami, with which General Zia had aligned himself in the course of his Islamization campaign. Throughout the 1970s, Jamaat, like many other conservative Muslim groups, had received generous support from oil-rich Saudi Arabia. Now the Jamaat students, inspired by the embassy takeover in Tehran and inflamed by the mosque seizure in Mecca, were ready to show what they could do.[65]

At noon on November 21, the day after the Grand Mosque takeover began, several hundred angry Jamaat students from Quaid-i-Azam arrived by bus at the gated American embassy in Islamabad and began a demonstration. After speaking with embassy officials, the protesters got back on their buses and drove off, seemingly having made their point. Just a few minutes later, however, the students returned, accompanied by many more busloads of demonstrators. Thousands of people now surrounded the embassy compound. The mood grew ugly — then violent. The crowd pushed past police and swarmed through the gates. The demonstration became a riot. Embassy officials called for help. One hundred thirty seven people took shelter inside the steel-reinforced code room on the building's second floor. Others fled to the nearby British embassy compound. The mob ran wild, rampaging through the grounds, wrecking everything it could, burning cars, and setting fire to the embassy itself. As smoke and flames filled the building, armed rioters climbed to the roof and began shooting down into the code-room vault, killing a young Marine corporal.[66]

Throughout the afternoon, US officials pleaded with the Pakistani government to intervene. President Carter even put in an "impassioned" call for help to General Zia. And yet it would be nearly four hours from the time the rioting began until the Pakistani military responded. At 5:30 in the afternoon, a single Pakistani army unit arrived at the embassy. The rioters slowly dispersed. In the meantime, US installations in Rawalpindi,

Lahore, and Karachi had also been attacked. In the end, one American Marine had been killed and the embassy compound had suffered twenty-three million dollars in damage.[67]

That night, Zia went on television to address the nation. Realizing that he could not afford to alienate the Islamic fundamentalists on whom his political life depended, the general offered only a mild reproach to the rioters. While saying that he understood the "anger and grief" over the seizure of the Mecca mosque that had sparked the attack on the Islamabad embassy, he went on to suggest that the rioters' actions had not been in keeping with "the lofty Islamic traditions of discipline and forbearance."[68]

In Washington, the response was quite different. American officials fumed at the failure of Zia's troops to respond more quickly to the attack. The CIA later determined that Zia, believing those inside the embassy had perished in the flames, decided to let the riot run its course. The diplomatic relationship between the two countries was in tatters. As one National Security Council (NSC) staff member put it, US relations with Pakistan were "about as bad as with any country in the world, except perhaps Albania or North Korea."[69]

But that would soon change, as the Carter administration would shortly find it convenient to exercise the Christian tradition of forgiveness.

Sowing Shit

A little more than four weeks after the Islamabad embassy attack, on Christmas Eve 1979, the Soviet army marched across the border into Afghanistan. The Russians had taken the bait and invaded, just as Zbigniew Brzezinski claims to have predicted. The Soviet action sent shock waves around the world. All hopes for a thaw in Cold War relations evaporated. New fears now emerged that the East–West contest would turn hot.

While the Soviets' primary objective in launching the attack had, at least in the beginning, been to remove the pro-Moscow but intractable Afghan government and install a more pliable client, Brzezinski and much of official Washington viewed the invasion as part of a possible Russian push toward the Indian Ocean. In that context, Pakistan's status in Washington improved markedly. No longer was Zia's military government a pariah to be shunned. Suddenly it was an ally to be embraced and protected. As one of Brzezinski's staff members recalled, the administration's attitude toward Pakistan, "overnight, literally . . . changed dramatically."[70] Small matters such as human rights, Islamic extremism, attacks on embassies, and nuclear weapons would no longer be allowed to stand in the way of good relations between the two countries.

In its newly rehabilitated guise, Pakistan would play a central role in Brzezinski's plan for countering the Soviet invasion. Brzezinski, the eager Cold Warrior, hoped to punish the Russians for their action. The invasion, as he saw it, offered an opportunity "to finally sow shit in [the Soviets'] backyard."[71] It would, however, have to be done covertly. The United States and its allies would pay for the shit, the Pakistanis would deliver it, and the Afghanis would do the actual sowing. Years later, the blowback from the operation would result in a worldwide shit storm.

Brzezinski laid out his scheme in a secret memo to Carter the day after Christmas, just two days after Soviet tanks first rolled into Afghanistan. The national security advisor cast the issue as a "regional crisis." "If the Soviets succeed in Afghanistan," Brzezinski wrote, "and if Pakistan acquiesces, the age-long dream of Moscow to have direct access to the Indian Ocean will have been fulfilled." As such, the situation posed "an extremely grave challenge" for the United States. Unless the US somehow managed to "project both confidence and power into the region," Pakistan would likely be intimidated and might eventually succumb to "some form of external Soviet domination." With Iran already "destabilized," there would no longer be a "firm bulwark" in the region against a "Soviet drive to the Indian Ocean."[72]

On the other hand, Brzezinski argued, if the United States *could* project power into the region, there was a chance that Afghanistan would become "a Soviet Vietnam," a quagmire from which the Soviet Union could not extract itself. As to what might be done to bring that about, Brzezinski offered some "preliminary thoughts." The plan he outlined was, in essence, a beefed-up version of the strategy he had promoted earlier in the year that had helped lure the Soviets into Afghanistan in the first place. There were, however, two important new features. One was that the United States would no longer be limited to providing only nonlethal assistance. The other was that the administration would have to put aside its concerns about Pakistan's nuclear weapons program. "It is essential," Brzezinski wrote, "that the Afghanistani resistance continues. This means more money as well as arms shipments to the rebels, and some technical advice." Toward that end, he added, the United States should "concert with Islamic countries both in a propaganda campaign and in a covert action campaign to help the rebels." It should also "encourage the Chinese" to assist. But most important, for the plan to work, the administration would have to "both reassure Pakistan and encourage it to help the rebels." That, Brzezinski wrote, "will require a review of our policy toward Pakistan, more guarantees to it, more arms aid, and, alas, a decision that our security policy toward Pakistan cannot be dictated by our nonproliferation policy."[73]

With that, Brzezinski had laid out the strategy that would guide American action in South Asia for the next decade. The United States and its allies would covertly enable a proxy Islamic holy war against the Soviet Union in Afghanistan. To do so, it would turn a blind eye to Pakistan's nuclear weapons program. The full implications of those actions would be realized only many years later.*

"Peanuts!"

President Carter quickly got behind Brzezinski's plan. The president viewed the invasion of Afghanistan as a major shift in Kremlin policy and an "extremely serious" threat to world peace.[74] He declared the move "the greatest foreign policy crisis confronting the United States since World War II."[75] Already under attack in some quarters for his seemingly timid response to the Iran hostage crisis, Carter believed strong, forceful action was needed to counter the Soviet move. In the days ahead, Carter would embargo wheat sales to Russia, order a boycott of the Olympic games scheduled to be held in Moscow in the summer of 1980, and withdraw from Senate consideration the second Strategic Arms Limitation Treaty (SALT II), painstakingly worked out with Moscow.[76] But those public gestures were just for show. The real action would be covert. And bringing Pakistan on board, Carter agreed, was key to the plan. Doing so would involve dropping large amounts of cash and all talk of nuclear weapons.

At an NSC meeting on December 28, two days after receiving Brzezinski's memo, Carter declared that ensuring Pakistan's participation

---

*Asked in a 2004 on-camera interview whether he still believed the policy outlined in his memo to Carter was appropriate, Brzezinski responded, "In general, yes." After further elaboration, he then ended the interview (authors' interview with Zbigniew Brzezinski, May 26, 2004). A clip from the interview can be seen at www.islamicbomb.org.

in the Afghan enterprise was even more important than bringing Egypt and Saudi Arabia — with whom Brzezinski's staff and State Department officials were already holding discussions — on board as financiers of the operation.[77] The following day, the president approved a covert action program that, in line with Brzezinski's proposal, instructed the CIA to provide the mujahedin with weapons and ammunition as well as nonlethal supplies and support.[78] With the plan formally approved, the wooing of Pakistan began.

On New Year's Day 1980, Assistant Secretary of State for Near Eastern and South Asian Affairs Harold Saunders sent a memo to Secretary Vance outlining a possible aid package for Pakistan. Saunders proposed that the administration submit a request to Congress for $150 million in aid and credits for Islamabad, along with debt relief, which the Pakistanis wanted. But providing that aid would require lifting the ban on assistance to Pakistan imposed eight months earlier because of its nuclear weapons program. Saunders therefore recommended that the administration seek the necessary Congressional waiver. Yet even with nuclear weapons off the table and the lure of cash, Saunders noted that Islamabad would find this "an inadequate basis for a closer, more supportive relationship." He cautioned, however, against offering more, such as F-16s or tanks, because of the predictable negative reactions from India.[79] This would be the starting point from which the bidding for Zia's affections would rise.

In a nationally televised speech on the Afghan crisis on January 4, President Carter pledged to bolster Pakistan, now a frontline state in the Cold War. He announced that the United States and its allies would provide military equipment, food, and other assistance to help Pakistan "defend its independence and national security against the seriously increased threat" posed by Soviet forces in Afghanistan.[80] But Carter gave no indication of the level of that support. He did, however, begin a concerted push to get allied Western and Arab states to significantly increase their assistance to Pakistan.

Zia took full advantage of the situation. The Soviet invasion of Afghanistan was like a Christmas present to the Islamic Republic of Pakistan. Suddenly, everyone was lining up to give it money. Realizing his much-improved negotiating position, Zia played hard to get, making clear that he would hold out for the best possible offer.

The week after President Carter's speech, Zia's foreign affairs advisor Agha Shahi arrived in Washington for talks with the White House and

State Department on the aid package. In meetings with Shahi, Carter and Secretary of State Vance offered the Pakistanis $400 million in US aid over two years. They also pledged to keep in place a 1959 security agreement in which the United States pledged to defend Pakistan against communist aggression. Moreover, Vance and Carter confirmed that they would seek authority from Congress to waive the sanctions imposed on Pakistan over its bomb program.[81]

To the Americans' great surprise, Shahi indicated that he found the offer insufficient.[82] He also reportedly gave no indication that there would be any change in Pakistan's position on its nuclear enrichment program.[83]

Zia's response was considerably less diplomatic. He subsequently dismissed the American aid proposal as "peanuts." Accepting such a paltry offer, he said, would simply "buy greater animosity from the Soviet Union" while doing little to improve Pakistan's security.[84]

Although the Carter administration was embarrassed and annoyed by Zia's public rebuff, it proceeded as if an agreement would still be reached. In his State of the Union address in late January, the president unveiled a new policy, soon dubbed the Carter Doctrine, under which an attack by the Soviets on the Persian Gulf region would be considered an attack on the United States. Carter made clear that US security guarantees would extend to Pakistan.[85]

In early February, Brzezinski and Deputy Secretary of State Christopher flew to Pakistan to ply Zia with more loot. During two days of talks, Brzezinski allowed the Pakistani leader to understand that the United States was now prepared to offer Islamabad nearly $500 million in economic aid over two years. Brzezinski also assured the general that there would be more funds available in the future — a prediction that would be more than amply borne out.[86] But the Americans also issued a warning. During the sessions, Christopher told the Pakistanis that a nuclear test by Islamabad "would foreclose increased military support from Washington," according to the *New York Times*. (He said nothing, apparently, about economic assistance.) The Pakistanis were said to have assured Christopher that they did not plan to build a nuclear weapon but refused to pledge that they would not build or test a "peaceful nuclear device."[87]

Unable to get a satisfactory response, the Americans gave ground. During the meetings, the *Christian Science Monitor* reported that the administration appeared to be "easing pressure" on Zia to abandon nuclear weapons development efforts. Brzezinski told the paper that the United

States remained "concerned" about nuclear proliferation. But when asked specifically whether the administration had "softened" its stance regarding Pakistan's bomb program, a "ranking US official" replied: "We will have to harmonize our goal of non-proliferation of nuclear weapons with the changes in the strategic situation in the area."[88] Zia presumably had no trouble interpreting the diplomatic code.

But even this was not enough to satisfy the Pakistanis. Zia personally called Carter to tell him that the American offer was offer was simply not good enough. Once again, the Americans were surprised by the rejection.[89] But Zia may have been looking past the Carter administration in anticipation of a better deal down the road. On the eve of the Brzezinski-Christopher mission to Islamabad, US presidential candidate Ronald Reagan told reporters on the campaign trail that he did not believe the United States should stand in the way of other countries developing nuclear weapons. "I just don't think it's any of our business," the future president said.[90] This must certainly have been welcome news to Zia.

Despite Zia's private rejection of the aid offer, the Carter administration continued to try to salvage a deal. Throughout February, the White House worked with Congress in a desperate effort to put together a package that Islamabad would find acceptable. At the same time, the administration also tried getting tough on the nuclear issue. In late February, the State Department revealed that the United States had received intelligence that Pakistan was continuing work on A.Q. Khan's uranium enrichment facility at Kahuta.[91] But if the administration believed Zia would be intimidated by the announcement, they were mistaken. Far from backing down, the Pakistanis became defiant. Just days after the administration floated the story, PAEC chairman Munir Khan publicly rejected what he described as the American thesis that technology that could be used to make nuclear weapons should be confined to the original members of the "nuclear club"— the United States, Britain, France, the USSR, and China.[92] Shortly afterward, on March 5, the Pakistanis publicly spurned the US aid offer and explicitly disassociated themselves from efforts to push a deal through Congress.[93] It was a severe slap in the face for the Carter administration, which had made such an obvious show of its eagerness to embrace Pakistan.

But the Pakistanis weren't done yet. To add insult to injury, after refusing to accept the administration's aid package, Zia announced that while he was no longer interested in US military assistance — which Christopher had warned him would be unavailable if Pakistan tested a bomb — he would still like economic aid and debt relief. Gluttons for punishment, the administration cravenly capitulated and pushed for a major debt-rescheduling package for its new Cold War partner.[94] Even as it did so, Zia continued to proclaim his determination to proceed with Pakistan's "peaceful" nuclear program.[95] In the end, it was the Carter administration, not Zia, that rolled over on the nuclear issue.

Seeds of Disaster

The Pakistanis had clearly come to realize that the United States would set aside its concerns about nuclear weapons in pursuit of Cold War objectives. As one of Zia's top aides, Gen. Khalid Mahmud Arif, later wrote, following the Soviet invasion of Afghanistan, "Pakistan's nuclear program remained irksome for the US administration but it no longer was an obstacle in developing a closer relationship with her. The other Western countries and agencies followed the US lead. It became expedient for the West to coexist with the ground reality in Pakistan."[96]

The reason it had become expedient to coexist with the reality of Pakistan's bomb program was, of course, the West's desperate desire to keep Islamabad on board the anti-Soviet Afghanistan operation. And in this, the Carter administration was successful. While the Pakistanis publicly embarrassed the administration on aid and refused to give up their pursuit of nuclear weapons or renounce testing, they enthusiastically joined in the effort to kill communists in Afghanistan. But the price for that cooperation would be higher than anybody could yet know.

Bzrezinski took up the issue of Pakistani participation in an expanded Afghan campaign with Zia during his trip to Islamabad in February.[97] Zia readily agreed — but once again there were conditions. First, Zia insisted that the CIA channel every dollar and every weapon through the ISI, which, in turn, would pass them on to the mujahedin. Second, the Pakistanis demanded that the Americans have no direct dealings with the Afghan rebels but instead work only through the ISI. Those two stipulations, which would remain in place throughout the Afghan operation, laid the basis for a long-term disaster.[98] Because the ISI was — and, some say, still is — dominated by Islamic fundamentalists, it funneled the vast majority of the weapons and resources supplied by the CIA to the most radical, anti-Western factions within the mujahedin. More moderate elements were largely frozen out. This suited Zia's political purposes, allowing him to promote his Islamist agenda in Afghanistan while currying favor with his fundamentalist allies at home. And because the CIA, in deference to Zia, kept its contacts with the Afghan holy warriors to a minimum in the early years, it exercised very little control over their activities. Those radical Muslim jihadists — armed, funded, and made powerful in large part by American taxpayer dollars — would go on to train legions of followers, who would eventually expand their holy war to other countries, including the United States.

Having set the wheels of this latent disaster in motion, Brzezinski headed off to seek additional funding for the jihad. An early stop was Saudi Arabia, where Brzezinski struck a deal in which the royal family would match US contributions to the mujahedin dollar for dollar.[99] Other US allies, including Britain, France, West Germany, Egypt, Turkey, the United Arab Emirates, and even China, also contributed to the cause. Although the program began on a relatively modest scale — $20 million in 1980 and $30 million the following year — it would expand exponentially, reaching some $630 million a year by 1987. Over the course of the 1980s, more than $3 billion would flow from the CIA, through the ISI, to the Afghan jihadis.[100] Much of that assistance disappeared along the way.[101] According to a top-level Defense Intelligence Agency source, a large portion of the missing money was siphoned off into Pakistan's bomb program.

While the Afghan operation succeeded in driving the Red Army back across the Russian border and contributed to the eventual collapse of the Soviet Union, those victories came at an unnecessarily high price to long-term world security. The great tragedy of the Afghan adventure is that the United States need not have sacrificed its principles on nuclear weapons or ceded its authority over the funding and control of the mujahedin in order to win Pakistani cooperation. The notion that Pakistan would not have participated in the effort without being bribed is a myth — one that Zia happily encouraged. In fact, national self-interest alone would dictate that Islamabad do everything in its power to resist the communist takeover of neighboring Afghanistan and a possible Soviet push into Pakistan or other states in the region. As Zbigniew Brzezinski and the CIA knew well, the Pakistanis had been doing just that well before the United States became involved. And yet the United States persisted in the false belief that all would be lost in its Cold War campaign in Afghanistan if it did not accede to Pakistan's every whim. Zia, fully aware of the American delusion, milked Washington for everything he could get.

The Reagan administration would adopt and massively expand the Carter-initiated Afghan campaign, complete with ISI control over the jihadi operation. It would also institutionalize the policy of winking at — and in some cases covering up — Pakistan's nuclear program to ensure Zia's cooperation. As a result, Islamabad's drive for an atomic bomb proceeded virtually unfettered. At the same time, the militant Islamic jihadist movement that would eventually give rise to al Qaeda and its ilk grew in power and influence.

6

Help from the West

IN THE SUMMER of 1979, A.Q. Khan's uranium enrichment program was in trouble. Although Khan and his team of scientists had managed to enrich small quantities of uranium the year before and had made additional progress since, the effort was now being slowed. Export controls put in place in the fall of 1978 by the British and later by the United States had begun to take effect. Khan's supply network had been interrupted, and he was now having difficulty obtaining critical centrifuge components and other equipment for the Kahuta facility. "The material that we were buying from the British and Americans has been stopped," Khan wrote to a Pakistani associate in Canada that summer. "They are even stopping screws and nails."[1] By October 1979, news reports would surface that work at Kahuta had come to a standstill.[2]

But Khan's troubles did not end there. In August, stories leaked out that the Carter administration was considering various options for dealing with Pakistan's nuclear program, including the possibility of a covert operation aimed at destroying Kahuta.[3] And although Khan could not know it, the State Department independently explored the possibility of launching covert actions or air strikes against Pakistani nuclear facilities.[4] To protect against such threats, the already heavily guarded Kahuta site was now ringed with antiaircraft missiles.[5] Khan may have, legitimately, felt under siege.

But Khan's luck was about to change. Even as these outward signs of pressure against Pakistan's nuclear effort mounted, the Carter administration had already begun collaborating with General Zia and Inter-Services Intelligence (ISI) in the covert operation that would eventually help lure the Soviets into Afghanistan. Once the Russians crossed the border and the United States turned its attention to engaging Islamabad's cooperation in the broader anti-Soviet Afghan jihad, the administration's policy toward Pakistan would, in the words of Zbigniew Brzezinski, no longer be "dictated" by its nonproliferation concerns.[6]

Pakistan's bomb program and its nuclear smuggling network would soon be running again at full tilt. And now there would be almost nothing to stop them. As A.Q. Khan himself noted years later, after Islamabad's successful nuclear tests, Pakistan's atomic weapons program was "indebted" to America's single-minded commitment to the Afghan campaign and the resultant willingness not to trouble Zia about his effort to build the bomb.[7]

"Why Shouldn't Pakistan Have a Nuclear Bomb Anyway?"

While the export controls belatedly imposed by the United States and Britain had slowed the flow of goods to Pakistan's nuclear program, the procurement network had never ceased operation. Once Islamabad became a partner in the anti-Soviet Afghan campaign and the Carter administration adopted a more lenient view of Pakistan's nuclear activities, the network expanded its operations dramatically. It would soon evolve into a truly global enterprise, obtaining the vast array of sophisticated equipment with which Pakistan would eventually build its bomb. A switch would later be thrown to reverse the flow and enable the network to send the nuclear technology Pakistan had been allowed to develop to other countries. That too would be tolerated by the United States.

Even after the uproar over the inverter sales in the summer of 1978, Britain remained a primary shopping location for Pakistan. And the British company at the center of the inverter controversy reportedly continued to serve as a supplier. In August 1979, reporter Simon Henderson of London's *Financial Times* revealed that Pakistan was still buying goods in Britain that were believed to be for use in Islamabad's nuclear program. Henderson reported that the goods, primarily machine tools, were being sent to Pakistan through Weargate Ltd., the company through which the inverter orders had been placed. As with the inverters, Henderson reported that most of the new orders had originated with Pakistan's Special Works Organization (SWO), the army engineering unit that was building the Kahuta enrichment facility. One of Weargate's principals, the Welsh engineer Peter Griffin, told Henderson that over the past eighteen months he had sold £800,000 (nearly $1.8 million) worth of equipment to the SWO and that an order to supply buses and ambulances was still outstanding.[8] Griffin insisted that none of the sales violated export regulations. "I am not helping Pakistan make a nuclear bomb," Henderson recalls Griffin saying, "but why shouldn't Pakistan have a nuclear bomb anyway?"[9]

In 1980, British Customs and Excise officers interviewed Griffin about his dealings with Pakistan. According to a confidential 2005 British Customs report that cites a contemporaneous Customs account of the interview, Griffin told the investigating agents that he had recently received an order from A.Q. Khan for six devices known as mandrels, rods or shafts around which metal or other materials are shaped on specialty lathes. They are used to produce hollow, high-precision cylindrical objects.The report states that Griffin told the Customs agents he understood that supplying the mandrels would be difficult because a previous order of equipment had been seized by Customs in February 1979. The report also indicates that Griffin said he had informed the head of the SWO that he would be unable to ship the mandrels because an export license would not be granted. Nevertheless, the report indicates that Griffin obtained the mandrels and moved them to the facilities of an export packager to prevent them from being damaged. The report says that Griffin told Customs investigators that the mandrels were the final consignment of equipment ordered by SWO for the production of bellows, which the 2005 Customs report describes as "centrifuge component parts." They are small, hollow cylinders used in high-speed centrifuges that allow the rotors to flex.[10] According to the report, Griffin also told the investigators in 1980 that he had not known the intended end use of the equipment but indicated that he now understood its sensitivity.

Griffin, who studied engineering at Oxford Brookes University in England, ran Weargate from his rambling Victorian manor house just outside of Swansea, in South Wales.[11] He claims that his business dealings with Pakistan began as the result of an odd mistake. Griffin, then forty-two years old, was working for another Swansea machine tool agency when he received a call from one of A.Q. Khan's friends and suppliers who was seeking equipment, according to published accounts and sources familiar with Griffin's version of events. Khan's friend, Abdus Salam, had been attempting to call another firm but dialed the wrong number from a business directory. Griffin, sensing an opportunity, convinced Salam to cut him in on part of the deal.[12] The two men eventually threw in together as business partners, forming Weargate in late 1977, followed by a string of about a half dozen other companies.[13] According to the book *Long Road to Chagai*, after the inverter deals with Pakistan became public, Salam was "kept under surveillance," presumably by British authorities. "A secret search of his office revealed documents and drawings which were traced to the Urenco plant in the Netherlands," the book reports.[14]

Griffin, the son of a Welsh railroad engineer, first met Khan over dinner in London. Khan was a frequent visitor to the city in those days and stayed at the house of an old college friend. At the dinner, which took place at a Pakistani restaurant not far from the House of Commons, Khan was accompanied by several men. One was a high-ranking Pakistani military officer. During the meal, Khan was reserved but nonetheless charming and amusing. Griffin has said he did not realize how important Khan was until the officer at one point referred to the scientist as "the boss."[15]

In the midst of Weargate's inverter deals with Pakistan in 1978, two retired army officers working for the Pakistan Atomic Energy Commission (PAEC) visited London. One was Maj. Mohammed Sadiq Malik, a procurement officer. The other was Capt. Fida Hussein Shah, an assistant administrative officer. According to the *Financial Times,* the two men told British officials that their trip had been arranged by their project director, Dr. Abdul Qadeer Khan. They also indicated that they would be visiting a company called Source Reliance International on Clovelly Avenue in London. Source Reliance, it turned out, was a sister company of Weargate. The directors and sole shareholders at that time were Peter Griffin's Weargate partner, Abdus Salam, and Salam's wife, Naseem. Griffin became a director, but not a shareholder, of Source Reliance on September 1, 1979. Like Weargate, Source Reliance also arranged to have engineering equipment sent to Pakistan. In late 1979, a British engineering firm revealed that it had earlier sent a metal finishing plant to Pakistan through Source Reliance that it subsequently came to believe was destined for use in Islamabad's uranium enrichment program, according to the *Financial Times.* An auditor's report later showed that Source Reliance's exports for fiscal year ending April 30, 1980, totaled £723,202 (about $1.66 million). Its net profit was £328 ($744).[16]

Source Reliance remained in existence for several more years. When it was dissolved, a liquidator's statement revealed that in the spring of 1983 the company received two payments, listed as "Book Debt," from "I.H. Khan." (This, of course, is the same name as the I.H. Khan who served as a procurement agent for Pakistan's SWO in West Germany and helped arrange the inverter deal in which Weargate was invovled.) One, in March 1983, was for £15,872.26 (roughly $36,000). The second, in April, was for £604.07 (approximately $1,370).[17]

The Canadian Caper

In the early 1980s, while Weargate, Source Reliance International, and the other Griffin–Salam enterprises continued to operate, the two principals parted company. Griffin remained based in Britain for a number of years and continued as a supplier to Pakistan. He shipped all manner of goods and equipment, every bit of it, he maintains, completely legal. "Anything that could be sent to Pakistan, I sent to Pakistan," he later said.[18]

In 1983, Griffin received a copy of the book *The Islamic Bomb,* an exposé that, among other things, details A.Q. Khan's role in Pakistan's quest for a nuclear capability and gives prominent mention to the roles Griffin and Salam played in the inverter affair. British Customs agents found the book years later during a search of Griffin's home. Inside the cover, according to a 2004 Customs report, the book was inscribed, "With kindest regards, my friend, Peter Griffin." It was signed, "Dr. AQ Khan."*

Griffin remained a steady source of supply for Pakistan in the years ahead. In the early 1980s, he reportedly began obtaining much of the gear he exported from Switzerland.[19] He eventually set up shop in Dubai, where, according to a 2005 British Customs report, he continued doing business with Pakistan. He later took up residence in a secluded villa in the south of France. By then he had long since faded from public view. Many who follow nuclear proliferation issues believed he had retired.[20] His name reemerged in 2004 amid allegations of his involvement in the so-called A.Q. Khan black market nuclear network. Griffin has repeatedly denied any wrongdoing and has not been charged with a crime.[21]

Abdus Salam would also eventually drop off the radar screen and land in sunny climes. In the short term, however, he quickly turned up in the midst of another inverter deal. In September 1980, reporter Simon Henderson traced the Indian-born British citizen to Dubai. There, Henderson discovered Salam serving as the director of a company known as Khalid Jassim General Trading. The company, named after Salam's local partner in the business, was, by all appearances, little more than a sign on a door. The office was a single room inside a small apartment located in the Arab Bank building in downtown Dubai. The office staff, Henderson discovered, consisted of just two people, a Mr. Amjad Ali and a young female secretary.[22]

---

*In the early 1980s, an unauthorized edition of *The Islamic Bomb* appeared in Pakistan. In it, all the unflattering references to Khan had been removed and negative comments about his rivals had beeen inserted. Khan himself reportedly had the book reprinted using government money, and sent it, free of charge, to prominent Pakistani military leaders, scientists, and government officials. (Greg Bearup, *South China Morning Post* [Hong Kong], February 11, 2004; Fahatullah Babar, "The Nuclear Sage of Pakistan," *The News* [Pakistan], April 22, 2005.)

The trail that led Henderson to Khalid Jassim General Trading began in Montreal, Canada. In July 1980, two officials of the PAEC arrived at Montreal's Mirabel Airport. They were traveling on diplomatic visas, but their mission was not one of diplomacy. Rather, the two men, Anwar Ali and Imtiaz Ahmad Bhatti, had gone to Canada to arrange the purchase of inverter components. Britain's tightening of export controls in the fall of 1978 had blocked the second order of inverters destined for Pakistan. As a result, Khan's enrichment facility was about one hundred inverters short of what it needed to run at full capacity. Although Khan and his team were well on their way to reverse engineering the inverters they had already obtained and would soon be building the devices themselves, there were some components they could not yet produce. Thus, Ali and Bhatti traveled to Canada carrying with them a high-priority shopping list.[23]

But Ali and Bhatti did not enter Canada unnoticed. The Royal Canadian Mounted Police, alerted by British authorities to the Pakistanis' impending arrival, kept the pair under close surveillance throughout their trip.[24] What the Mounties apparently did not know was that Ali, despite his diplomatic visa, was a centrifuge expert employed at Khan's uranium enrichment lab.[25] Nevertheless, the activities of the two Pakistanis quickly led the Canadians to an alarming discovery. Rather than doing diplomatic work at the Pakistani Consulate, Ali and Bhatti made contact with three naturalized Canadian citizens who would serve as a local purchasing network for inverter components. In the coming weeks, the Canadians, all of Arab or South Asian origin, would buy an array of resistors, capacitors, condensers, and other gear through two small electrical supply stores in Montreal. The equipment came from manufacturers in the United States, including General Electric, Westinghouse, RCA, and Motorola. Once in Canada, the parts were repackaged for shipment to Pakistan through one of two intermediary agents, one of which was Khalid Jassim General Trading.[26]

The Canadian purchasing team had been recruited before Ali and Bhatti arrived. One of the members, Salam Elmenyawi, was a native Egyptian who owned an electrical parts store in Montreal. Another was Mohammad Ahmad, a mechanical engineer of Indian extraction who worked for the province of Quebec. The third was Abdul Aziz Khan, the Pakistani-born electrical engineer with whom A.Q. Khan had maintained an extended correspondence. A.Q. Khan had originally encouraged the Canadian engineer to join his team in Pakistan. Although A.A. Khan chose to remain in Canada, he did spend vacations in Pakistan, helping to train engineers working in A.Q. Khan's program. He also sent A.Q. Khan technical literature from the United States and Canada.[27] Much of the correspondence between the two men, all of it in Urdu, contained cryptic references. In one letter, for example, A.Q. Khan wrote: "June 4 is a historic day for us. . . . We put 'air' in the machine and the first time we got the right product and its efficiency was the same as the theoretical. Everybody was quite happy."[28] June 4, 1978, was the day that Khan's team first produced enriched uranium.[29]

The two men also commiserated about the increasing difficulty of obtaining inverters and related components. In an October 1978 letter, A.Q. Khan complained about impending British efforts to tighten export controls following exposure in Parliament of the inverter deal. A.A. Khan wrote back that this was "mischievous news."[30] In a letter the following summer, A.Q. Khan lamented that materials from Britain and the United States had been cut off. "Now we will have to do some work ourselves," he wrote. A.A. Khan and his Canadian cohorts would help.[31]

The Canadian trio eventually made at least ten shipments of parts and equipment to Pakistan. The total value was estimated at $560,000 (Canadian). Then, on August 29, 1980, as the group prepared to send yet another consignment, Canadian Mounties, acting on a tip from the British, descended on the operation. They seized nineteen boxes of electrical components as they were being readied for shipment at Mirabel Airport. The goods, worth some $56,000 (Canadian), had been split into two consignments. One was labeled for delivery to a company called Tech Equipment in Islamabad. The other was addressed to Khalid Jassim General Trading in Dubai.[32]

The three Canadians were taken into custody. Throughout an all-night interrogation, Elmenyawi and Ahmad reportedly denied knowing the intended purpose of the equipment they were sending to Pakistan. A.A. Khan was released the following day. Mounties followed him to Montreal's Central Railway Station. There, Khan reportedly removed a suitcase from a storage locker, took out a pile of documents, tore them up, and dumped them in the trash. He then hopped a bus to the airport, where he intended to get on a flight to Pakistan. The Mounties nabbed him before he could get on the plane. The police later reassembled the documents Khan had shredded. Among them, they found a paper by an American scientist on the use of centrifuges for enriching uranium. Khan told investigators he was taking the article to another scientist.[33]

The Canadians were eventually charged on a variety of counts, including failure to obtain necessary export licenses, exporting goods originally imported from the United States without "adding value" (that is, without integrating them into larger items), and violating the Atomic Energy Materials and Equipment Act, which regulates nuclear sales to other countries.[34] After repeated delays, the case eventually went to trial. But even then there were problems. A strict prohibition on hearsay evidence prevented prosecutors from explaining to jurors the roles Ali, Bhatti, and A.Q. Khan played in Pakistan's nuclear program. As a result, Abul Aziz Khan was able to offer apparently convincing explanations for the cryptic references in his correspondence with A.Q. Khan. He suggested, for example, that A.Q. Khan was engaged in research on synthetic butter factories and that the mention of putting "air" into the machine concerned an effort to produce cooking gas.[35] Operation Butter Factory was reportedly a code name used by at least one of A.Q. Khan's longtime European associates to refer to his dealings with Pakistan.[36] A.A. Khan also testified that the inverter components were for use in textile and food-processing plants.[37]

Witnesses also proved difficult. Prosecutors sought testimony from a representative of the British arm of Emerson Electric. Controversy arose, however, over the witness's refusal to provide details about the sale of Emerson inverters to Pakistan through third parties. As a result, only portions of his testimony were admitted to the jury.[38] The Crown's chief prosecutor, Guy Gilbert, says that the testimony would have provided a "clear demonstration" that the exported parts were to be used for building inverters.[39]

At the end of a two-month trial, Salam Elmenyawi and Mohammad Ahmad were convicted on one count each of exporting goods without the proper license. The convictions carried a maximum penalty of five years in prison and a $25,000 (Canadian) fine. At sentencing the men were each fined $3,000. In handing down the sentence, Justice Gerald Ryan said Elmenyawi and Ahmad had been too causal in their actions and showed no concern for the "legal implications" of shipping the restricted parts to Pakistan. A.A. Khan was acquitted entirely.[40] It was a result that would become familiar in the years ahead.

Gilbert believes Justice Ryan, who died in 2002, was "partial" to the defense. "I remember the address by the judge to the jury," Gilbert said in a 2006 interview. "He almost gave them [the case] on a silver fork." Gilbert recalled crossing paths with Ryan several years after the case ended. "I said,

'Gerry, don't tell me that you think those guys did not export the parts to make the inverters.' He said, 'Of course, of course they did it.' 'Well,' I said, 'that's not the way it came out when you addressed the jury.'"[41]

Gilbert says he also taunts the defense lawyers in the case when he sees them. "I tease them once in a while," he said. "I say, 'You think these guys knew what they were doing? . . . I think you contributed to them [Pakistan] having the bomb today.'"[42]

The two PAEC officials, Anwar Ali and Imtiaz Bhatti, were never charged in the case. The Pakistani ambassador to Canada later acknowledged the two men had been involved in the equipment purchases but insisted their activities were inconsequential. "It's not nuclear material," the ambassador said in an interview broadcast on Canadian television. "What they bought was what anyone could pick up from a catalogue from the store, and that equipment was ordered. There is no ban on it."[43]

Imtiaz Bhatti went on to a long career as a finance administrator at A.Q. Khan's research facility at Kahuta. The government of Pakistan later awarded him one of its highest honors, the Sitara-i-Imtiaz or star of excellence, in recognition of his service to his country. He died in 2002.[44]

Anwar Ali enjoyed a long and distinguished career with the PAEC and played key roles in Pakistan's development of nuclear weapons and missile-guidance systems. In 2006, he was rewarded for his service by being named chairman of the commission, placing him at the head of Pakistan's entire nuclear program.[45]

That left only one loose end in the case.

Land of the Free

When reporter Simon Henderson visited the Dubai office of Khalid Jassim General Trading in September 1980, shortly after Mounties raided the Canadian purchasing ring, Abdus Salam was nowhere in sight. A representative of the firm refused to discuss the company's business. But Henderson found tantalizing clues in the company's files and in a shipment of goods outside the office door.

In the hallway, Henderson spotted four boxes labeled "Mikron infrared thermometers." An agent in the Gulf for the Mikron Instruments company of New Jersey told Henderson it had sold the thermometers to Khalid Jassim, ostensibly for use at a cement factory in nearby Sharjah. But Mikron itself informed Henderson that the instruments could be used to measure the temperature of "moving objects without making contact and in conditions of extreme radiation."[46] This, Henderson later wrote, suggested the thermometers would be "ideal" for measuring the temperature of uranium enrichment centrifuges spinning at high speed.[47]

But the question remained: Where was Abdus Salam, the man at the center of the British and Canadian inverter purchases? It would be more than twenty-five years before an answer emerged. During dinner with a reporter from the National Security News Service in Washington, DC, in 2006, Henderson, by then a fellow with the Washington Institute for Near East Policy, said he was told in the early 1980s that Salam had moved to Florida. Based on that lead, the News Service reporter started digging. What he and his colleagues discovered was disturbing.

The News Service quickly determined that Salam was, in fact, living in Tampa, and had been there openly for some time, under his own name. Further, he had established a long string of companies in Florida. These included at least three import-export firms, two trading companies, two communications outfits, a computer retailer, two hospitality companies, a financial services enterprise, and several companies involved in indeterminate business. None seemed to have had any noticeable presence in the area. A number of the companies overlapped in time with U.K. enterprises in which Salam was still involved with Peter Griffin. One, in fact, had a name remarkably similar to that of Source Reliance International, the Salam–Griffin outfit that had been visited by PAEC officials working for A.Q. Khan and that had reportedly arranged the shipment of engineering equipment to Pakistan in 1979. Salam's Florida company, International Reliance, was established on December 31, 1982, near the time, and possibly slightly before, Salam arrived in the United States. This was less than six months after Salam and Griffin began the long legal process of dissolving Source Reliance International in the UK.[48]

But that was just the beginning. As the National Security News Service continued to look into Salam's activities, an even stranger picture took shape. In June 2006, a reporter from the News Service traveled to Tampa in hopes of speaking with Salam. Salam lives, and works, in a simple one-story house on a quiet cul-de-sac in a nondescript housing development on the outskirts of Tampa. On the day the News Service reporter visited, a car with a flat tire sat in the driveway. When the reporter knocked, Salam, a small, almost tiny man in his late sixties, answered the door dressed in a sleeveless white T-shirt and a traditional Indian wrap-around garment known as a dhoti. He made it clear he did not wish to discuss his affairs. "Why should I tell things to you," he said when asked about his dealings with Peter Griffin and his business activities in Florida. "Are you my mother-in-law?"

Some of Salam's Florida business associates proved more communicative. Salam's most public profile during the 1980s was as a partner in a group of Curtis-Mathes home-entertainment equipment stores. Salam operated an outlet in Saint Petersburg, across the bay from Tampa, while his partners ran stores elsewhere around the state. One of the partners, David Reed, was also involved in several other enterprises with Salam. The men first met and went into business together shortly after Salam, whom Reed and the other partners knew as Abdul, arrived in the United States from Dubai around 1983. Reed recalls Salam telling him that there had been trouble in Dubai. As Salam told the story, he had been involved in a very successful import-export and trading business in Dubai. He had a local Arab partner from whom the company took its name. (Reed could not recall the name of the partner.) At some point, the partner took Salam to court claiming, falsely, according to Salam, that it had been his idea to start the company, that he had put up all the money, and that he had put more time into it than Salam. The Dubai court, Salam told Reed, had sided with the Arab partner. Salam, according to Reed, ended up going to jail and losing all his money. "The partner ripped him off," Reed said. That, at least, was the story Salam told.[49]

While the information Reed provided offered new insight into Salam's activities in Dubai around the time of the Canadian inverter deal, it did little to explain what he had been doing in the United States for the past quarter century. But another partner in the Curtis-Mathes franchises, Julio Bagiardi, offered intriguing information on that score. Bagiardi remembered his former partner well. He described Salam as a "very shrewd" businessman. "That guy was pretty slick," Bagiardi told the National Security News Service reporter. And yet, there were problems. "We [Bagiardi and the other partners] didn't like the way things were being run," Bagiardi said. Salam did a good job of managing his store, according to Bagiardi, but he was seldom there. And then there was the matter of the bank. "All he wanted to do was make a lot of deposits at the same bank," Bagiardi said. "He insisted on keeping [all his deposits] at BCCI."[50] BCCI. The Bank of Credit and Commerce International. The notorious, scandal-ridden financial institution that would explode in a blizzard of criminality and corruption in 1991. The bank through which the CIA and Saudi intelligence financed covert operations. The bank used to channel money to the Nicaraguan contras and the Afghan mujahedin. A bank that supported terrorism. And the bank that helped Pakistan build the Islamic bomb. That was where Abdus, aka Abdul, Salam insisted on stashing his company's money.

Moreover, Salam wanted his partners to do the same. Bagiardi says that Salam was "very forceful" in trying to convince the others to deposit all the funds from their stores at BCCI. The payoff, Salam told them, would be access to "unlimited" cash. Bagiardi says Salam told the partners that as long as they made daily deposits at BCCI, "it would never be a problem if we needed to get money." Bagiardi says Salam told them they could "get any kind of money we wanted from that bank as along as we made our deposits there." And, according to Bagiardi, Salam explained that the best part was that they "never had to pay nothing back." There would be "no payments," Salam told his partners. "I never understood that situation," Bagiardi said.[51]

Bagiardi said that Salam had been doing business with BCCI since he arrived in the United States. He also said that Salam "knew somebody" at the bank.[52]

David Reed confirmed that Salam kept an account at BCCI. He also made clear that he and the other partners in the Curtis-Mathes franchises had nothing to do with the bank. "That was Abdul's thing," Reed said.[53]

But Salam's "thing," it turned out, involved more than just depositing money at BCCI. A search of Florida court records revealed that in 1990, BCCI won a nearly $50,000 judgment against Salam and one of his import-export companies. That seemed odd, since by that time the outlaw bank had already begun to unravel. Why, at that stage, would it go after one of its most enthusiastic patrons? The answer proved surprisingly difficult to obtain. The court had destroyed almost all of the records of the case in 1995. The lawyers involved claimed they too had destroyed their files, or they refused to answer questions. Several denied having any recollection of the case. BCCI liquidators were unable to provide any information. Prosecutors and investigators involved in the criminal proceedings against BCCI in Tampa said they knew nothing about Salam. And the various managers and assistant managers of the bank where Salam deposited his money every day for years either would not return phone calls, denied knowing who Salam was, or claimed that, while the name did "ring a bell," they could recall no specifics about their former client.

A breakthrough came when the News Service located sources familiar with the records of BCCI's liquidation in Europe and the United States. The information provided by those sources shed considerable light on Salam's ties to the bank. According to the sources, BCCI's liquidation records show that arrangements for the loan were made through a manager of the bank's Park Lane branch in London, the branch where the Abu Nidal terrorist organization kept a $50 million account used to finance its nefarious activities.[54] The records show that the Park Lane manager recommended Salam for a loan from BCCI's Tampa branch in early 1989, according to the sources. The previous fall, a federal grand jury had indicted ten current and former BCCI employees, including a former manager of the Tampa branch, along with four of the bank's corporate entities, on charges of money laundering and conspiracy to import and distribute huge quantities of cocaine from South America. (The charges against the Tampa branch manager were later dropped due to a procedural error by investigators.)[55] According to the sources familiar with BCCI's liquidation records, for reasons that are unknown, instead of issuing Salam a loan, the Tampa branch granted him what the sources describe as a "virtually unsecured" $120,000 overdraft line. Ordinarily, overdraft lines are reserved for customers with very substantial account balances, which they are then allowed to overdraw. Salam's account, according to the sources, was "very skimpy." The records show that the line was later shifted to one of his import-export companies, Centaur Impex, backed only by a personal guarantee from Salam and the meager assets of Centaur Impex. By the end of the year, the line had been paid down to $35,000. But in the meantime, as BCCI began to implode, the Tampa branch had been shut down. And at some point the Salam–Centaur Impex overdraft line went into default. The overdraft debt was shifted to BCCI's Miami branch. According to the sources, BCCI's liquidation records show that the Miami office, desperate to convince Florida banking regulators that they were running a legitimate operation, decided to try to collect from Salam, along with a handful of other nonpaying debtors. The Miami branch filed suit against Salam, and by the time a judgment was rendered the debt had jumped to nearly $50,000. The sources familiar with BCCI's liquidation records describe the fact that Salam "had the influence to obtain such a loan without any meaningful collateral" as "very curious." They say it "suggests that he may have had influence among highly placed people within BCCI and/or Pakistan."*

The records helped explain how the man who allegedly told business associates that they could borrow money from BCCI without having to pay it back had ended up on the wrong end of a $50,000 judgment from the bank. It also suggested even closer ties to high-ups in the bank and possibly in Pakistan than had previously been understood. But it still left open the question of what Salam has been up to since then. A closer look at Centaur Impex's activities suggested that Salam remained involved in unusual financal practices.

Salam formed Centaur Impex in August 1983, around the time he arrived in Florida.[56] David Reed, one of the partners in the Curtis-Mathes franchises, says he worked with Salam at Centaur during the mid-to-late 1980s, although he was not an officer or owner of the company. According to Reed, Centaur imported computer components and built computers for sale in the United States. Reed recalled that Centaur also did business with BCCI and that Salam had "some loans" from the bank, none of which Reed was involved with.[57]

Publicly available records yield almost no information about Centaur's activities during the 1980s or 1990s. One of the few indications of its existence is the judgment BCCI won against the company and Salam in 1990. Then, in 2000, Centaur Impex was reborn under the name of Natasha International. Natasha was, in fact, a company Salam had established in May 1990, around the time the trouble began with BCCI over the Centaur Impex debt. Like Salam's other Florida enterprises, Natasha left little trace of its activities during the 1990s. In 2000, Salam appears to have allowed its corporate existence to lapse briefly. But in October of that year, he reinstated the company and at the same time reincorporated Centaur Impex as Natasha International.[58] In this new incarnation, the company would adopt a much higher public profile than its predecessor enterprises.

By no later than 2005, Natasha began advertising on international commerce Web sites as a business consulting and financial services firm.

---

*Abdus Salam did not respond to a letter from the authors seeking comment on his dealings with BCCI and other matters.

Its ads offer a wide array of loans and financing, all with "no up front pay." Services advertised as available in the United States include loans for construction of apartment complexes starting at a minimum of $1.5 million and with "no maximum," short-term bridge loans, and financing and refinancing of commercial property for borrowers with less than perfect credit. But it is the services Natasha advertises as being available outside the United States that are most interesting. Under the heading of loans available "worldwide," Natasha's ads state, "Any who will bring BG [bank guarantee] we may get from USD 50 million to USD 500 million as loan." Moreover, "Borrower who are not able provide BG we shall [be] able to get from USD 50 million to USD 500 million as JV [joint venture] Funding." And for the truly adventurous, under the heading of worldwide "Private Placement, HYIP [High Yield Investment Programs] or Trading," the ads state: "I have program with USD 100 million, it pays 400%, with split of 40:40:20." The ad instructs interested parties to call "Abdul Salam," and lists Salam's home telephone number and a post office box in Tampa.[59]

In addition to offering $500 million dollar loans and 400 percent returns on investments, Natasha's ads also state that through an "affiliate program" it is able to provide such services as "Lease[d] Proof of Fund[s]," "Lease[d] Reserve Cash Certificate[s]" and "Lease[d] Standby Letter[s] of Credit." For further details about these offerings, the ads provide a link to the Web site of an entity known as EllaFinance of Barbados Leasing. At the top of the page, "Abdus Salam" is listed as a "Certified Agent of EllaFinance." Salam's home address and phone number are also listed. Directly below, the page lists a variety of services offered by EllaFinance. Under the heading, "Leasing of a Proof of Funds," the page states:

> EllaFinance can proof the availability of large sums of money, or a credit line available to you, or, cash funds reserved in your name and for a certain period of time. Funds can be reserve or blocked for a week, a month, or a year. . . . [Y]ou can show a significant increase of your financial capacity. . . . Available funds are clear, clean, unencumbered and of non-criminal origin and legally earned. Using the services of EllaFinance, you will be able to come up with a history of funds.[60]

Similarly, under the heading of "Asset Leasing," the site states:

> If you require to increase your financial capacity, there are possibilities that you lease a title to real estate or tangible property, [or other goods and assets]. Such can be made available through the services of EllaFinance at a fixed leasing rate for a certain period of time. Assets can be leased as collateral of value to put on account or on a business balance sheet.

Other services listed are: "Leasing of a Reserved Cash Certificate," "Leasing of Investment Power," and "Leasing a Standby Letter of Credit."[61] According to the company's Web site, "Up to US $100,000,000 are immediately available to be reserved by our funders."[62] There is even an area where those interested in leasing their assets to others can sign up.[63]

Details of how these opportunities work are available by signing into the Web site. As former Congressional investigator and financial fraud expert Jack Blum notes, in each case the routine is roughly the same. "EllaFinance offers to rent its clients other people's assets for fixed periods in order to create the appearance of resources that clients can use as collateral to obtain loans or to create the appearance of wealth and resources," Blum says. Blum, who was a lead investigator in the US Senate inquiry into BCCI in the early 1990s, describes this activity as "very unusual."[64]

Who or what exactly is behind EllaFinance is not clear. The company identifies itself as being based in "Bridgetown, Barbados."[65] Yet the Central Bank of Barbados has issued a warning that EllaFinance is not incorporated in Barbados or licensed to engage in activities regulated by the Central Bank.[66] In fact, EllaFinance does not even operate out of Barbados.[67] Determining where it *is* based is a difficult task. The company does not list a physical address on its Web site, and nobody answered repeated phone calls. Nor did anybody at the company respond to e-mail requests for an interview. Abdus Salam refused a request from the authors to discuss his involvement in EllaFinance. In a July 2006 e-mail message, he indicated that he was no longer associated with the company.[68] Yet an EllaFinance Web page still lists him as a "Certified Agent."[69]

Welcome to America

Perhaps more surprising than Abdus Salam's activities in the United States is just how little US officials seem to know about them. Given

Salam's involvement with the inverter deals in Britain and Dubai, it seems reasonable to assume that the US authorities would have kept tabs on him once he arrived. Under what circumstances had he come to the United States and what is his current status? With those questions in mind, the National Security News Service contacted US Customs and Immigration sources to find out what they knew about Salam. The answer was not much. But it was worse than that.

It is not clear from immigration records exactly when Salam first took up residence in the United States. The earliest indication of his presence came in September 1995, when he applied for and was granted work authorization. This was more than a decade after he first arrived and began conducting business in the United States. In December 2005, he became a naturalized US citizen. His wife was granted citizenship in May 2006. Until he became an American citizen, Salam traveled in and out of the United States either on his alien-residency card or on his British passport.

That is as much as the immigration record reveals. There are no red flags. And the Customs records contain nothing about Salam's business activities.

It seemed odd. Given Salam's past, why was the file on him so thin? Part of the answer may have to do with faulty record-keeping. The computer system used by the US Customs and Immigration services, now jointly known as Immigration and Customs Enforcement, or ICE, came on line in 1987. But when it did, old case information was not added into the new database. As a result, everything prior to 1987 is more or less lost in a void. So even if the United States began keeping track of Salam when he arrived, that information would effectively have disappeared in 1987. Investigators coming along later would be unaware of anything that had turned up in the earlier period.

But there is also another, more disturbing problem. The birth date and Social Security number that appear on the records for virtually every one of Salam's known businesses and addresses do not match those on file for him with Immigration and Customs authorities. The date of birth is just a few days off, but still enough to potentially trip up efforts to make a positive identification. The Social Security number, on the other hand, is way off. In fact, the number that ICE has on file for Salam actually tracks to a thirty-two-year-old African American man with an entirely different name who has a long history of arrests and incarceration on a wide range of theft, drugs, and weapons charges dating back to 1993. He is currently in prison in Florida on charges of possession, sale, manufacture, and delivery of cocaine.

Yet somehow, the US government granted Salam citizenship, despite this rather glaring discrepancy.

Or perhaps there is another explanation?

7

Aiding and Abetting

THE US GOVERNMENT'S apparent disinterest in Abdus Salam once he arrived in this country is emblematic of American policy toward Pakistan during the 1980s and beyond. The Carter administration elected to downplay concerns about nuclear weapons in an effort to ensure Islamabad's cooperation in the anti-Soviet Afghan jihad. In doing so, it opened the door to Pakistan's virtually unfettered pursuit of an atomic bomb. But it was the Reagan administration that kicked down the door. Reagan and his team arrived in Washington in 1981 convinced that Carter had not done enough to confront Soviet advances in the Third World. In particular, they believed that much more could be done to punish the Soviets in Afghanistan. But that, in turn, would require shoring up relations with Pakistan. As Secretary of State George Shultz wrote in a secret memo to President Reagan on the eve of a visit to Washington by General Zia, "We must remember that without Zia's support, the Afghan resistance, key to making the Soviets pay a heavy price for their Afghan adventure, is effectively dead."[1]

One of the keys to keeping Zia on board, the argument went, was money. Lots of it. Anxious to avoid a repeat of the embarrassing aid debacle suffered by the previous administration, the Reagan White House quickly arranged a five-year, $3.2 billion assistance package that Zia found more to his liking than the "peanuts" offered by Jimmy Carter. The Reagan administration also pushed through the sale of forty advanced F-16 fighter-bombers, previously available only to NATO allies and Japan, and pledged increased support for the mujahedin.[2]

But even then, Zia imposed strict conditions on his cooperation. Well aware of the leverage he enjoyed as a result of the administration's overriding interest in keeping him happy, Zia extracted several key concessions from the Reagan team. One involved the US stance on human rights. After nearly four years of haranguing by Jimmy Carter about Pakistan's dismal human rights record and continued military rule, Zia wanted the complaining to stop. During a meeting in April 1981 with Secretary of State Alexander Haig in Washington to discuss the proposed aid package, Agha Shahi, by then Zia's foreign minister, and General Khalid Arif made clear that there should be no more harping on these matters. "We would not like to hear from you on the type of government we should have," General Arif bluntly told Haig. The usually gruff secretary of state compliantly responded, "General, your internal situation is your problem."[3]

Another demand concerned the procedures for supplying covert US assistance to the mujahedin. The Pakistanis insisted that the United States recommit to the ground rules established during the Carter administration whereby all money, guns, and equipment would be funneled to the Afghan jihadis though Pakistan's Inter-Service Intelligence, and the CIA would deal with the insurgents only though the intelligence service. The Reagan administration, anxious to appease Zia, readily agreed, thereby keeping in place a practice that would have disastrous long-term consequences.

And finally, there was the nuclear issue. While the Carter administration had made clear its willingness to tolerate Pakistan's nuclear program in the wake of the Soviet invasion of Afghanistan, the issue remained an underlying source of contention in US–Pakistani relations. Before signing on as a partner in the Reagan administration's ramped-up anti-Soviet crusade, therefore, Zia sought assurances that there would be no more trouble on this point. Former State Department South Asia specialist Dennis Kux reports that when Shahi and Arif met with Alexander Haig in April 1981, they flatly told the secretary that Pakistan "would not compromise" on its nuclear program. Haig replied that if Pakistan conducted a nuclear test, the resulting uproar in Congress would "make it difficult to cooperate with Pakistan in the way that the Reagan administration hoped." If, however, Islamabad refrained from testing, Haig explained, the nuclear issue "need not become a centerpiece of the US–Pakistani relationship." As State Department South Asia specialist Jane Coon saw it, there was "a tacit understanding that the Reagan administration could live with Pakistan's nuclear program as long as Islamabad did not explode a bomb."[4]

It was a truly Faustian bargain. In exchange for assistance in pursuing a short-term Cold War objective, the United States had formed a partnership with a despotic military regime and agreed not to interfere in its "internal" affairs. In addition, the partner's Islamist-dominated intelligence service would be allowed to continue channeling US funds, now in ever-increasing amounts, to the most radical, anti-Western factions within the mujahedin, bolstering militant Islamic fundamentalism that would eventually give rise to groups such as the Taliban and al Qaeda. And the United States would turn a blind eye to its partner's development of the world's most deadly technology — technology that A.Q. Khan and his associates would later sell to rogue states and sponsors of terrorism. Thus, in an effort to "punish" the Soviet Union over the Afghan invasion — a move provoked, at least in part, by the United States — the Reagan administration unleashed a collection of forces that would eventually emerge as the gravest security challenges facing the world.

A primary enabler of this mad scheme was Saudi Arabia. Under an arrangement worked out with President Reagan, the Saudi royal family agreed to continue the practice begun during the Carter administration of matching US covert assistance to the mujahedin dollar for dollar.[5] Over the course of the 1980s, billions of dollars in Saudi and American funds would flow to the holy warriors in Afghanistan, much of it through the coffers of Bank of Credit and Commerce International (BCCI). But even as the Saudis joined with the United States in backing the mutually beneficial anti-Soviet jihad, it was also contributing to Pakistan's development of atomic weapons. On the very eve of Ronald Reagan's inauguration, London's *Sunday Times* reported that Riyadh, which had been an early backer of Ali Bhutto's Islamic bomb program, had offered $800 million to help Islamabad build a nuclear device.[6] And while continuing to assist Pakistan's nuclear weapons program both directly and indirectly in the years ahead,[7] at the conclusion of the Afghan campaign, Saudi Arabia would emerge as a major funder of al Qaeda and other terrorist spawn of the mujahedin as they set their sights on the world's lone remaining superpower. Such were the partners the United States engaged to confront the Evil Empire.

But the Reagan administration had another important partner in facilitating this slow-motion train wreck — the US Congress. Because of the aid cutoff imposed by President Carter in 1979 in response to revelations that Pakistan had acquired unsafeguarded uranium enrichment technology, any US assistance to Islamabad required a congressional waiver. Although longtime nonproliferation advocates on Capitol Hill strongly opposed lifting the sanctions, the Senate Foreign Relations Committee ultimately approved the White House request for a six-year waiver of the aid ban by a vote of 10 to 7 in mid-May 1981. Amid mounting intelligence reports that Pakistan would soon be capable of conducting a nuclear test, Democratic congressman Stephen Solarz of New York managed to push through an amendment in the House of Representatives barring aid to any country that detonated a nuclear device. When, as Dennis Kux notes, the Senate adopted a similar measure at the end of the year, "the 'tacit' understanding about Pakistan's not testing became a legal requirement for US aid."[8] But throughout the 1980s, Congress repeatedly allowed its attempts to sanction Pakistan for its nuclear program to be steamrolled by the White House.

The Reagan administration trotted out a variety of arguments to justify US assistance to Pakistan. One was that the threat of losing the multibillion-dollar aid package would serve as a deterrent to Islamabad conducting a nuclear test.[9] The administration might alternately have made aid contingent on Pakistan giving up its nuclear program altogether, but that was not a tack the White House chose to adopt. The Pakistanis would later admit that they would have continued to support the Afghan jihad even if American aid had been cut off.[10]

Administration officials also claimed that the intelligence about Pakistan's nuclear program remained "rudimentary." As Reagan's ambassador to Islamabad, Ronald Spiers, put it, "It was very clear that there was a group in Pakistan that was working towards development of a nuclear capability . . . but we didn't know how far it had advanced." Moreover, according to Spiers, the United States remained skeptical that Pakistan "could actually achieve" a nuclear capability.[11]

But such assertions run counter to the facts. The CIA station in Islamabad had explicit orders to monitor Pakistan's bomb program, and it collected detailed information about the nation's nuclear capabilities and progress toward development of an atomic weapon.[12] In December 1981, on the eve of the Senate vote to waive the aid ban on Pakistan, the *Washington Post* reported that intelligence estimates indicated that Islamabad could detonate an atomic device by the end of the following year.[13] And in June 1983, a secret State Department report, based on intelligence gathered by the CIA, concluded that there was "unambiguous evidence" that Pakistan was "actively pursing a nuclear weapons development program."[14] And yet, US aid continued.

The Reagan administration willingly accepted Zia's false "assurances" that Pakistan was not building a bomb. In private meetings with President Reagan, Vice President George H.W. Bush, and other top US officials, Zia repeatedly lied about what his government was up to.[15] When, on one occasion, a pair of high-ranking State Department officials had the temerity to confront Zia about one of his many misrepresentations, the Pakistani dictator responded, "It is permissible to lie for Islam."[16]

The Reagan administration knowingly participated in this charade out of fear that openly challenging Zia over nuclear weapons would jeopardize the all-important anti-Soviet campaign in Afghanistan. As Secretary of State Shultz explained in a secret memo to President Reagan in November 1982, "How we handle the nuclear issue can have a profound effect on our ability to continue to cooperate with Pakistan in supporting the Afghan freedom fighters."[17] How the administration handled it was by looking the other way, clinging to the flimsy rationale that Pakistan had not yet actually tested a nuclear weapon and continuing to pump money into Zia's government. As one administration official put it, "We have two motherhood issues clashing here — Afghanistan and nonproliferation."[18] One of those issues, however, had been orphaned.

With the Reagan administration's willingness to look the other way, Pakistan's bomb program flourished during the 1980s. In 1981, construction of the Kahuta enrichment facility was completed, and rechristened as the Khan Research Laboratories (KRL).[19] Although the CIA was soon receiving detailed, firsthand information about activities inside the plant, work there went on uninterrupted.[20] In 1983, Pakistan carried out a successful test of a nuclear bomb design, minus the fissionable core, an exercise known as a cold test.[21] And in early 1984, A.Q. Khan told an Urdu-language newspaper that Pakistan had "broken the Western countries' monopoly" on uranium enrichment and could, if necessary, produce an atomic bomb.[22] All throughout this period, moreover, Pakistan continued its pursuit of technology that could be used to develop a plutonium weapon, despite the earlier setback over the French reprocessing facility.[23] Still, the administration did not move to cut off funding of its Cold War ally.

Pakistani nuclear cooperation with China got underway in earnest during this period as well. Ali Bhutto had been instrumental in helping to establish Islamabad's close ties to China during the 1960s, and as prime minister during the 1970s secretly gained Beijing's commitment to help Pakistan build a bomb.[24] But it was under Zia that real cooperation began. In 1983, the secret State Department report on Islamabad's nuclear effort found that China, which was helping Washington and Islamabad supply the anti-Soviet forces in Afghanistan, had "provided assistance to Pakistan's program to develop a nuclear weapons capability." On an ominous note, the report added, "We now believe the cooperation has taken place in the area of fissile material production and possibly also nuclear device

The true father of the Islamic bomb, Pakistani Prime Minister Zulfikar Ali Bhutto (center, in hat) attending Friday prayers with Libyan leader Colonel Muammar el-Qaddafi (to Bhutto's left) during an Islamic summit in Lahore in February 1974. It was during this visit to Pakistan that Bhutto and Qaddafi reportedly finalized a deal in which Libya agreed to help finance Pakistan's bomb program. Also in attendance at the prayer session were Palestinian leader Yasser Arafat (third from left) and King Faisal of Saudi Arabia (far right). (Associated Press)

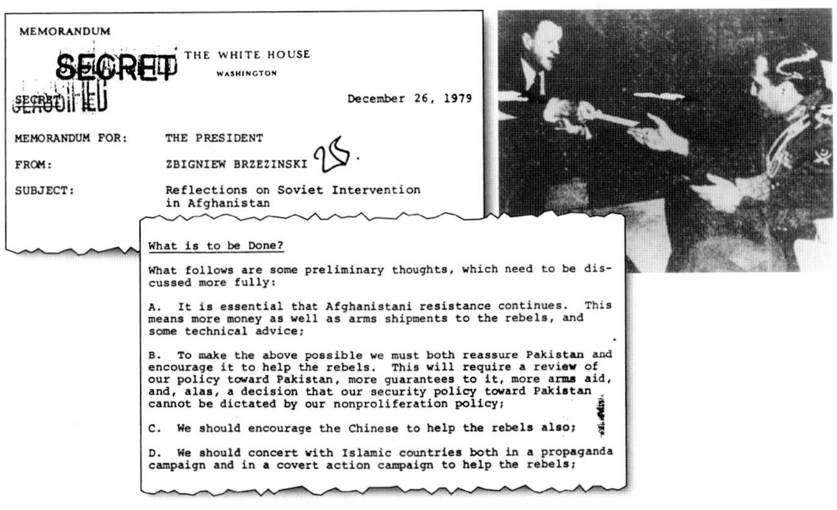

MEMORANDUM

SECRET

THE WHITE HOUSE

WASHINGTON

SECRET

December 26, 1979

MEMORANDUM FOR: THE PRESIDENT

FROM: ZBIGNIEW BRZEZINSKI ZS.

SUBJECT: Reflections on Soviet Intervention in Afghanistan

What is to be Done?

What follows are some preliminary thoughts, which need to be discussed more fully:

A. It is essential that Afghanistani resistance continues. This means more money as well as arms shipments to the rebels, and some technical advice;

B. To make the above possible we must both reassure Pakistan and encourage it to help the rebels. This will require a review of our policy toward Pakistan, more guarantees to it, more arms aid, and, alas, a decision that our security policy toward Pakistan cannot be dictated by our nonproliferation policy;

C. We should encourage the Chinese to help the rebels also;

D. We should concert with Islamic countries both in a propaganda campaign and in a covert action campaign to help the rebels;

Zbigniew Brzezinski's secret 1979 memo urged President Carter to set aside concerns about Pakistani nuclear proliferation to help win Islamabad's support for the Islamic forces resisting the Soviet invasion of Afghanistan. (Cold War International History Project) In Islamabad, Brzezinski presented Pakistani military ruler Gen. Mohammad Zia ul-Haq with a letter from Carter in February 1980. Brezinski won Zia's agreement to assist the Afghan campaign, but at a price.

President Ronald Reagan meets with Pakistani leader Gen. Zia ul-Haq at the White House in December 1982. The Reagan administration expanded the Afghan campaign, as well as the practice of winking at Pakistan's nuclear program. (Courtesy Ronald Reagan Library)

Vice President George H.W. Bush and General Zia wave from the steps of Air Force Two in Islamabad during Bush's mission to Pakistan in May 1984. (Courtesy George Bush Presidential Library)

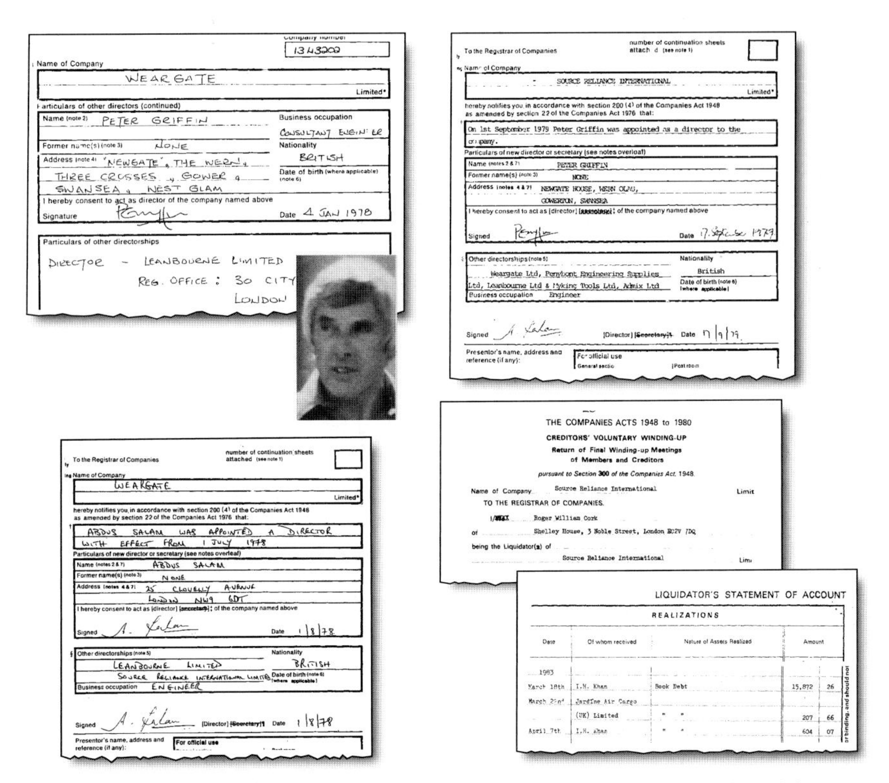

Company number 1343202

Name of Company
WEARGATE Limited*

Particulars of other directors (continued)
Name (note 2) PETER GRIFFIN
Business occupation CONSULTANT ENGIN: ER
Former name(s) (note 3) NONE
Nationality BRITISH
Address (note 4) 'NEWGATE', THE WERN, THREE CROSSES, GOWER, SWANSEA, WEST GLAM
Date of birth (where applicable) (note 6)
I hereby consent to act as director of the company named above
Signature
Date 4 JAN 1978

Particulars of other directorships
DIRECTOR - LEANBOURNE LIMITED
REG. OFFICE: 30 CITY
LONDON

To the Registrar of Companies
number of continuation sheets attached (see note 1)
Name of Company
SOURCE RELIANCE INTERNATIONAL Limited*
hereby notifies you in accordance with section 200 (4) of the Companies Act 1948 as amended by section 22 of the Companies Act 1976 that:
On 1st September 1979 Peter Griffin was appointed as a director to the company.
Particulars of new director or secretary (see notes overleaf)
Name (notes 2 & 7) PETER GRIFFIN
Former name(s) (note 3) NONE
Address (notes 4 & 7) NEWGATE HOUSE, WERN OLAU, GOWERTON, SWANSEA
I hereby consent to act as [director] [secretary] of the company named above
Signed
Date 17 September 1979
Other directorships (note 5)
Weargate Ltd, Penybont Engineering Supplies Ltd, Leanbourne Ltd & Hyking Tools Ltd, Admix Ltd
Nationality British
Date of birth (note 6) (where applicable)
Business occupation Engineer
Signed
[Director] [Secretary] Date 17/9/79
Presentor's name, address and reference (if any):
For official use
General section | Post room

To the Registrar of Companies
number of continuation sheets attached (see note 1)
Name of Company
WEARGATE Limited*
hereby notifies you in accordance with section 200 (4) of the Companies Act 1948 as amended by section 22 of the Companies Act 1976 that:
ABDUS SALAM WAS APPOINTED A DIRECTOR WITH EFFECT FROM 1 JULY 1978
Particulars of new director or secretary (see notes overleaf)
Name (notes 2 & 7) ABDUS SALAM
Former name(s) (note 3) NONE
Address (notes 4 & 7) 25 CLOVELLY AVENUE LONDON NW9 6DT
I hereby consent to act as [director] [secretary] of the company named above
Signed
Date 1/8/78
Other directorships (note 5)
LEANBOURNE LIMITED
SOURCE RELIANCE INTERNATIONAL LIMITED
Nationality BRITISH
Date of birth (note 6) (where applicable)
Business occupation ENGINEER
Signed
[Director] [Secretary] Date 1/8/78
Presentor's name, address and reference (if any):
For official use

THE COMPANIES ACTS 1948 to 1980
CREDITORS' VOLUNTARY WINDING-UP
Return of Final Winding-up Meetings of Members and Creditors
pursuant to Section 300 of the Companies Act, 1948
Name of Company Source Reliance International Limited
TO THE REGISTRAR OF COMPANIES.
I/We Roger William Cork
of Shelley House, 3 Noble Street, London EC2V 7DQ
being the Liquidator(s) of Source Reliance International Limited

LIQUIDATOR'S STATEMENT OF ACCOUNT
REALIZATIONS

Date	Of whom received	Nature of Assets Realized	Amount	
1983				
March 18th	I.H. Khan	Book Debt	15,872	26
March 2?nd	Jardine Air Cargo (UK) Limited	" "	207	66
April 7th	I.H. Khan	" "	604	07

Corporate records for Weargate Ltd. and Source Reliance International Ltd. list Abdus Salam and Peter Griffin as directors. A liquidator's statement shows payments to SRI from "I.H. Khan." Pictured is Peter Griffin, circa the late 1970s.

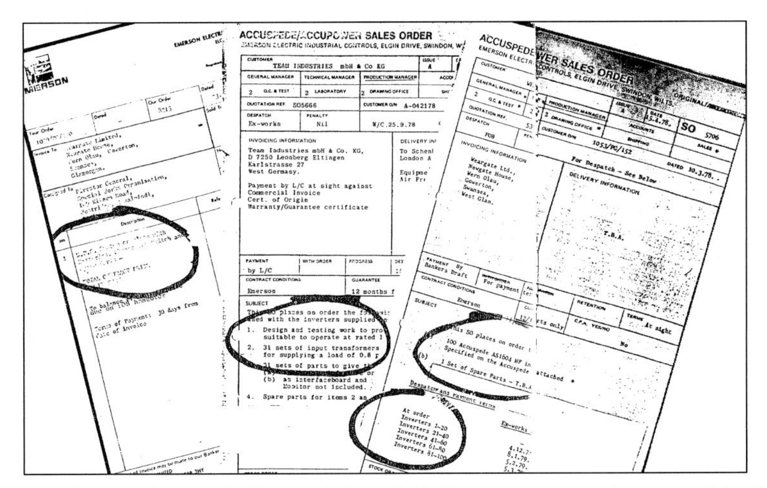

EMERSON
Weargate Limited, Newgate House, Wern Olau, Gowerton, Swansea, Glamorgan.
Consigned to Director General, Special Works Organisation
Terms of Payment: 30 days from date of invoice

ACCUSPEDE/ACCUPOWER SALES ORDER
EMERSON ELECTRIC INDUSTRIAL CONTROLS, ELGIN DRIVE, SWINDON, WILTS.
CUSTOMER TEAM INDUSTRIES mbH & Co KG
QUOTATION REF S05666
CUSTOMER O/N A-042178
DESPATCH Ex-works
PENALTY Nil
W/C.25.9.78
INVOICING INFORMATION
Team Industries mbH & Co. KG, D 7250 Leonberg Eltingen, Karlstrasse 27, West Germany.
Payment by L/C at sight against Commercial Invoice, Cert. of Origin, Warranty/Guarantee certificate
PAYMENT by L/C
CONTRACT CONDITIONS Emerson
GUARANTEE 12 months
SUBJECT
1. Design and testing work to pro... suitable to operate at rated l...
2. 31 sets of input transformers for supplying a load of 0.8 ...
3. 31 sets of parts to give ...
(b) as interfaceboard and Monitor not included.
4. Spare parts for items 2 an...

ACCUSPEDE/ACCUPOWER SALES ORDER
EMERSON ELECTRIC INDUSTRIAL CONTROLS, ELGIN DRIVE, SWINDON, WILTS.
DATED 10.3.78.
SO 5706
INVOICING INFORMATION
Weargate Ltd., Newgate House, Wern Olau, Gowerton, Swansea, West Glam.
CUSTOMER O/N 1053/PG/152
For Despatch – See Below
PAYMENT By Bankers Draft
CONTRACT CONDITIONS Emerson
TERMS At sight
SUBJECT
This 50 places on order ...
100 Accuspede AS1604 HF in... Specified on the Accuspede attached
(b) 1 Set of Spare Parts – T.B.A.
Despatch and Payment terms
At order
Inverters 1-20
Inverters 21-40
Inverters 41-60
Inverters 61-80
Inverters 81-100
Ex-works

Copies of the orders at the heart of the British inverter controversy appeared in the publication *8 Days.* From left, a Weargate order (consigned to Pakistan's Special Works Organization) for thirty-one inverters and uninterrupted power supply units, Team Industries' order for thirty-one sets of input transformers, and Weargate's order for one hundred inverters and a set of spare parts.

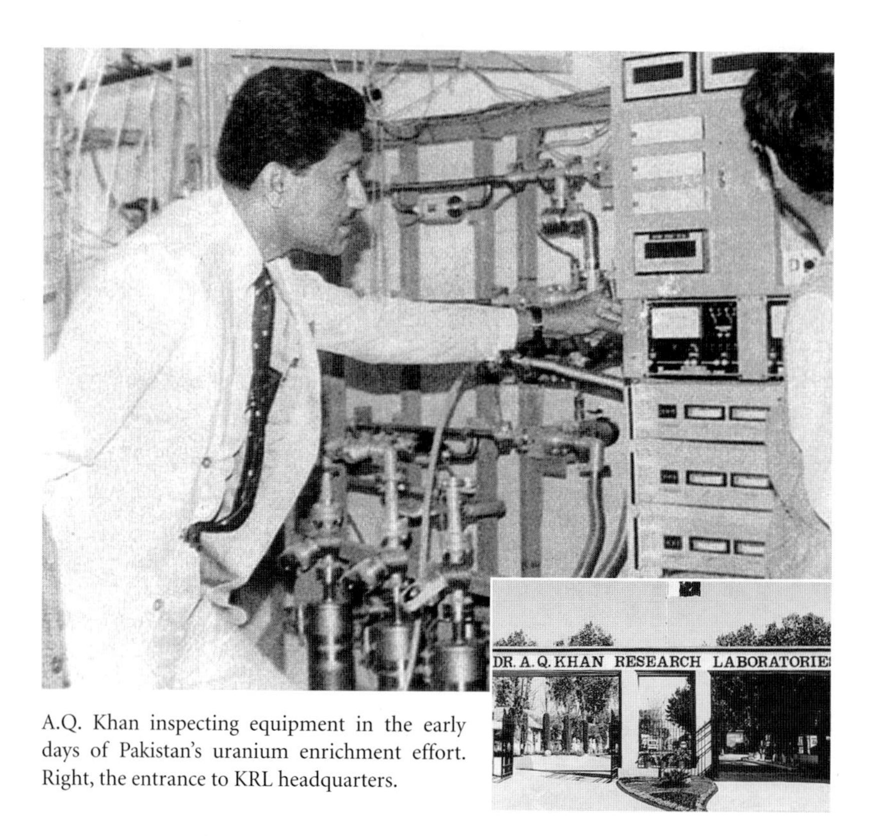

A.Q. Khan inspecting equipment in the early days of Pakistan's uranium enrichment effort. Right, the entrance to KRL headquarters.

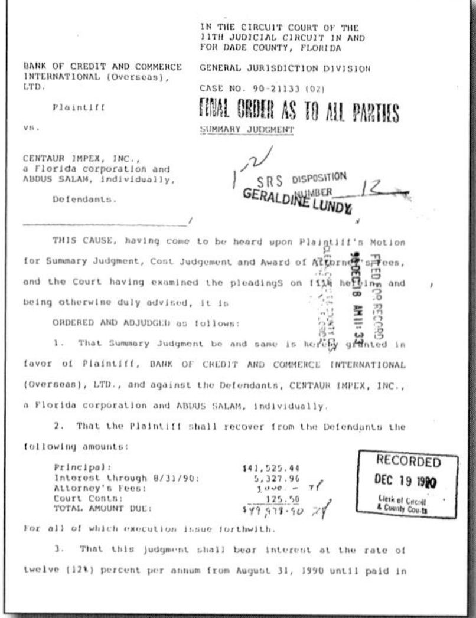

IN THE CIRCUIT COURT OF THE
11TH JUDICIAL CIRCUIT IN AND
FOR DADE COUNTY, FLORIDA

GENERAL JURISDICTION DIVISION

CASE NO. 90-21133 (02)

BANK OF CREDIT AND COMMERCE INTERNATIONAL (Overseas), LTD.

Plaintiff

vs.

CENTAUR IMPEX, INC., a Florida corporation and ABDUS SALAM, individually,

Defendants.

FINAL ORDER AS TO ALL PARTIES

SUMMARY JUDGMENT

SRS DISPOSITION NUMBER 12
GERALDINE LUNDY

THIS CAUSE, having come to be heard upon Plaintiff's Motion for Summary Judgment, Cost Judgement and Award of Attorney's Fees, and the Court having examined the pleadingS on file herein and being otherwise duly advised, it is

FILED FOR RECORD
90 DEC 18 AM 11:33

ORDERED AND ADJUDGED as follows:

1. That Summary Judgment be and same is hereby granted in favor of Plaintiff, BANK OF CREDIT AND COMMERCE INTERNATIONAL (Overseas), LTD., and against the Defendants, CENTAUR IMPEX, INC., a Florida corporation and ABDUS SALAM, individually.

2. That the Plaintiff shall recover from the Defendants the following amounts:

Principal:	$41,525.44
Interest through 8/31/90:	5,327.96
Attorney's Fees:	3,000.—
Court Costs:	125.50
TOTAL AMOUNT DUE:	$49,978.90

RECORDED
DEC 19 1990
Clerk of Circuit & County Courts

For all of which execution issue forthwith.

3. That this judgment shall bear interest at the rate of twelve (12%) percent per annum from August 31, 1990 until paid in

Former Weargate director Abdus Salam in Tampa, Florida, circa 1994 (courtesy *Tampa Tribune*), and a copy of the judgment the Bank of Credit and Commerce International (BCCI) won against Salam and his company, Centaur Impex.

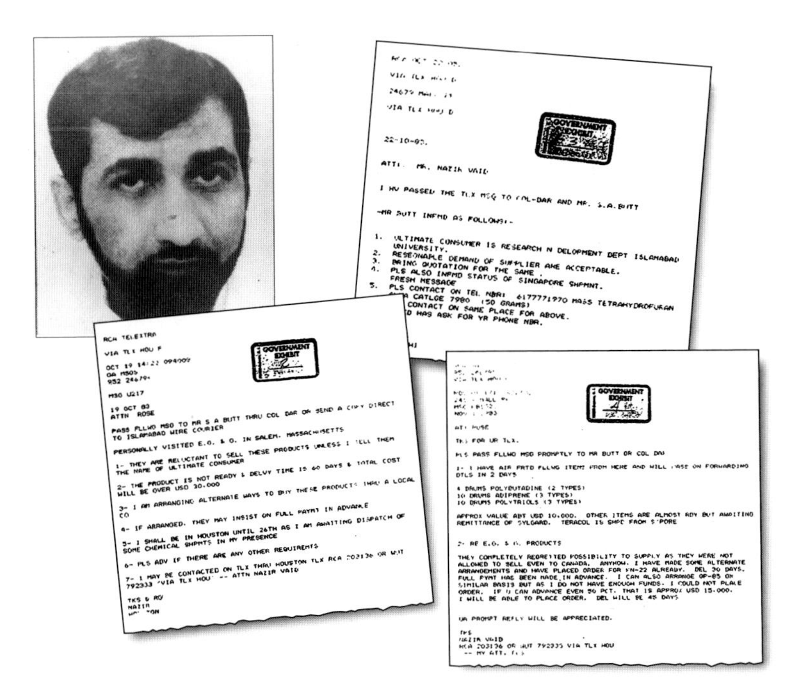

22-10-83.

ATTN. MR. NAZIR VAID

I HV PASSED THE TLX MSG TO COL-DAR AND MR. S.A. BUTT

-MR BUTT INFMD AS FOLLOWS:-

1. ULTIMATE CONSUMER IS RESEARCH N DELOPMENT DEPT ISLAMABAD UNIVERSITY.
2. RESEONAPLE DEMAND OF SUPPLIER ARE ACCEPTABLE.
3. BRING QUOTATION FOR THE SAME .
4. PLS ALSO INFMD STATUS OF SINGAPORE SHPMNT. FRESH MESSAGE
5. PLS CONTACT ON TEL NBR: 6177771970 MASS TETRAHYDROFURAN ... CATLGE 7980 (50 GRAMS) ... CONTACT ON SAME PLACE FOR ABOVE. ... HAS ASK FOR YR PHONE NBR.

RCA TELEXTRA

VIA TLX HOU F

OCT 19 14:22 094007
OA MSO5
952 24679

MSO U217

19 OCT 83
ATTN ROSE

PASS FLLWG MSG TO MR S A BUTT THRU COL DAR OR SEND A COPY DIRECT TO ISLAMABAD WIRE COURIER

PERSONALLY VISITED E.G. & G. IN SALEM. MASSACHUSETTS

1- THEY ARE RELUCTANT TO SELL THESE PRODUCTS UNLESS I TELL THEM THE NAME OF ULTIMATE CONSUMER

2- THE PRODUCT IS NOT READY & DELVY TIME IS 60 DAYS & TOTAL COST WILL BE OVER USD 30.000

3- I AM ARRANGING ALTERNATE WAYS TO BUY THESE PRODUCTS THRU A LOCAL CO

4- IF ARRANGED. THEY MAY INSIST ON FULL PAYMT IN ADVANCE

5- I SHALL BE IN HOUSTON UNTIL 26TH AS I AM AWAITING DISPATCH OF SOME CHEMICAL SHPMTS IN MY PRESENCE

6- PLS ADV IF THERE ARE ANY OTHER REQUIREMTS

7- I MAY BE CONTACTED ON TLX THRU HOUSTON TLX RCA 203136 OR WUT 792333 VIA TLX HOU -- ATTN NAZIR VAID

TKS & RGDS
NAZIR
HOUSTON

ATT: ROSE

TKS FOR UR TLX.

PLS PASS FLLWG MSG PROMPTLY TO MR BUTT OR COL DAR

1- I HAVE AIR FRTD FLLWG ITEMS FROM HERE AND WILL PASS ON FORWARDING DTLS IN 2 DAYS

4 DRUMS POLYBUTADINE (2 TYPES)
10 DRUMS ADIPRENE (3 TYPES)
10 DRUMS POLYTRIOLS (3 TYPES)

APPROX VALUE ABT USD 10.000. OTHER ITEMS ARE ALMOST RDY BUT AWAITING REMITTANCE OF SYLGARD. TERACOL IS SHPD FROM S'PORE

2- RE E.G. & G. PRODUCTS

THEY COMPLETELY REGRETTED POSSIBILITY TO SUPPLY AS THEY WERE NOT ALLOWED TO SELL EVEN TO CANADA. ANYHOW. I HAVE MADE SOME ALTERNATE ARRANGEMENTS AND HAVE PLACED ORDER FOR KN-22 ALREADY. DEL 30 DAYS. FULL PYMT HAS BEEN MADE IN ADVANCE. I CAN ALSO ARRANGE GP-85 ON SIMILAR BASIS BUT AS I DO NOT HAVE ENOUGH FUNDS. I COULD NOT PLACE ORDER. IF U CAN ADVANCE EVEN 50 PCT. THAT IS APPROX USD 15.000. I WILL BE ABLE TO PLACE ORDER. DEL WILL BE 45 DAYS

UR PROMPT REPLY WILL BE APPRECIATED.

TKS
NAZIR VAID
RCA 203136 OR WUT 792333 VIA TLX HOU
-- MY ATT. ...

Nazir Ahmed Vaid in 1984, about the time he was convicted of violating US export law by attempting to ship devices to Pakistan that could be used to trigger nuclear weapons. (Courtesy Mark Obenhaus) Telexes indicate that Vaid had purchased the devices on behalf of S.A. Butt and Col. Umar Din Dar, officials of the Pakistan Atomic Energy Commission.

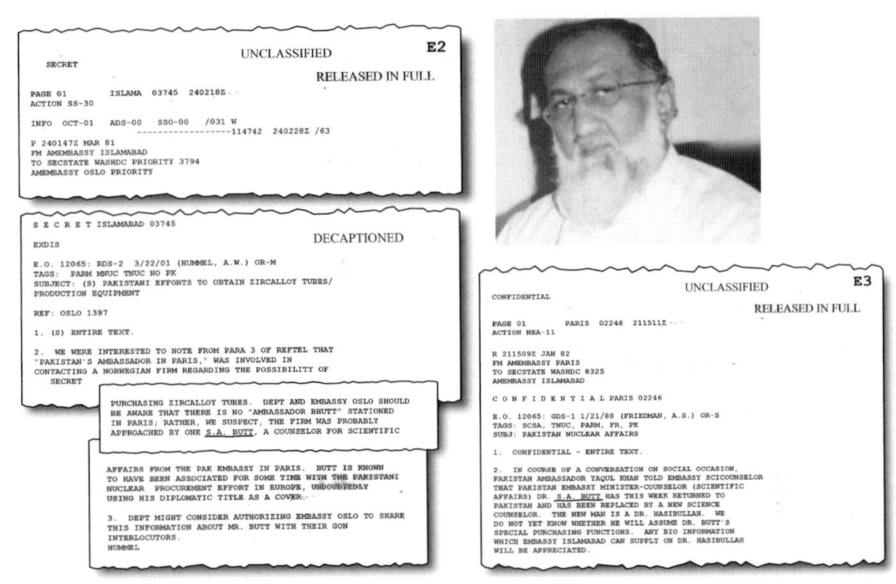

E2

UNCLASSIFIED

SECRET

RELEASED IN FULL

PAGE 01 ISLAMA 03745 240218Z
ACTION SS-30

INFO OCT-01 ADS-00 SSO-00 /031 W
------------------114742 240228Z /63

P 240147Z MAR 81
FM AMEMBASSY ISLAMABAD
TO SECSTATE WASHDC PRIORITY 3794
AMEMBASSY OSLO PRIORITY

S E C R E T ISLAMABAD 03745

DECAPTIONED

EXDIS

E.O. 12065: RDS-2 3/22/01 (HUMMEL, A.W.) OR-M
TAGS: PARM MNUC TNUC NO PK
SUBJECT: (S) PAKISTANI EFFORTS TO OBTAIN ZIRCALLOY TUBES/ PRODUCTION EQUIPMENT

REF: OSLO 1397

1. (S) ENTIRE TEXT.

2. WE WERE INTERESTED TO NOTE FROM PARA 3 OF REFTEL THAT "PAKISTAN'S AMBASSADOR IN PARIS," WAS INVOLVED IN CONTACTING A NORWEGIAN FIRM REGARDING THE POSSIBILITY OF
SECRET

PURCHASING ZIRCALLOY TUBES. DEPT AND EMBASSY OSLO SHOULD BE AWARE THAT THERE IS NO "AMBASSADOR BHUTT" STATIONED IN PARIS; RATHER, WE SUSPECT, THE FIRM WAS PROBABLY APPROACHED BY ONE S.A. BUTT, A COUNSELOR FOR SCIENTIFIC

AFFAIRS FROM THE PAK EMBASSY IN PARIS. BUTT IS KNOWN TO HAVE BEEN ASSOCIATED FOR SOME TIME WITH THE PAKISTANI NUCLEAR PROCUREMENT EFFORT IN EUROPE, UNDOUBTEDLY USING HIS DIPLOMATIC TITLE AS A COVER.

3. DEPT MIGHT CONSIDER AUTHORIZING EMBASSY OSLO TO SHARE THIS INFORMATION ABOUT MR. BUTT WITH THEIR GON INTERLOCUTORS.
HUMMEL

E3

UNCLASSIFIED

CONFIDENTIAL

RELEASED IN FULL

PAGE 01 PARIS 02246 211511Z
ACTION NEA-11

R 211509Z JAN 82
FM AMEMBASSY PARIS
TO SECSTATE WASHDC 8325
AMEMBASSY ISLAMABAD

C O N F I D E N T I A L PARIS 02246

E.O. 12065: GDS-1 1/21/88 (FRIEDMAN, A.S.) OR-S
TAGS: SCSA, TNUC, PARM, FR, PK
SUBJ: PAKISTAN NUCLEAR AFFAIRS

1. CONFIDENTIAL - ENTIRE TEXT.

2. IN COURSE OF A CONVERSATION ON SOCIAL OCCASION, PAKISTAN AMBASSADOR YAQUL KHAN TOLD EMBASSY SCICOUNSELOR THAT PAKISTAN EMBASSY MINISTER-COUNSELOR (SCIENTIFIC AFFAIRS) DR. S.A. BUTT HAS THIS WEEK RETURNED TO PAKISTAN AND HAS BEEN REPLACED BY A NEW SCIENCE COUNSELOR. THE NEW MAN IS A DR. HASIBULLAH. WE DO NOT YET KNOW WHETHER HE WILL ASSUME DR. BUTT'S SPECIAL PURCHASING FUNCTIONS. ANY BIO INFORMATION WHICH EMBASSY ISLAMABAD CAN SUPPLY ON DR. HASIBULLAH WILL BE APPRECIATED.

State Department cables from the early 1980s demonstrate US knowledge of S.A. Butt's involvement in Pakistan's nuclear procurement effort well before the Vaid affair. Right, Nazir Vaid in more recent years.

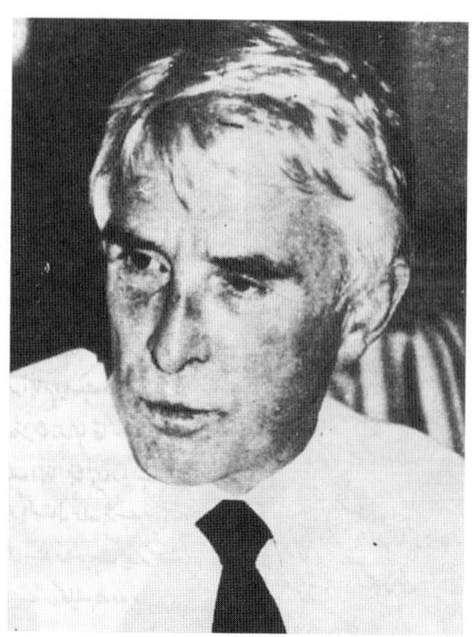

Left, former CIA nonproliferation analyst Richard Barlow. (Courtesy Richard Barlow) Right, Peter Griffin. This picture appeared in a 1986 Pakistani magazine article about AQ Khan in which Griffin was interviewed.

A.Q. Khan in an undated photograph. A caption accompanying the picture read: "'Don't worry if I am assassinated by the enemy. There are dozens of A.Q. Khans, fully groomed, waiting in the wings,' says Dr. Khan with the smile that could launch any number of bombs and missiles in defence of Pakistan."

A.Q. Khan and associates at the Shah Hassan II Mosque in Casablanca in February 1998. Pictured with Khan (center) are B.S.A. Tahir (far left), Henk Slebos (second from left), Abdul Mabood Siddiqui (fourth from right), Pakistan's honorary consul general in Morocco, Hussain Bin Jeloun (third from right), and Lieutenant General (Retired) Riaz Ahmed Chowhan (second from right).

SCOMI PRECISION ENGINEERING SDN BHD

Lot 28 & 30
Jalan Beliong 15/11
Seksyen 15
40200 Shah Alam
Selangor.

Tel : 03-55118295
Fax : 03-55118301

DELIVERY NOTE/PACKING LIST

DATE: 1st August 2002

TO : Gulf Technical Industries L.L.C.
P.O. Box 29576
Dubai, United Arab Emirates

Attn :

Tel : [00 971] 4 394 0342/333 2226
Fax: [00 971] 4 394 0352/333 1626

ITEM	DESCRIPTION	SIZE TUBE	QTY
1	Aluminium Al Mg Si Cu/6061 / EN AW 6061	OD212 x WT22mm, ID=168mm, App 2300mm length	305 pcs/ 26118.0 kgs
2	Aluminium Al Mg Si Cu/6061 / EN AW 6061	OD212 x WT22mm, ID=168mm, App 2300mm length	314 pcs/ 26446.0 kgs
3	Aluminium Al Mg Si Cu/6061 / EN AW 6061	OD212 x WT22mm, ID=168mm, App 2300mm length	213 pcs/ 17423.0 kgs

SCOMI PRECISION ENGINEERING SDN BHD

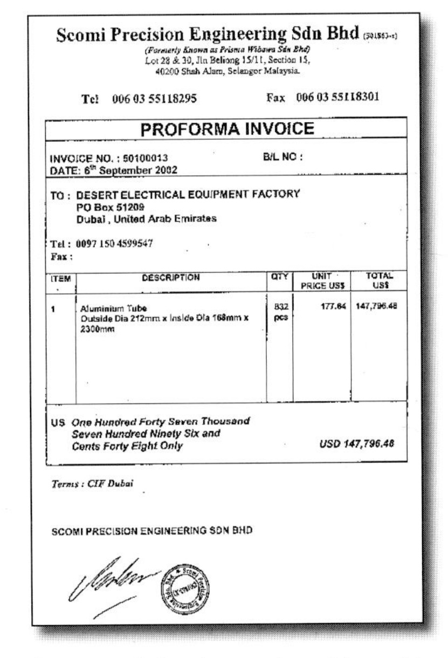

Scomi Precision Engineering Sdn Bhd
(Formerly Known as Prisma Wibawa Sdn Bhd)
Lot 28 & 30, Jln Beliong 15/11, Section 15,
40200 Shah Alam, Selangor Malaysia.

Tel 006 03 55118295 Fax 006 03 55118301

PROFORMA INVOICE

INVOICE NO. : 50100013 B/L NO :
DATE: 6th September 2002

TO : DESERT ELECTRICAL EQUIPMENT FACTORY
PO Box 51209
Dubai, United Arab Emirates

Tel : 0097 150 4599547
Fax :

ITEM	DESCRIPTION	QTY	UNIT PRICE US$	TOTAL US$
1	Aluminium Tube Outside Dia 212mm x Inside Dia 168mm x 2300mm	832 pcs	177.64	147,796.48

US *One Hundred Forty Seven Thousand Seven Hundred Ninety Six and Cents Forty Eight Only* USD 147,796.48

Terms : CIF Dubai

SCOMI PRECISION ENGINEERING SDN BHD

A delivery note/packing slip (left) shows a consignment of aluminum tubes addressed to Gulf Technical Industries in Dubai. A proforma invoice (right) for the same number of tubes is addressed to Desert Electrical Equipment Factory, also in Dubai. The documents were included in a 2004 Malaysian police report.

A.Q. Khan receiving an award from Pakistani President Ghulam Ishaq Khan in 1990 (top left).

Khan and Army Chief of Staff Gen. Aslam Beg admiring a new addition to Pakistan's military arsenal (top right).

Khan meeting with Saudi Defense Minister Prince Sultan bin Abdul Aziz and Pakistani Prime Minister Nawaz Sharif in Khan's office at KRL (center).

Khan with Pakistani President Farooq Leghari and Prime Minister Benazir Bhutto (above left).

Khan with Pakistani President and Army Chief of Staff General Pervez Musharraf (bottom left). Khan has long boasted of having the full support of Pakistani leaders.

design."[25] In fact, during one of his visits to Beijing in the early 1980s, A.Q. Khan had traded Pakistani centrifuge technology for the blueprints to a bomb China had successfully tested in the mid-1960s. The proven design, while relatively simple, was compact enough to fit atop a missile, and gave Pakistan the ability, in theory, to build a reliable nuclear weapon that did not require testing, rendering the Reagan administration's tacit no-test policy effectively meaningless. Copies of the Chinese blueprints would wind up in Libya years later when Khan tossed them in as a token of his appreciation for Tripoli's purchase of $100 million worth of uranium enrichment equipment from his nuclear supermarket. During the 1980s, China would supply Pakistan with highly enriched uranium, tritium, scientific advisors, and tools and components for building a nuclear weapons production facility.[26]

As Islamabad's bomb program took off during the early years of the Reagan administration, the nuclear procurement network shifted into high gear. US intelligence agencies monitored its activities closely. The 1983 State Department report noted that "Pakistan's procurement agents" were "seeking from commercial companies items which are unambiguously identified as major components of a nuclear explosive device." Much of this activity centered around Khan's lab. "In late 1981 through to 1982," according to the report, procurement agents who had "long been associated" with Khan's operation at Kahuta ordered metal components from European companies that had been "unambiguously identified as those of a nuclear explosive device." In addition, the report noted, the Pakistanis were also seeking to acquire metal-working equipment, such as precision lathes and other machines tools "intended specifically for the manufacture of these components."[27]

Confronted with this evidence that Pakistan was, despite Zia's assurances, building a bomb, the Reagan administration did not cut off funding. Instead, according to the report, the United States approached the Pakistani government "at the highest levels" and communicated its "extreme concern." It didn't work. Islamabad simply went to Plan B. "After the initial discussion with the Pakistanis," the report states, "we noted a shift in emphasis from procurement of weapons components themselves to procurement of machinery necessary for their manufacture."[28] Pakistan would now make what it needed to build its bomb.

But there were some materials and components that Pakistan could not yet produce on its own. Among them were items that were produced only in the United States. And when Pakistani agents got caught trying to acquire them, the Reagan administration shifted from merely winking at Pakistan's nuclear bomb program to actively aiding and abetting in its cover-up.

Operation EXODUS

On October 18, 1983, a man with a distinct foreign accent called an electronics firm in Salem, Massachusetts, with an unusual request: according to court records and published accounts, he wanted to purchase fifty high-speed electronic switches known as krytrons — tiny, odd-shaped bulbs used in such things as oil-exploration equipment, lasers, and industrial strobe lights. The particular model the caller asked for could also be used to trigger nuclear bombs.[29]

The employees of the company, EG&G Electro-Optics, were stunned. The firm's parent, EG&G Inc., was a Fortune 500 company that had been involved in America's nuclear weapons program since the 1940s and was the world's lone manufacturer of krytrons. No individual had ever attempted to purchase the switches directly.[30] But their astonishment would only grow deeper.

The caller was directed to EG&G Electro-Optics' customer service manager, John McClafferty. The caller explained to McClafferty that he was in Salem and asked if he could stop by. He arrived dressed in a knee-length shirt and flowing trousers, traditional Pakistani attire. He identified himself as N.A. Vaid (his full name was Nazir Ahmed Vaid, a thirty-two-year-old native of Lahore, Pakistan) and said he was in the import-export business in Karachi and Houston. He then repeated his request for the krytrons, known as model KN-22s, indicating they were to be sent to Pakistan, according to court records.[31] "He offered to pay more than the going rate, pay in cash — I'm pretty sure he offered to pay in gold," an EG&G official recalled. "He made it clear he was willing to pay almost anything."[32]

McClafferty informed Vaid that sending krytrons overseas legally would require a federal export license. Vaid requested that EG&G take care of the licensing paperwork. When McClafferty asked who the "end users" of the krytrons would be, Vaid said he did not wish to identify them, according to an affidavit filed in the case. At that, McClafferty abruptly ended the meeting.[33]

Alarmed by Vaid's behavior and the large number of krytrons involved, EG&G alerted the FBI. The FBI, in turn, brought in the Customs Service, which had jurisdiction over illegal foreign commerce.[34] All agreed Vaid was up to something.

The following week, Vaid called McClafferty from Houston. This time, according to court records, Vaid said he was calling on behalf of the University of Islamabad and wanted to purchase fifty spark gaps, another type of electrical switch that can be used to trigger nuclear bombs, as well as fifty krytrons. McClafferty said he would not accept the order until he received proper end-user documentation. Later that day, EG&G received a request for fifty krytrons and fifty spark gaps from a technical services company in Rawalpindi, Pakistan, the military and nuclear hub that abuts Islamabad. EG&G ignored the request.[35]

Three days later, on Halloween, Vaid showed up in a neighborhood electrical supply shop in Houston dressed, once again, in South Asian robes. The store, called Electrotex, was a distributor of EG&G products. Vaid said he wanted to buy fifty EG&G KN-22 krytrons, according to court records and published reports. One of the store's owners, Jerry Simon, called EG&G for a price quote. EG&G, which by then was cooperating with the Customs Service, asked for the name of the customer and whether the switches would be shipped overseas. Simon told them he'd checked with his customer, who said they would not be exported, but he declined to provide Vaid's name. EG&G agreed to accept the order and quoted a price of $3,950 plus tax.[36]

Vaid returned to Electrotex a few days later and made a cash down payment of one thousand dollars on the krytrons. He was told to expect delivery in sixty to ninety days.[37] Vaid then returned to Pakistan, where he awaited word of his order.

Three months later, two undercover Customs agents approached Simon and his brother Don, the co-owner of Electrotex. The agents, assigned to a Customs program known as Operation EXODUS, which investigated possible arms exports violations, asked for the Simons' assistance in nabbing Vaid. The Simons agreed to cooperate. Over the next several months, they stayed in close touch with Customs as they waited for Vaid to return and claim his order.[38]

Vaid had asked that Electrotex deliver the krytrons to a Houston photocopy shop operated by two of his relatives. On April 16, 1984, an undercover Customs agent posing as an Electrotex employee delivered a box containing fifty disabled krytrons to the shop. One of Vaid's relatives signed for the package. Customs then placed the shop under twenty-four-hour surveillance. By law, the government could not make an arrest until an attempt was made to export the krytrons without the appropriate license.[39]

Vaid finally returned to Houston from Pakistan on June 19. Two days later he contacted a freight-forwarding company that he used regularly and arranged pick-up of a package bound for Karachi, according to court records and published accounts. Vaid asked casually whether Customs would be able to open and examine the contents of the parcel and was told that they could. The shipping company, which was also cooperating with the authorities, then called Customs and informed them that Vaid was preparing a shipment.[40]

The following day, June 22, a driver for the shipping company arrived at the copy shop. Vaid handed him a sealed package. The shipping manifest listed the contents as "printed material and office supplies," including "50 bulbs/switches," according to media reports and an affidavit filed in the case. The stated value was less than $250.[41]

The driver delivered Vaid's package to Houston Intercontinental Airport. Customs agents there opened it and found the krytrons. Shortly afterward, Vaid and two employees of the copy shop were taken into custody. Bond for Vaid was set at two hundred thousand dollars.[42]

So far, the case appeared to be a success. Sensitive technology had been kept out of Pakistani hands, and the main suspects were in custody. But the investigation would soon take an unusual turn.

Missing Link

At the time of Vaid's arrest, Customs agents seized documents directly linking him to officials of the Pakistan Atomic Energy Commission. Among Vaid's papers, the agents found copies of letters between Vaid and PAEC officials S.A. Butt and Colonel Umar Din Dar dating back to 1983. The letters identified Butt as the procurement director for the PAEC. He was, of course, well known to American and European arms-control officials as the point man in Pakistan's successful campaign to obtain uranium enrichment and plutonium reprocessing technology from Europe, and his activities had been extensively detailed in the 1981 book *The Islamic Bomb*.[43] Documents obtained through the Freedom of Information Act by the National Security News Service in 2007 show that by 1981 the US State Department knew Butt had been "associated for some time with the Pakistani nuclear procurement effort."[44] Dar was a senior executive officer of the PAEC. Yet Customs investigators and prosecutors involved in the Vaid case apparently remained unaware of the men's involvement in Pakistan's nuclear program.[45] Although the government would subsequently turn up evidence showing that Vaid's efforts to buy krytrons had been directed by Dar and Butt, the letters seized by Customs concerned Vaid's purchase of unregulated chemicals on behalf of the PAEC. They nevertheless provided clear evidence that Vaid had operated at the behest of the top officials of Pakistan's nuclear program. The US government, however, claims to have missed the link.

A week after his arrest, Vaid appeared at a bond hearing before a federal magistrate in Houston. Government prosecutors emphasized the serious nature of the case, saying they were convinced Vaid was a Pakistani agent. The lead prosecutor, Assistant US Attorney Samuel Longoria, noted that the krytrons Vaid had attempted to export could be used as nuclear triggers. "We strongly suspect that Mr. Vaid is operating at the instructions of the Pakistani government and that the purchase of krytrons was for Pakistani use in obtaining a nuclear bomb," Longoria told the court. He gave no indication, however, of the basis for that suspicion and made no mention of the letters seized by Customs. He would later back away from his statement, saying there was no evidence to support it, only to acknowledge its accuracy when confronted with the evidence after the trial. Years later he would tell a reporter that the comment was based more on "supposition than direct knowledge." At the bond hearing, however, Longoria made clear that the allegations in the complaint against Vaid were "not simply technical violations of United States export laws." To emphasize the seriousness of the matter, Longoria even referenced a case in the Netherlands the year before in which A.Q. Khan had been convicted in absentia of attempting to obtain components from Holland for Pakistan's nuclear program. Longoria also highlighted the fact that the State Department had taken a strong interest in the case due to "serious concern regarding the nature and direction of Pakistan's nuclear program."[46]

Based on Longoria's presentation, Magistrate Calvin Botley rejected defense motions to reduce Vaid's bond. Botley also ruled that Vaid would have to put up the full two hundred thousand dollars in cash, rather than the 10 percent that is customary — an indication of the judge's awareness of the seriousness of the charges. Vaid's attorneys indicated their client was unable to raise the money.[47]

As Longoria noted in court, the State Department had taken an unusually strong interest in the Vaid case. Documents obtained through the Freedom of Information Act show that US and Pakistani diplomats feared that the case could jeopardize continuation of US aid to Pakistan, a development that would undermine cooperation in the anti-Soviet war in Afghanistan. The State Department's Bureau of Near Eastern and South Asian Affairs (NEA), which was responsible for Pakistan policy, was especially vigilant in protecting the Afghan program. As Vaid awaited trial, NEA Senior Deputy Assistant Secretary Arnold Raphel sent a memo to a senior State Department political-affairs official warning that Vaid's case "could have an influence on how Senators vote on [an upcoming renewal of] security assistance to Pakistan." Raphel, who would later become ambassador to Pakistan, added that the NEA would factor that consideration into its "Congressional strategy on the Pakistan nuclear program."[48] In the weeks following Vaid's arrest, the State Department monitored the case closely, keeping a sharp eye on media coverage and obtaining "relevant intelligence" on the matter from the CIA and the National Security Agency.[49]

Pakistani officials also worried about repercussions from the Vaid case. As Vaid's trial approached, the Pakistani Embassy Minister in Washington, Mujahid Hussein, complained to a US diplomat that the timing of the case "could not be worse," given the upcoming congressional vote on continued aid to Pakistan.[50] As it turned out, Islamabad had nothing to worry about.

While the diplomats fretted, Sam Longoria and his team took their case against Vaid to a federal grand jury. Had the prosecutors made the connection between Vaid and the Pakistani government, they could have obtained an indictment under the Atomic Energy Act or the Export Administration Act. Violation of either act carried a maximum sentence of twenty years in prison, if the licensable goods being exported could be used to enhance the national security of a foreign country.[51] But the prosecution failed to make the link. As a result, on July 16, the grand jury handed down an indictment against Vaid and his associates from the copy shop on the far less serious charges of conspiracy, making false statements to federal officials, and illegal export of licensable munitions.[52] Those charges carried a maximum penalty of two years in prison.[53] Although the indictment noted that the krytrons Vaid had attempted to export could be used "as part of a firing set for a nuclear explosive devise," all reference to nuclear weapons would soon disappear from the case.[54]

Vaid's good fortune extended beyond the charges in the indictment. In a flurry of legal challenges, his lawyers won two important concessions from the court and the prosecutors. First, Vaid's criminal attorney, William Burge, petitioned the court to issue a gag order prohibiting participants from discussing the case with the media. Burge later said he requested the order to avoid adverse publicity for his client. To his surprise, the prosecutors readily agreed to the request. "Their reaction," Burge recalled, "was, 'Hey, that sounds okay. We're not interested in having a lot of publicity about this case either.'"[55] With no objection from the prosecutors, the gag order was issued, effectively keeping the case out of public view.[56]

Vaid's lawyers also asked the court to remove any mention of the potential use of krytrons as nuclear triggers from the indictment. After initially petitioning the court to reject the motion, the prosecutors unexpectedly reversed course and agreed to the request. The court granted the defense motion and prosecutors rewrote the indictment with the offending language removed.[57]

Longoria later denied that the government's position in the case had been weakened by acceding to the defense requests. Burge later said he believed the government agreed to the gag order and to rewriting the indictment because of the US relationship with Pakistan. "The case, apparently, was embarrassing to the government of Pakistan, and they didn't want any more publicity than was necessary," Burge said.[58]

Another of Vaid's attorneys says the US government actively interfered in the case. Peter Williamson, who was one Vaid's immigration lawyers, told a reporter for the National Security News Service in 2004 that around the time the grand jury handed down its indictment he was summoned to the courthouse for a meeting with Sam Longoria and the judge hearing the case, James DeAnda. The conversation, Williamson recalls, was brief and one-sided. Williamson says DeAnda stated, "I have received a call from Washington," then added, "I have been instructed to get rid of this case." According to Williamson, DeAnda then said, "This case is going to be dismissed," and told the two attorneys, "There's not going to be any complaints about it."[59]

Williamson says he was stunned. "I was certainly not going to complain that my client was not going to be sent to jail for a long time," Williamson said. "But I had never before even heard of a federal judge saying that he'd been called by Washington — and acknowledging it. It was quite unbelievable." Williamson says DeAnda, a former trial lawyer appointed to the federal bench by President Carter in 1979, did not indicate from whom the call had come. The incident was particularly mystifying, says Williamson, because federal judges are appointed for life and, therefore, are presumably immune from political pressure.[60]

DeAnda, who died in 2006, flatly denied Williamson's assertions. "I can categorically tell you that that didn't happen in this case," DeAnda told the National Security News Service in 2004. In fact, DeAnda said he did not recall the government showing any special interest in Vaid's case. He did acknowledge that for reasons of national security, or to protect an investigation, the government might, on occasion, try to withhold information about a case from the public. But DeAnda said that if anyone in Washington had wanted to have information withheld in the Vaid case there was a "much more efficient" way of doing so. "They had a way of handling that that would be a lot easier and would not involve any risk," DeAnda said. "That would be telling . . . the US Attorney not to indict him."[61]

Sam Longoria says he does not recall an incident such as the one described by Williamson. "I would have remembered that, I would think," he said. He did, however, acknowledge in a 1985 PBS documentary about the Vaid case that both the State Department and the Department of Justice took "a very great interest" in the matter.[62] Asked by a reporter for the National Security News Service in 2004 whether he had sensed any pressure or interference from the government, Longoria said, "Not in the least." Like DeAnda, Longoria expressed doubt that anyone would have tried to manipulate a federal judge. "I just can't imagine that that would have happened," he said. "Federal judges don't take orders from anybody."[63]

There are, however, additional factors that bolster the possibility of government interference in Vaid's case. In 2004, reporter Seymour Hersh, who, in 1985, had revealed the existence of the correspondence between Vaid and the PAEC, told the National Security News Service of a conversation he had with the State Department's Arnold Raphel. According to Hersh, after the legal proceeding against Vaid had ended, Raphel told him that the "fix was in" on the case. In addition, Hersh said that Raphel, who served as the US ambassador to Pakistan from 1987 until his death in the mysterious 1988 plane crash that also took the life of Pakistani leader Zia ul-Haq, told him that the CIA had taken a special interest in the Vaid matter. In fact, top-level CIA officials were deeply worried about what might happen to the anti-Soviet Afghan campaign in the event that the Vaid case triggered a cut-off of US aid to Pakistan.[64]

Texas Congressman Charlie Wilson, a leading advocate of the covert Afghan program and a staunch supporter of Pakistan, shared that concern. Fearing that congressional opponents of Pakistan's nuclear program would use Vaid's case to call for an end to US aid to Zia's government and thus endanger Islamabad's cooperation in the anti-Soviet jihad, Wilson took his concerns to the Reagan administration. Wilson told the National Security News Service in 2005 that after discussing the Vaid case with administration officials he was "pretty well comforted that nothing was going to happen" that would disrupt US assistance to Pakistan. Wilson said he came away from the discussions with the sense that the administration had the Vaid case under control.[65]

"Smoking Gun"

Amid the behind-the-scenes maneuvering, Judge DeAnda set a trial date of September 17 for Vaid and his codefendants.[66] But just as the proceedings were to get underway, the prosecution obtained dramatic new evidence. A subpoena of the company Vaid used to send his messages to Colonel Dar and S.A. Butt yielded a series of telexes showing that Vaid's krytron purchase had been orchestrated by the PAEC.[67]

In a telex to Dar and Butt dated October 19, 1983, the day after Vaid's visit to EG&G Electro-Optics in Salem, Vaid explained that the firm was "reluctant" to sell him the krytrons unless he could tell them "the name of the ultimate consumer." He added that he was "arranging alternate ways to buy these products" through another company. On October 22, Butt replied, stating that the "ultimate consumer" was the Research and Development Department of the University of Islamabad, the same institution Vaid identified as the end user in his second conversation with EG&G Electro-Optics' customer service manager just days later. In a message to Dar and Butt on November 1, 1983, the day after the visit to Electrotex in Houston, Vaid reported that EG&G "completely regretted" being unable to supply the krytrons due to export restrictions, but added that he had made "alternate arrangements" and placed the order.[68]

Prosecutors later said they considered the telexes a "smoking gun," conclusively demonstrating that Vaid had knowingly attempted to skirt export restrictions. Yet they claim still not to have recognized the larger significance of the evidence they held. Although the letters seized by Customs in June had been addressed to Butt as the PAEC's procurement director and the telexes obtained in September showed that Vaid had purchased the krytrons on behalf of Dar and Butt, Longoria and his associates said they did not know Vaid was supplying Pakistan's nuclear weapons program. A Customs agent involved in the case told Seymour Hersh in 1985 that the names Dar, Butt, and Vaid had been run through a Treasury Department database but came up negative. "There was no record," the agent said. Customs officials in Washington also told Hersh that they found nothing in their files about the three men.[69] Yet State Department cables from the early 1980s, later obtained by the National Security News Service, make clear the US government was well aware of Butt's "special purchasing functions" on behalf of Pakistan's nuclear program.[70]

But prosecutors said at the time that because Customs investigators supplied them with no information about Dar and Butt, they did not know of Vaid's ties to Islamabad's atomic weapons program. "We did not know that Dar and Butt were officials [involved] with nuclear bombs in Pakistan," Longoria told Hersh. "We didn't know who they were." Longoria acknowledged that had they known the role Dar and Butt played in the PAEC, prosecutors had all the evidence they needed to connect Vaid to Pakistan's bomb program.[71] Yet inexplicably, given Butt's well-known history and the letters addressed to him at the PAEC, Longoria and his team did not connect the dots.

While prosecutors were seemingly unable to recognize the significance of the evidence in their possession, Vaid was not. Confronted with the telexes on the eve of his trial, Vaid, according to his attorneys, realized he could not avoid conviction and quickly struck a deal with the prosecution. In exchange for prosecutors dropping the conspiracy and false statements charges, Vaid agreed to plead guilty to one count of attempting to illegally export licensable defense articles. He was also to be deported from the United States as a convicted felon. The charges against Vaid's codefendants were dropped outright. Judge DeAnda signed off on the plea bargain and set sentencing for October 22.[72]

On the day of sentencing, Vaid arrived in court wearing what DeAnda later described as a South Asian "frock."[73] As the proceeding began, DeAnda announced that he was "left with the impression" that Vaid "apparently had no malicious intent beyond trying to expedite what he thought was a business deal and trying to accommodate a customer," a striking foreshadowing of a view expressed by the judge in a future nuclear smuggling case. Vaid "did all the things the government charges," DeAnda added, "but there's nothing in the evidence aggravating beyond that." He then asked Longoria whether the prosecution had "anything to the contrary."[74] DeAnda told the National Security News Service years later that he asked the question because he vaguely recalled Longoria's earlier statement about Vaid being a Pakistani agent.[75] It was a perfect opening for the prosecution to voice any concerns about the larger significance of Vaid's activities, particularly in light of the discovery of the "smoking gun" telexes. But Longoria did not take advantage of that opportunity. Instead, he responded to DeAnda's question by saying that the prosecution believed Vaid was guilty of more than simply trying "to accommodate a customer" — he had attempted "to accommodate himself to a profit that other people who obey the law wouldn't ordinarily get." Curiously, Longoria added that the prosecution had "never claimed Mr. Vaid was an enemy of the country or Pakistan was an enemy, far from it." Nevertheless, Longoria urged DeAnda to "consider an appropriate term of imprisonment" for Vaid.[76]

Judge DeAnda sentenced Vaid to two years in prison, but suspended the sentence, saying the four months Vaid spent in jail awaiting trail was "adequate." DeAnda also placed Vaid on five years probation, without supervision. As a "special condition," DeAnda prohibited Vaid from engaging in "any business activities" while on probation. "I know you have suffered severe financial loss already, but you brought all this on yourself," DeAnda told Vaid. "You can go back home. I hope you don't have to see me again."[77]

Peter Williamson believes the entire exchange was disingenuous. "That's what DeAnda said was going to happen," Williamson said. "This case was going to go away."[78] It was an outcome that, like the sentences handed down in the Canadian inverter-components case, would become increasingly familiar.

As Vaid awaited deportation, Seymour Hersh interviewed him at the federal detention center. Vaid protested his innocence, saying he was merely a businessman involved in a commercial transaction — an odd contention from someone who'd been caught with orders signed by PAEC officials. He also insisted he was unaware of the krytrons' potential use in nuclear weapons. In the course of the interview, Vaid encouraged Hersh to speak with his lawyers and waived his attorney-client privilege.[79]

Taking advantage of the opportunity, Hersh contacted Peter Williamson and his law partner, Spencer Gardner, who also represented Vaid. The two, freed of their obligation to protect their client's information and incensed by the government's handling of their client's case, provided Hersh with the documents showing that Vaid had purchased the krytrons at the behest of Colonel Dar and S.A. Butt. Williamson told Hersh he simply wanted the government to follow the law. "The government started off this case going into court . . . saying this man is a big spy," Williamson said. "At the end of the case they quietly folded their books and went away." Williamson added that there was no reason Vaid "shouldn't have done a lot of time in the penitentiary."[80] Years later Williamson explained to the National Security News Service that he had not told Hersh about DeAnda saying the case would "go away" because he feared that angering the judge could have negative repercussions for Vaid, who was still in federal custody.[81] Nevertheless, he felt strongly enough about the prosecution's mishandling of the case to share his concerns with a reporter.

Armed with the evidence that Vaid had purchased the krytrons on behalf of PAEC officials, Hersh returned to the federal detention center. Shown the communiqués, Vaid suggested they were fabrications. He acknowledged writing the text, but denied addressing the messages to Dar and Butt. He offered no explanation as to who might have forged the documents.[82]

Sam Longoria, on the other hand, had no doubts about the authenticity of the messages. He insisted, however, that he did not know of Dar and Butt's connection to the PAEC until Hersh informed him. "The evidence was not there to support a theory, which I personally espoused early on, that this person was a spy or an agent of the Pakistani Government," Longoria told Hersh.[83] Longoria elaborated on this point in an interview with National Security News Service in 2004. "What I was relying on was what the Customs Service gave me," Longoria said. "They didn't have that information at the time."[84] As a result, he contended, the government did not understand that Vaid had purchased the krytrons for Pakistan's nuclear weapons program.

Judge DeAnda said he was never informed of Vaid's letters and telexes to the PAEC and learned about them only as a result of questions from a reporter at the National Security News Service in 2004. "This was never raised at all that I can remember," DeAnda said.[85]

Operation Clusterf#@k

Three weeks after sentencing, federal marshals transported Vaid — dressed in his familiar Pakistani garments — to Houston Intercontinental Airport. There, the marshals placed Vaid on a plane bound for Pakistan, and he was deported without incident.[86]

By all rights, the Vaid saga should have ended there — albeit imperfectly. Vaid had been captured and deported. Pakistan had been denied at least one shipment of components that could be used to build a nuclear weapon. Publicity had largely been avoided. Islamabad had been spared embarrassment. Pakistani cooperation with the United States in the Afghan effort had been preserved. And questions about the wisdom of ignoring the threat posed by Pakistan's continuing quest for nuclear weapons had been set aside for consideration at some future date.*

But Vaid's story was far from over. In 2004, as part of an investigation of the Pakistan bomb program, the National Security News Service took another look at the Vaid case. What they found was alarming: Vaid was back in business in the United States.

Not surprisingly, the News Service found that the Houston branch of Vaid's old import-export company had long since been dissolved. But in searching Texas's corporate records, News Service reporters discovered that since 1996 Vaid had set up no fewer than seven new businesses — some of them seemingly noteworthy given his past. Six of those companies, moreover, had been established after 9/11, when tough, new immigration

---

*While Washington and Islamabad were undoubtedly relieved by the outcome of the Vaid trial, the Reagan administration realized the result could be a possible irritant to US–Indian relations. With that in mind, on the eve of a visit to Washington by Prime Minister Rajiv Gandhi in mid-1985, the State Department issued a press "guidance" that, among other matters, offered suggestions on how to deal with the Vaid matter. The guidance, which was approved by Arnold Raphel, stated: "The key points are that Vaid was arrested, tried and convicted, and that his attempt to export krytrons was foiled. This is evidence of how serious an effort we are making to stop the export of any item to Pakistan that could contribute to the development of a nuclear weapons capacity. . . . His [Vaid's] sentencing was the perogative of the court, as it is in our system generally." ("NEA Guidance for Press Spokesman on India," June 5, 1985, available from the National Security Archive, Washington, DC.)

regulations were ostensibly in place and should have prevented a convicted felon such as Vaid from entering the country. Yet Vaid's notarized signature on the corporate records suggested he had, in fact, been in the United States.

Several of Vaid's new companies had foreign corollaries. Three in particular stood out. One was the US arm of an Islamic investment firm based in India. Another was the American subsidiary of a Pakistani metal engineering company. And the third was the Texas branch of a Dubai-based trading company. Vaid and his brother, a former doctor who now lives in Houston, set up all three entities on the same day in July 2002.[87] Together, the companies bear what a former US intelligence nonproliferation analyst calls a "striking similarity" to the structure of the A.Q. Khan nuclear smuggling network, in which financial entities provided funds for the purchase of specialty metals and other strategic components by engineering front companies that then shipped raw materials and finished products to Pakistan, Libya, and beyond via Dubai.

Surprised by this discovery, the News Service contacted the Bureau of Immigration and Customs Enforcement (ICE), the post-9/11 incarnation of the US Customs and Immigration services, to determine the circumstances under which Vaid had entered the country. It quickly became clear that ICE officials in Washington were unaware that Vaid had returned to the United States. The News Service also contacted ICE and FBI sources in Houston to find out what they knew about Vaid's recent activities. The phone calls resulted in a scene reminiscent of a bad Keystone Cops parody. The call to the FBI came on a day that Attorney General John Ashcroft had announced an increased risk of terrorist attacks within the United States. As a result, the FBI field office in Houston reacted to the news about Vaid as if responding to a potential terrorist strike. The Houston field office ordered agents pulled off the security detail for the upcoming Major League Baseball All-Star game at Houston's Minute Maid Park and reassigned them to the Vaid case. Both the FBI and ICE began round-the-clock surveillance of Vaid's known addresses. At one point, FBI agents reported seeing a suspicious vehicle near one of Vaid's residences only to discover, after hours of surveillance, that the vehicle was part of an ICE stakeout of the same location. ICE agents, meanwhile, maintained twenty-four-hour, "eyes-on" surveillance for days of a man they believed to be Vaid for days before realizing they were watching his brother, Mohammad.

After tripping over each other for several days, both the FBI and ICE determined that Vaid was no longer in the United States. He had returned to Pakistan in November 2002, and, as far as they could tell, had not returned. According to an ICE official, however, Vaid had been in and out of the United States repeatedly in recent years. ICE sources told the National Security News Service that Vaid had "fraudulently" obtained visas from the State Department under the name Nazir Ahmed. Had he applied under his full name, according to ICE sources, he presumably would have been denied entry to the United States, because "Nazir Ahmed Vaid" appears on both ICE and State Department watch lists. By simply dropping his last name, however, Vaid was apparently able to beat the system. This is all the more remarkable for the fact that "Nazir Ahmed" was listed as a known alias for Vaid in court records from his 1984 nuclear smuggling case and presumably, therefore, should have raised red flags.[88] It is also the name he used in incorporating several of his new US companies.[89] The name Nazir Ahmed, however, was not on the watch lists.

Having satisfied itself that Vaid was not currently in the United States, the FBI lost interest and dropped its investigation. ICE, on the other hand, kept its investigation open, as did the National Security News Service. Following an unannounced visit by a News Service reporter to Vaid's brother in Houston in June 2004, Vaid himself called the reporter from Pakistan. Vaid told the reporter he visited the United States "frequently" on work-transfer visas known as L1s. He also explained that he'd set up his foreign-affiliated Texas companies to do business in the United States, but insisted he was no longer involved in international trade. Vaid said his life had been "ruined" by his 1984 legal troubles, and he "categorically" denied that he currently dealt in dual-use or nuclear materials. He acknowledged having met A.Q. Khan "socially" but denied ever doing business with the disgraced scientist. Vaid also indicated that he was well acquainted with "a lot" of Pakistani generals, intelligence operatives, and political elites.[90]

In fact, Vaid has done quite well for himself in Pakistan since his deportation from the United States. Among his many business ventures, he heads a Karachi-based enterprise known as the Finatra Group of Companies. Finatra controls several businesses, including cybercafés, a Web hosting service, an energy generation company, phone and cell phone rental agencies, and a pre-paid calling card dealer called Finatra Communication Private Limited.[91] In 1998, Finatra Communication signed a contract with Pakistan's official phone company, Pakistan Telecommunication Company Ltd., to provide pre-paid calling card service in Pakistan. The service also allows direct international dialing.[92] In 1996, Vaid and his brother established a branch of Finatra in Texas.[93]

In another conversation with the National Security News Service in July 2004, Vaid indicated that he was in the process of liquidating his assets in Pakistan and planned to return to the United States with the money and get his businesses up and running. Vaid said he expected to be in Houston toward the end of summer.[94] ICE, meanwhile, had obtained warrants for Vaid's arrest on charges of misleading the federal government and making false statements to obtain a visa, according to a source familiar with the case. ICE planned to detain Vaid the next time he attempted to enter the United States, the source said.

Vaid's travel plans were delayed several times. Finally, in October 2004 he was set to return to the United States. His son, Samad, who is a director of one of Vaid's Texas companies, made the trip a week ahead of him. According to Nazir Vaid, as Samad passed through customs in Houston, an immigration officer pulled him out of line and informed him that there was a warrant out for his father's arrest. Samad, who was allowed to enter the country, informed his father. Vaid canceled his planned return to the United States, thereby avoiding arrest. As far as is known, he has not returned to the United States. The Department of Homeland Security conducted an internal investigation of the incident but has refused requests to release its findings.

Most of Vaid's Texas businesses have since folded. One, however, Catalyst Communications USA, remains in operation. In 2001, Catalyst, of which both Vaid and his son were directors, received a license from the Federal Communications Commission to provide international telecommunications services.[95] As of late 2006, Vaid's son was still listed as a Catalyst director and vice president.[96] Vaid himself was listed in 2005 as the chief executive officer of a related company in Pakistan known as Catalyst Communication (Private) Ltd.[97]

In 2005, one of the authors of this book, Joseph Trento, published an abbreviated account of the Vaid affair in his book *Prelude to Terror.* After a pirated version of that account appeared on the Internet, Vaid sent an e-mail message to the other author of this book, David Armstrong, with whom he had previously been in communication. Vaid's message read: "I have read the material you have posted about me on the web and this pack [of] lies will be accounted for [by] Almighty God suitably. For such unethical acts you will npt ne [sic] spared by GOD in this world and the world hereafter."[98]

The Vaid saga is remarkable on several fronts. Vaid's ability to obtain US visas and return to the United States in recent years despite his 1984 conviction and deportation demonstrates the glaring holes in US security that remain despite the measures put in place since 9/11. More broadly, the handling of Vaid's nuclear smuggling case provides vivid evidence of the Reagan administration's willingness to help cover up Pakistan's bomb program in order to ensure Islamabad's continued cooperation in the anti-Soviet Afghan campaign. Once again, short-term Cold War policy goals trumped concerns about the spread of nuclear weapons. And in many ways, Vaid's case set the stage for much of what was yet to come.

8

A Turn of the Screw

THE REAGAN ADMINISTRATION dodged a bullet in the Nazir Vaid affair. The case had been handled relatively quietly, the threatened disruption of US aid to Pakistan had been avoided, and Islamabad's cooperation in the Afghan war had been preserved. It was as good an outcome as the White House could have hoped for. But the matter did not go away entirely.

In late February 1985, five months after Vaid's sentencing, the first news stories reporting that the government had overlooked crucial evidence linking Vaid to Pakistan's nuclear program appeared. The revelations set off new consternation both in Washington and in Islamabad. The State Department quickly tried to defuse the controversy by telling reporters that the United States had received assurances from Pakistan that its nuclear program was purely "peaceful in intent." Department spokesman Bernard Kalb added that the US did not believe Islamabad had a nuclear device or that it was in position to build one. Kalb noted, however, that Washington had made it clear to Islamabad that the US would be "unable to sustain" its assistance unless Pakistan showed some "restraint in the nuclear area."[1]

While Pakistan offered no signs of restraint, Zia did make an odd attempt at explaining away the krytron caper. In an unusual interview with the *Christian Science Monitor* on March 1, Zia insisted that although Vaid and his compatriots had been arrested, they were ultimately "let off." "So, you see," he said, "they were not stealing anything." In any case, Zia added, the incident "wasn't important." Pakistan was merely seeking "timing switches for revolving lights." The krytrons, Zia said, were for "search lights, ambulances, [and] lights affixed to the tops of buildings to warn aircraft of the buildings' height."[2] Why Pakistan's Atomic Energy Commission would need switches for such items was neither asked nor explained.

On Capitol Hill, longtime critics of Pakistan's nuclear effort apparently found Zia's explanation unconvincing. When news of Vaid's ties to Islamabad's bomb program broke, Rep. Stephen Solarz, chairman of a House subcommittee on Asian affairs, began calling for hearings on the matter. With another major vote on aid to Pakistan just weeks away, it appeared that the assistance package was once again in trouble, and with it, the all-important Afghan campaign.[3]

Alarmed by the implications, the CIA and its most committed ally on the anti-Soviet jihad, Rep. Charlie Wilson, quickly swung into action. Wilson understood that, unlike Vaid's criminal case in Houston, the administration had limited options for handling Solarz. The Texas Democrat therefore took it upon himself to rein in his fellow congressman and party member. His method of doing so was cynically ingenious. Solarz was widely known as India's best friend in Congress. Wilson is said to have taken advantage of this by presenting Solarz with a dossier of classified intelligence on India's nuclear weapons program. Wilson reportedly inferred that on the issue of the bomb, New Delhi might be more vulnerable than Pakistan.[4] Solarz presumably got the message, yet pressed ahead.

Wilson's close collaborator in the Afghan cause, Gust Avrakotos, the CIA officer in charge of supplying arms to the mujahedin, also sought to intimidate Solarz. Avrakotos, who had earned the nickname "Dr. Dirty" for his willingness to engage in ethically questionable operations, claimed to have shown up at a private briefing for Solarz and put on a show for the congressman's benefit. According to the book *Charlie Wilson's War*, which chronicles the efforts of Wilson and Avrakotos to support the Afghan program, Avrakotos suggested to Solarz that Congress could find itself in a "terrible position" if it cut off aid to Pakistan. Zia might start charging the United States for the "services of the ISI [Inter-Services Intelligence]" and related logistical facilities. If he did, the bill would come to "many billions of dollars a year." Worse yet, Avrakotos suggested, Zia might simply pull out of the program altogether.[5] It was a highly dubious argument, but it, again, sent a message.

In the end, neither Solarz nor any other nonproliferation advocates in Congress mounted a serious bid to cut off funding for Pakistan. The best they could manage was to fend off the administration's request for a $25 million increase in the aid package. In late March, Solarz's subcommittee and the Senate Foreign Relations Committee approved another $626 million in military and economic assistance to Islamabad, holding the allocation at the previous year's level. Both bodies did, however, pass measures ostensibly aimed at restraining Pakistan's bomb building efforts. In response to the new disclosures about Vaid's activities, Solarz pushed through a proposal banning aid to any country that smuggled or attempted to smuggle nuclear-related equipment or technology out of the United

States. But the measure also contained a major loophole: a provision granting the president authority to waive its application. On that basis the White House found the measure acceptable.[6] President Reagan would take full advantage of the waiver provision.

While the White House could reconcile itself to the Solarz measure, it had deep concerns about a proposal introduced on the other side of the Capitol by Democratic Senator John Glenn of Ohio. Glenn, another persistent critic of Pakistan's nuclear program, believed that the six-year sanction waiver granted by Congress in 1981 had effectively given Islamabad free rein to develop a bomb as long as it did not conduct a nuclear test. Glenn, therefore, proposed that continued aid be made conditional on the president certifying each year that Pakistan neither possessed nor was developing a bomb. Unlike the Solarz measure, Glenn's proposal did not include a waiver provision. The Reagan administration found Glenn's concept wholly unacceptable and mounted a vigorous campaign to quash it. It ultimately convinced the Senate Foreign Relations Committee, which had initially approved the proposal, to substitute a less objectionable version. The White House arranged for Senator Larry Pressler of South Dakota, a Republican who had not previously inserted himself into the Pakistani bomb fracas, to sponsor a measure requiring only that the president certify that Pakistan did not a have nuclear weapon and that American assistance would reduce the odds of it obtaining one in order for aid to continue.[7] But even this watered-down version would be effectively ignored by both the Pakistanis and the Reagan administration.

Even before these measures went into effect, new evidence of Pakistan's bomb making capability emerged. In an ironic twist, on July 11, ABC news reported that, according to intelligence sources, Pakistan had recently conducted a successful cold test of a nuclear triggering package using American-made krytrons.[8] How Islamabad obtained the triggers was not explained. Just a day earlier, the Reagan administration had announced that it was expediting delivery of sophisticated antiaircraft missiles to Islamabad to bolster defenses against an increasing number of Soviet and Afghan incursions into Pakistani territory.[9] The administration gave no indication it planned to cancel or even slow the delivery in light of this newest indication of Pakistan's determination to build a bomb.

Early the following month, on August 8, 1985, President Reagan signed the Solarz and Pressler amendments to the Foreign Assistance Act into law. It would not be long before their effectiveness was put to the test.

The Bomb in the Basement

The inadequacies of the Pressler amendment were apparent from the start. By 1984, the White House had begun receiving intelligence reports showing that A.Q. Khan and his team were producing significant quantities of enriched uranium. Concerned that news of this activity would spark new calls in Congress for an aid cutoff, the Reagan administration attempted to establish a new line of demarcation for Pakistan. In August, the White House drafted a letter for Reagan's signature cautioning Zia not to enrich uranium above 5 percent — a critical threshold above which it becomes much easier to enrich all the way to the 90-percent, weapons-grade level. Although some in the administration urged the president to issue an unambiguous warning, the final draft, sent to Zia on September 12, merely stated that crossing the 5-percent "red line" would result in "grave" but unspecified consequences. Zia issued a noncommittal response, saying only that the American-imposed limit would be respected, but even that would quickly prove meaningless.[10]

On March 14, 1985, in the midst of the uproar over the Vaid revelations and the jockeying over the nascent Solarz and Pressler amendments, A.Q. Khan gave another provocative interview to an Urdu-language publication, this time declaring that Pakistan could carry out "an atomic explosion in a very short time, if required, without conducting any test."[11] If true, it implicitly meant that Pakistan not only had crossed Reagan's "red line," but had also produced weapons-grade uranium. Within months, US intelligence was reporting that Pakistan was, at a minimum, enriching uranium beyond the 5-percent limit. Yet when Reagan met with Zia at the United Nations in late October, the president reportedly did not confront the Pakistani leader about the violation. There were no "grave consequences."[12] Instead, one month later, President Reagan signed the first in a series of annual Pressler amendment certifications declaring that Pakistan did not possess a nuclear bomb and that continued US aid reduced the odds of it acquiring one.[13]

The process became even more absurd the following year. By mid-1986, the administration had received a Special National Intelligence Estimate confirming that Pakistan had, in fact, produced weapons-grade uranium. In addition, press reports during the year revealed that Islamabad had conducted at least two more cold tests. Yet in June, when the Soviet Union charged that Pakistan had developed the capability to build a bomb, the Reagan administration rushed to its Cold War partner's defense, effectively telling Moscow to lay off. And in July, when Pakistani Prime Minister Mohammed Khan Junejo visited Washington, the White House went through the familiar ritual of warning that Islamabad would have to show "restraint" in its nuclear program if US assistance was to continue.[14]

But no restraint was actually required. In October, President Reagan signed another Pressler amendment certification letter. Just days later, the *Washington Post* quoted a well-informed source as saying that Islamabad was only "two screwdriver turns" from having a fully assembled weapon. But preventing Pakistan from making those final screwdriver turns was simply a lower priority than ensuring Islamabad's continued help in the Afghan war. As one senior US official responsible for monitoring the Pakistani bomb program put it, the Reagan administration "wouldn't come down on Pakistan if we found a bomb in Zia's basement."[15]

Presidents Reagan and Bush continued to issue certifications declaring Pakistan to be nuclear-weapons free for the next three years. During that time, much of the intelligence community and many State Department officials had concluded that Pakistan had produced everything it needed to make a bomb and had likely already done so. According to former CIA analyst Richard Barlow, who monitored Pakistan's bomb program closely during the mid-to-late 1980s, by the time President Bush issued the last Pressler amendment certification in 1989, Pakistan had already had a completed bomb for at least one and possibly two years or more. "Pakistan possessed nuclear weapons from 1987–88 at the latest," Barlow says.[16] In fact, by the end, the United States had already confronted Zia with a copy of the designs for the Pakistani bomb and his successor, Benazir Bhutto, with a full-scale model of the weapon.[17] As the CIA's former Islamabad station chief Milt Bearden put it, "Reagan and Bush said it ain't a bomb until they turn that last screw and paint B-O-M-B on the side."[18] As a result of that attitude, the American aid that was advertised as deterring Pakistan from acquiring a nuclear weapon continued to flow.

Kept in the Dark

While the Reagan administration was busy certifying Pakistan bomb-free, Islamabad went right on buying material in the United States and elsewhere for its nuclear weapons program. Most of the illegal purchases were simply ignored. One case, however, became too big to overlook and forced the implementation of the Solarz amendment. But when put to the test, the measure proved no more effective than the Pressler amendment in halting Pakistan's nuclear drive.

The challenge came in July 1987, when a Pakistani-born Canadian citizen was arrested in Philadelphia on charges of attempting to illegally export to Pakistan large quantities of two exotic metals that can be used to manufacture nuclear weapons. The arrest was the culmination of a twenty-month sting operation carried out by US Customs agents assigned to the Operation EXODUS program in conjunction with Canadian authorities and the CIA. The culprit, Arshad Z. Pervez, a forty-two-year-old Toronto-based import-exporter, had been nabbed trying to send a shipment of specially formulated beryllium and twenty-five tons of high-strength steel alloy to Pakistan, in violation of US law. Beryllium is used as casing for the fissile core of atomic bombs. The steel alloy, known as maraging 350 steel, is used to produce rotors and other components of high-speed uranium enrichment centrifuges.[19]

The person orchestrating the purchase on the Pakistani end was a sixty-two-year-old retired military engineer named Brigadier Inam ul-Haq, who ran an import-export business in Lahore.[20] US nonproliferation officials in the intelligence community were well acquainted with ul-Haq as a procurement agent for Pakistan's bomb program and for the Khan Research Laboratories (KRL) in particular. In a letter to Pervez, ul-Haq urged the Pakistani expatriate to carry out the metals purchase in the "national interest." Pervez's notebooks contained an entry about the deal that stated, "my expert is a procurement manager for nuclear plant." There were also references to the project being "atom" and "military."[21] And during negotiations with an undercover Customs agent posing as a metals salesman, Pervez had at one point acknowledged that the maraging steel was intended for "the Kahuta client."[22] Financing for the $333,000 deal came through the London and Toronto branches of the Bank of Credit and Commerce International (BCCI).[23]

Pervez's arrest ignited a furor in Congress. The Solarz amendment, adopted in response to the Vaid case, prohibited US aid to any country that attempted to export American goods or equipment that could be used in a nuclear weapons program. The existing Pakistani aid package was then in its final year. The White House and Congress had negotiated a new, six-year, $4 billion assistance program that was set to begin in October. In April, following a rancorous debate, House and Senate committees had adopted a proposal offered by Representative Solarz calling for a new, two-year waiver of antiproliferation laws that would allow the aid to continue.[24] Now, all of that was once again in jeopardy.

As soon as news of the Pervez case broke, Solarz demanded that the Reagan administration take action under the legislation that bore his name and cut off aid to Pakistan. Senator Glenn and other nonproliferation proponents in Congress also called on the White House to impose sanctions on Islamabad. But the administration countered that it was "premature" to charge that the Pakistani government was behind the attempted purchase and urged a go-slow approach. The State Department insisted that there should be "no rush to judgment" as to whether an aid cutoff was required. In the meantime, State Department officials pledged that the United States would engage in an "intensive dialogue" with Pakistan and would demand an explanation.[25]

Solarz decided to get some explanations of his own. Shortly after Pervez's arrest, the congressman convened a closed-door session of his Asian affairs subcommittee to learn what the CIA and State Department knew about Pakistan's nuclear procurement activities and to determine whether Pakistan had violated the terms of his amendment. One of those sent to testify was CIA director William Casey's top aide on proliferation issues, retired army major general David Einsel. Casey had designated Einsel as the lone intelligence official responsible for briefing Congress on Pakistan's nuclear program. This diverged from the standard practice. Typically, the senior CIA officers charged with overseeing the collection and analysis of intelligence on foreign nuclear programs kept Congress apprised of the agency's findings. As a result of Casey's new arrangement, the CIA's nonproliferation experts did not know what Congress had been told about Pakistan's extensive nuclear procurement efforts within the United States. Some senior agency officers were deeply suspicious of Casey's motives in implementing the scheme. They believed that he and Einsel were more interested in protecting the anti-Soviet Afghan campaign than in halting Pakistani violations of American law. Although Einsel's official title was National Intelligence Officer for Proliferation, he was known within certain circles of the CIA as the National Intelligence Officer for the Mujahedin.

Because Solarz had called for a formal hearing, the CIA's nonproliferation office had an opportunity to bypass Casey's Einsel-only policy and present Congress with a full accounting of Pakistan's nuclear procurement activity in the United States. To make that presentation, senior CIA officials assigned nonproliferation specialist Richard Barlow to appear at the hearing. Although still in his early thirties, Barlow, who played a lead role in the undercover operation that led to Pervez's arrest, had already earned a reputation as one of the intelligence community's top analysts. He tracked Islamabad's nuclear procurement program closely and had worked on many unpublicized cases of illegal Pakistani nuclear exports from the United States, Europe, and elsewhere. This would be his first time testifying before Congress.

In preparation for the hearing, Barlow says he compiled ten pages of talking points detailing a litany of Pakistani efforts to obtain American nuclear goods and technology. Each item on the list, he says, had been cleared throughout the intelligence community for presentation to Solarz's subcommittee. According to Barlow, his superiors and the CIA general counsel's office instructed him to hold back nothing from the panel. "You go down there and you tell the truth," Barlow recalls being told. "You don't mislead, you don't evade. You be very direct no matter what happens."[26] Barlow would do just that, but he would pay a heavy price.

The hearing took place on July 22, 1987, in the Rayburn House Office Building. Just down the hall, the Iran–Contra hearings were in full swing. While much of the country remained riveted on the arms-for-hostages scandal and the activities of Lt. Col. Oliver North, for some in Washington the real action was playing out before Solarz's subcommittee. Charlie Wilson showed up in what he described as an effort to "intimidate" Solarz.[27] Also in attendance was the State Department's new Deputy Assistant Secretary for Near Eastern Affairs, Robert Peck, who had taken over the post the year before when Arnold Raphel became ambassador to Pakistan. Peck, like Wilson, was an ardent defender of the mujahedin.

Solarz began the hearing by explaining that he and the members of his subcommittee wanted to know about any nuclear export cases that might trigger implementation of his amendment. He made it clear that he considered any transfer of American nuclear-related products or equipment, whether purchased directly from the United States or through an intermediary country, to be a violation as long as the transaction was carried out on behalf of a foreign government. The first question, therefore, was whether Pervez and ul-Haq were agents of the Pakistani government. Barlow recalls Einsel telling the panel that the CIA did not know whether the pair were Pakistani agents or simply freelancers acting on their own. Barlow was astonished. "We absolutely knew," he says. "I knew that ul-Haq was a fairly major agent of A.Q. Khan before this case even started. We knew he was a major agent of the Pakistani government." Moreover, according to Barlow, there was "absolute smoking gun evidence" that the metals Pervez had attempted to purchase on ul-Haq's instructions were "going to Kahuta and that KRL knew it and was ordering this activity." When Barlow was asked what the CIA knew about Pervez and ul-Haq, he says he responded that there was "no doubt" they were Pakistani government agents. But before he could go any further, he says, Einsel cut him off, preventing him from getting into details.[28]

While Barlow was disturbed by what he viewed as Einsel's attempt to mislead the subcommittee, it was what happened next that he found truly outrageous. The panel's second question was whether there had been other instances of Pakistan attempting to purchase US nuclear technology. It was at that moment, Barlow says, that he realized Congress had been kept in the dark about Islamabad's ongoing procurement efforts in the United States. "We didn't know that they didn't know what was going on," Barlow says. "At that point, everything became clear." According to Barlow, Einsel's response to the question was, "No, there haven't been other cases." But Barlow says he knew otherwise and that Einsel did too. "I knew exactly what Einsel knew," Barlow says. "He'd seen all the stuff I had. I knew what his state of knowledge was. I knew he was lying." Barlow says that when the subcommittee asked to hear from him he testified that there had been "scores" of other attempts by Pakistan to illegally purchase American nuclear technology. With that, Barlow says, the hearing room erupted in commotion. Congressmen were screaming and yelling, demanding to know why they had not been informed of the other cases.[29]

Einsel, according to Barlow, tried to regain control of the situation, telling the subcommittee that Barlow didn't know what he was talking about and that the cases he had referred to were just rumors based on unreliable information.[30] With the atmosphere growing increasingly combative, the State Department's Robert Peck rushed from the hearing room and grabbed a senior State Department official who had been standing by in the hallway. Peck implored the official to step in and try to minimize the damage Barlow had done. "That s.o.b. Barlow is telling them all sorts of things and you've got to straighten it out," Peck reportedly told the official. Brought before Solarz's panel, the official reiterated Einsel's argument, emphasizing the imprecise nature of intelligence reports and generally downplaying Barlow's testimony.[31]

But the subcommittee, Barlow says, wasn't buying it. "They thought there had been nothing going on since the Vaid case," Barlow explains. "Suddenly, right after they approve a $4 billion aid package, they hear that Pakistan has been routinely violating American export laws and that people have been lying to them or withholding information with the intent to mislead them. That was very clear in their minds."[32]

Einsel, who died in 2006, rejected Barlow's assertion that he misled Congress or withheld information. "There is always information that is in process, that's evolving," the former general told the *New Yorker* in 1993. "You generally give Congress evaluated answers — not all the rumors that are flying."[33] But Barlow insists the information the CIA had collected on Pakistan's procurement activity was rock solid. "This was not the kind of intelligence that was subject to interpretation," Barlow says.[34] Both Solarz and Senator Glenn have since acknowledged that with the exception of the Pervez case, they were never formally briefed, as required by law, about Pakistan's efforts to purchase nuclear technology in the United States.[35]

At the conclusion of the closed-door session, Solarz's subcommittee held an open hearing. During the public testimony, panel members continued to press for information about possible violations of the Solarz amendment but got largely evasive responses. Attempting to follow up on the issue that had caused such an uproar in the closed hearing, Democratic Congressman Howard Wolpe of Michigan asked Robert Peck whether the State Department was aware of any instances other than the Pervez case in which Pakistan had attempted to skirt American law to obtain nuclear materials. Peck offered only one example, that of an American couple and a Hong Kong national who had recently been indicted in California for illegally exporting high-tech equipment to Pakistan several years earlier that could be used to analyze the triggering mechanism of a nuclear bomb. When pressed further by Wolpe, Peck would say only that the State Department had received information that had "caused us concern" but insisted there were no "current cases" that fell under the "purview of the Solarz amendment."[36]

Richard Barlow was not asked to testify at the public hearing. Immediately after his executive-session testimony he was whisked away from the Capitol in a CIA staff car. By the time he reached CIA headquarters in nearby Langley, Virginia, Barlow says, "the phones were ringing off the hook at the highest levels" of the agency. "The shit really hit the fan," Barlow says. "You wouldn't believe how bad it was."[37] In fact, it was only the beginning of Barlow's troubles.

"You Can Make All the Bombs You Want"

As Solarz and others in Congress continued to call on the administration to cut funding to Pakistan in the wake of Pervez's arrest, the defenders of the anti-Soviet Afghan jihad went all out to save their prize program. Solarz became a particular target of their campaign. During a dinner at the Pakistani embassy, Jimmy Carter's former national security advisor, Zbigniew Brzezinski, the man who helped initiate both the covert Afghan proxy war and the blind-eye approach to Pakistan's nuclear program, confronted Solarz about his call to halt US assistance to Islamabad. "Steve, what are your objectives in cutting off aid to Pakistan?" Brzezinski asked. "Because if you do, I foresee the following things happening: one, the Afghan resistance collapsing and the Soviets triumphing; two, the present government of Pakistan will disappear; and three, you'll have an anti-American government in Pakistan in possession of the bomb. Is that what you want?"[38] It was an argument Solarz would hear repeatedly.

Arnold Raphel, the State Department's lead protector of the Afghan cause, adopted a different strategy for dealing with Solarz. In October, Raphel, now the ambassador to Pakistan, drafted a letter to Solarz in which he attempted to appeal to the congressman's interest in nonproliferation. In the process he provided a remarkable insight. Raphel began his letter by zeroing in on a recent statement by Solarz suggesting that the aid issue was a struggle between "whether we attach more importance to our nuclear nonproliferation objectives or to the support of our Afghan policies." Raphel explained that he was convinced the choice the United States faced was "not whether nonproliferation is more important than Afghanistan, or vice versa." Rather, Raphel wrote, "the central issue is whether the possibility of cutting off US assistance, in spite of what such a step may do to our Afghanistan policy, is an effective means of encouraging a positive change in Pakistan's nuclear program."[39] Raphel noted that he "firmly" believed that "if we cut off assistance the most likely result would not be a slowing of Pakistan's nuclear program but just the opposite." He added, "I believe that the most likely result of a cut-off of US assistance would be to undercut our objectives regarding both Afghanistan and nonproliferation."

Raphel insisted that the United States did not face an "either-or" choice on Pakistan. "Rather," he wrote, "through our assistance and close ties to Pakistan we can continue to keep a cap on the Pakistan nuclear program — although not at the point we would want — and continue to pursue a successful Afghanistan policy, or we will suffer reverses on both fronts." While acknowledging that the United States "should and must do more" to corral Islamabad's bomb program, Raphel also argued that US assistance was having a positive effect. "We have heard the contention that Pakistan could manufacture a device on short notice," Raphel wrote. "If this assessment is correct, it is an important sign of restraint that Pakistan has chosen not to do so. . . . I believe that a major impediment to crossing that threshold is our assistance program."* Raphel failed to note that all US assistance had done so far was allow Pakistan to come within two screwdriver turns of building a bomb.

Raphel then offered a remarkable revelation. On the question of how an aid suspension would affect Pakistan's support for the Afghan war, Raphel wrote, "the government of Pakistan has already made clear that Pakistan would continue its support even if US aid to Pakistan is cut off." That, of course, flew in the face of years of executive branch arguments that American aid — along with a tolerant policy toward Pakistan's bomb program — was crucial to maintaining Islamabad's cooperation in the anti-Soviet Afghan jihad. Now, it turned out, that simply wasn't true. But that was all billions and billions of dollars and untold numbers of kilograms of enriched uranium under the bridge.

Finally, on the issue of the Pervez case, Raphel informed Solarz that the United States had made some "useful headway" on the matter, having won several new "commitments" from Zia and Prime Minister Junejo. "First, they have agreed to pursue the Pervez case, no matter where it leads," Raphel wrote. "Second, they will design and implement new and more rigorous procedures to guard against any further illegal procurement in the [United] States. Finally they have agreed to cooperate fully." But neither Pakistan's latest assurances nor Raphel's strained appeals to Solarz's proliferation concerns would be sufficient to mollify the congressman. Customs investigators would later reveal that they could detect no letup in Pakistan's illegal acquisition activities.[40]

While Arnold Raphel and Zbigniew Brzezinski performed their good-cop–bad-cop routine for Solarz, Charlie Wilson launched a separate

*In testimony before Solarz's subcommittee the following March, Robert Peck would take up Raphel's argument that aid acted as a deterrent to Pakistan's nuclear program, saying imposing sanctions "would make it significantly more likely that Pakistan would proceed to acquire nuclear weapons." (Kux, *The United States and Pakistan 1947–2000*, p. 286.)

effort to preserve the Afghan campaign. Instead of trying to convert Solarz and the other congressional sticklers for nonproliferation, Wilson decided to go around them. Amid mounting calls to cut aid to Pakistan, the Texas Democrat concluded that the only way to save the anti-Soviet jihad was to market it. Earlier in the year, he had convinced Zia to hire Washington foreign lobbyist Denis Neill, a friend and ally of Wilson, to promote Pakistan's cause in Congress. Now Neill and Wilson would turn to selling the mujahedin. Neill had made generous contributions to the campaigns of members of the House Foreign Affairs Committee as well as the powerful Appropriations Committee, on which Wilson sat. Wilson convinced Zia to host a congressional delegation, then began cajoling colleagues into making the trek to Pakistan to see up close what Americans were really getting for their tax dollars. During the 1987 Thanksgiving break, Wilson led a politically diverse contingent of seven congressmen on a highly scripted tour of the Afghan operation. Wilson believed that by witnessing the tremendous fighting spirit of the Islamic holy warriors and experiencing the great anticommunist struggle in which they were involved, his colleagues could be won over to the cause. Accordingly, Wilson arranged for the delegation, decked out in full rebel regalia, to be taken to the secret training camps where the mujahedin learned how to kill Soviets. The congressmen also visited American-funded aid stations where wounded jihadis were treated. Some members of the delegation even donated blood as a show of support.[41]

But for all the appeals to the honor and glory of the fight, Wilson knew that the great obstacle to keeping alive the Afghan campaign was the bomb. In the midst of the Pakistan junket, news stories surfaced indicating that recent US intelligence reports had concluded that Islamabad had completed a "workable" nuclear weapon. One way or another, it was an issue that would have to be addressed. And in typically idiosyncratic fashion, Wilson tackled it head on. At a gala state dinner in Islamabad, Wilson rose before the assembled delegation to offer praise for Zia and to provide him an opportunity to make his best case for continued support. "Mr. President," Wilson said, "in history I have three heroes. Winston Churchill, President Lincoln, and Zia ul-Haq." Had it not been for Zia's leadership, Wilson declared, the Soviets would long since have rolled through Pakistan to the Indian Ocean in their quest for world domination. And as for Pakistan building nuclear weapons, Wilson made clear that he had no objections. "Mr. President," Wilson said, "as far as I'm concerned you can make all the bombs you want because you are our friends and they, the Indians, are our enemies." But, Wilson noted, not everyone in the United States shared that view. The issue was "getting hot" and there were questions that had to be answered.[42] That was Zia's cue.

Zia stepped to the front of the room to address his audience. He began by insisting that Pakistan's nuclear program was purely peaceful in nature and asked his guests to accept his word on the matter. Besides, he added, Pakistan had no plans to build a delivery system. A delivery system for *what* he did not say.[43]

On the question of aid, Zia spoke of the common struggle in which the United States and Pakistan were engaged in Afghanistan. With the Soviets on the verge of defeat, cutting aid now would be "a betrayal of history," Zia said, "and the judgment of history would be very severe on those who take this decision." But having no sooner said that, the general confirmed what Raphel had told Solarz the month before. "Whether there is American aid or not we will continue to fight," Zia said. "So please go back and assure my American friends and all those who are insisting that we should succumb to American pressure that Pakistan is not ready to accept any conditional elements."[44]

So there it was. Pakistan would not back down on the bomb and it would continue to fight in Afghanistan even if the United States cut off assistance. It was an odd argument for continuing the flow of US aid. And it might logically have prompted the question of why, under the circumstances, the United States should continue providing aid to Pakistan, much less tolerate its bomb program. But somehow, that didn't happen.

Once back in Washington, Charlie Wilson lobbied frantically to save the Pakistani aid package. He pulled out all the stops, twisting arms and calling in every favor that was owed to him.[45] But even as he did so, two new developments placed additional strain on the rationale for continuing to provide assistance to Pakistan. During a summit meeting with President Reagan in early December, Soviet leader Mikhail Gorbachev officially announced that the Red Army would pull out of Afghanistan over the next twelve months, based on an agreement with the United States to end external aid to the mujahedin.[46] Then on December 17, a jury in Philadelphia convicted Arshad Pervez on five counts of conspiring to ship nuclear-related materials to Pakistan.[47] At that point there could be no more arguments about whether or not the Solarz amendment applied.

And yet, the Reagan administration managed to salvage the aid package. Secretary of State George Shultz made a personal appeal to Congress to continue supplying assistance to Islamabad. Echoing Zia's earlier argument, Shultz contended that with the Soviets on the verge of withdrawing from Afghanistan, it was essential that the United States not undermine Pakistan at that crucial moment.[48] Thus, the issue was framed as a choice between continuing aid to Pakistan as a means of ensuring the Soviet ouster from Afghanistan or cutting off aid as required by law to prevent the spread of nuclear weapons. It was, as some in Congress pointed out, a false argument. But that didn't seem to matter. Despite the fact that the Soviet Union had agreed to pull out of Afghanistan, that Pakistan had announced it would continue to support the Afghan cause even without American aid and refused to give up its bomb program, and that US law required a cutoff of aid to Islamabad, on the very *day* of Pervez's conviction, congressional negotiators worked out a deal that allowed a massive new infusion of US assistance to go forward. Under an agreement worked out by a congressional conference committee, there would be a two-and-a-half-year extension on the aid-ban waiver, and Pakistan would receive a one-year, $480 million package of economic and military aid.[49] The following day, President Reagan issued his third Pressler amendment certification, providing the necessary declaration that Pakistan did not have a nuclear bomb.[50]

In January, as required by law, Reagan finally invoked the Solarz amendment in response to Pervez's conviction. The president's formal certification, which implemented the aid suspension, confirmed that the material in question was to be used "in the manufacture of a nuclear explosive device." But with another stroke of his pen, Reagan invoked the waiver provision of the Solarz amendment, thereby lifting the aid cutoff. In doing so, the president affirmed that continuing US assistance to Pakistan was "in the national interest."[51] As a result, US aid to Islamabad was preserved. American support for the Afghan jihadis — who would soon finish off the Soviets and begin looking for new targets against which to turn their talents — continued. The Russian pullout from Afghanistan went forward. And Pakistan continued building its bomb.

Six months earlier, at the time of Arshad Pervez's arrest, as the White House and State Department resisted implementing the Solarz amendment, Senator Glenn had stated, "We have short-term interests in Afghanistan, but we have a long-term interest in trying to keep our finger in the dyke to prevent the spread of nuclear weapons. That is more important than anything else."[52] But clearly the senator was mistaken. As the Reagan administration had once again demonstrated, Cold War objectives trumped concerns about nuclear proliferation.

In the meantime, the Pervez case continued to play out. Facing up to twenty-five years in prison, Pervez was sentenced to five. He appealed his conviction and was granted a new trial in 1990. But before that case went to court, Pervez's lawyers reached a plea-bargain agreement with prosecutors. Five of the charges against Pervez were dropped. He pleaded no contest to a sixth. The judge determined that Pervez had already received sufficient punishment and sentenced him to time served.[53] Pervez returned to Toronto.

In 1991, Brig. Inam ul-Haq was arrested in Germany and extradited to the United States. While awaiting trial in Philadelphia on charges of conspiring with Pervez to send nuclear-related goods to Pakistan, ul-Haq was released on $150,000 bail. In July 1992, a jury convicted him of conspiracy to defraud the US government and making false statements to a US government agency. He faced up to ten years in prison and a $500,000 fine. In September, the judge in the case sentenced ul-Haq to time served and imposed a $10,000 fine. Afterward, the former brigadier proclaimed his innocence. "I am a victim of a conspiracy," ul-Haq said. "I have been a victim of the pro-India, pro-Israeli, anti-Pakistan lobby."[54] Ul-Haq returned to Pakistan, where he ran several companies, including an automotive components manufacturer. He now lives in retirement in Lahore.[55]

For Richard Barlow, who had played the lead role on the intelligence side of the operation that led to the convictions of Pervez and ul-Haq, the affair marked the beginning of a long nightmare. After his appearance before Solarz's panel, senior CIA officials chastised him for his testimony and accused him of undermining the Afghan campaign. "I was called a traitor to my country," Barlow says. He quickly became a pariah within some quarters of the CIA. David Einsel and the State Department's Near East bureau reportedly urged the CIA's nonproliferation office to fire Barlow or, at a minimum, bar him from doing work on Pakistan's nuclear program. Within weeks, the CIA rewrote Barlow's job description, eliminating his responsibility for the Pakistan account. A year after his congressional testimony, Barlow left the CIA. He eventually landed a job as a proliferation analyst in the Department of Defense. But once again there was trouble. His persistent digging into Pakistani nuclear smuggling and his public contradictions of what he considered to be misleading testimony by his superiors about Pakistan's nuclear capabilities lead to his forced resignation under threat of firing. Barlow continued to work as a consultant to the intelligence community but was never again able to find work as a federal employee and lost his pension. "They ruined my life," he says.[56] Although derided by some for what they consider overzealousness, Barlow's efforts to expose Pakistani nuclear smuggling twenty years ago can today be seen clearly as a lost opportunity. He would not be the last government investigator to discover that there are those who are not interested in having Pakistan's nuclear activities revealed.

9

Bombs 'R' Us

IN LATE JANUARY 1987, as the undercover operation that would snare Arshad Pervez and Brigadier Inam ul-Haq took shape, A.Q. Khan granted a rare interview to a foreign journalist — and an Indian, no less. In the tranquil surroundings of his well-guarded suburban Islamabad home, Khan sat down for an hour with veteran Indian reporter Kuldip Nayar to discuss the state of Pakistan's nuclear weapons program. "What the CIA has been saying about our possessing the bomb is correct," Khan told his Indian guest. "And so is the speculation of some foreign newspapers."[1] It was a remarkable declaration, given Pakistan's repeated assurances that its nuclear program was for strictly peaceful purposes.

The notoriously egocentric Khan made clear, moreover, that he took great pride in the accomplishment. "They told us that Pakistan could never produce the bomb and they doubted my capabilities, but they now know we have done it," Khan said. And unlike Zia and other Pakistani officials, Khan made no pretense about Islamabad's intentions. "The word 'peaceful' associated with the nuclear program is humbug." Khan told Nayar bluntly. "There is no peaceful bomb."[2]

When Nayar's article appeared at the beginning of March, Khan immediately issued a statement denying that Pakistan possessed nuclear weapons. Khan claimed that some of the comments attributed to him were "concocted," while others had been "taken out of context."[3] Yet just a few weeks later Zia himself confirmed the substance of what Khan had said. In an interview with *Time* magazine, Zia stated, "Pakistan has the capability of building the bomb. You can write today that Pakistan can build a bomb whenever it wishes." Zia insisted, however, that Pakistan had no plans to produce nuclear weapons.[4]

At the time, both Khan's comments and Zia's follow-up were generally understood as a response to a recent flare-up in tensions between India and Pakistan. In late 1986, the Indian army began a huge military exercise in the Rajasthan desert near the Pakistan border. The exercise, dubbed Operation Brasstacks, was the largest mobilization of armed forces on the subcontinent since World War II. Fearing that the maneuvers might be a pretext for an invasion, Pakistan deployed reinforcements to the frontier. Soon, nearly four hundred thousand Indian and Pakistani troops stood facing off across the border. The crisis peaked in January, just as Khan gave his interview with Nayar. Tensions eased following the opening of diplomatic negotiations, and both nations began withdrawing forces in late February, a few weeks before Zia's interview with *Time*.[5]

While the statements of Zia and Khan posed a significant problem for the Reagan administration, struggling, as it was, to maintain the fiction that Pakistan did not possess nuclear weapons, most commentators regarded the disclosures as a little more than thinly veiled warnings to India. Pakistan had, after all, gone into the bomb business to ward off possible attacks by its much larger and more powerful neighbor. Having succeeded — with America's tacit consent — in developing a nuclear deterrent, Islamabad was now telling New Delhi to back off. There can be little doubt that this was a primary consideration in the timing of the statements.

But in retrospect it seems clear that there were additional factors driving Islamabad's decision to reveal its nuclear capability, and for Khan's bold declaration in particular. The Reagan administration had repeatedly demonstrated, most recently in the Vaid case, that it would go to great lengths to cover up Pakistan's nuclear program in order to preserve the anti-Soviet Afghan campaign. Islamabad, therefore, had much to gain in its long-running contest with India by declaring its nuclear capability and had little to fear in the way of retribution from the United States, at least in the short term. But Pakistan also understood that because of developments in Afghanistan, the grace period on the bomb program was running out. Just a year earlier, the Soviets had been ascendant in Afghanistan, making significant gains against the mujahedin and becoming increasingly aggressive in their use of helicopter gunships. When the rebels sought sophisticated American antiaircraft missiles, known as Stingers, to counter the helicopter attacks, the Reagan administration balked, fearing further exposure of the already well-known US role in the "covert" Afghan campaign. But under pressure from the Islamic holy warriors' friends in Congress, the administration relented. The jihadis got their missiles and in September 1986 brought down their first Soviet helicopter. More takedowns quickly followed. Soviet helicopter losses mounted rapidly, and pilots became much more cautious in their operations. Suddenly, the Soviets were on the defensive. By early 1987, Zia was suggesting that a "miracle" might be possible. Instead of simply punishing the Soviets and blocking the Red Army's advance farther south, it now seemed the Russians might actually be driven from Afghanistan altogether.[6]

But Zia knew that such a miracle would be a mixed blessing for Pakistan. On one hand, the ouster of the Soviets held the prospect of a truly Pakistani-friendly regime emerging in Kabul. Such a development would not only greatly decrease the odds of future Russian adventures in Afghanistan but would also provide Islamabad with an important ally in its ongoing contest with India. In addition, Zia hoped that a new Afghan government would share his own Islamist vision, providing a further buffer against Soviet interference and strengthening the bond to Pakistan.[7] Although Zia would not live to see it, when such a government eventually came to power in Kabul, with considerable help from Pakistan's Inter-Services Intelligence (ISI), the consequences would be profound. Under the rule of the Taliban during the 1990s, Afghanistan became a haven for militant Islamic fundamentalists, including Osama bin Laden and his band of terrorists.

On the other hand, Zia also knew that a Soviet pullout from Afghanistan would mean an end to the free ride Pakistan had enjoyed with the United States. American interest in Afghanistan was strictly limited to countering Moscow. Pakistan knew from bitter experience during its wars with India how fickle an ally the United States could be. Zia realized that once the Russians withdrew from Afghanistan, Pakistan would likely be on its own. Islamabad faced losing its powerful ally and a major source of support. It would also likely come under sanctions, due to pressure from American nonproliferation advocates — although it would be too late to rein in Pakistan's nuclear program.

During the 1980s, Pakistan had gone all-out to build a bomb while it had the chance. As Washington looked the other way, and even provided cover, Islamabad spent millions of dollars purchasing equipment from the United States, Europe, and Asia with which it created its weapon. Now that episode was coming to an end. Pakistan had built its bomb but would soon lose much of the political protection and financial support with which it had done so. Yet as the Brasstacks standoff had shown, tensions with India still ran high. From that point forward, every confrontation had the potential to go nuclear. Pakistan was not about to give up its bomb and would, in fact, expand its arsenal. The United States had allowed Pakistan's nuclear genie out of the bottle, and it could not be coaxed back in.

And now that technology would start to spread. The bomb that Pakistan had developed with America's acquiescence would be sold and traded as a commodity. The network that had become so adept at procuring equipment for Pakistan's bomb program would now, for a price, provide other countries everything they needed to build their own nuclear weapons. A.Q. Khan's pronouncement at the beginning of 1987 was in many ways a declaration that Pakistan's nuclear bomb shop was open for business.

Just days after Khan's interview appeared in print, Sen. John Glenn, testifying before the Senate Foreign Relations Committee, stated, "Pakistani nuclear weapons production will, sooner or later, whether by design or by espionage, result in the wider transfer of nuclear weapons technology to countries in the Middle East."[8] Although Glenn could not have known it, that process had already begun.

The Dubai Deal

Pakistan's entry into the retail atomic bomb business began with a pair of meetings in 1987. The first is believed to have taken place early in the year in Switzerland at the Zurich-Kloten International Airport. There, a member of A.Q. Khan's network reportedly presented a top official of Iran's Atomic Energy Organization with what amounted to an order form for a nuclear weapons starter kit.[9] The one-page, handwritten document listed everything necessary to attain a full-fledged uranium enrichment capability. Among the items on the list were dismantled centrifuges and components; auxiliary equipment, such as vacuum pumps and electrical drives; and drawings, plans, and specifications for a complete centrifuge plant complete with a workshop for manufacturing additional parts. The list also included equipment for casting enriched uranium into the hemispherical forms used as the cores of atomic bombs as well as technical reports on the process — normally, one of the most difficult to master and closely guarded secrets of the bomb-making arts.[10] Yet Khan's network was offering all of this to Iran in a straight cash deal. Prices ranged from millions of dollars for individual items to hundreds of millions for the complete package.[11]

The offer came against a backdrop of increased nuclear cooperation between Iran and Pakistan. Iran's nuclear ambitions stretched back to the mid-1970s. As money poured into the country following the 1973 oil crisis, Shah Mohammad Reza Pahlavi not only doled out large sums to Ali Bhutto, who was then raising funds for his Islamic bomb, but also announced plans to launch his own large-scale nuclear power program.[12] Work soon began on four French- and German-designed reactors with four more on the drawing board.[13] By the late 1970s, US intelligence had determined that in addition to this ostensibly peaceful program, the shah had also established a secret nuclear weapons development effort.[14] But in 1979, the shah's nuclear ambitions fell victim to the Islamic revolution that drove the monarch himself from power. Many top scientists fled the country, and the nuclear effort languished.[15] But in the mid-1980s, with the country bogged down in a long, bloody war with Iraq, Iran's clerical leaders took a renewed interest in developing a nuclear option.[16] In 1985, they revived a scaled-back version of the shah's nuclear power program.[17] That same year, Iran also launched a secret uranium enrichment effort. Procurement agents quickly obtained critical equipment from a West German firm, and Iranian scientists and engineers began the daunting task of mastering the highly complex technology.[18] They would soon get significant help from outside.

In February 1986, Tehran entered into an agreement with Islamabad to send Iranian engineers to Pakistan for training in nuclear energy.[19] That same month, A.Q. Khan traveled to Iran and paid a secret visit to the nuclear reactor facility in Bushehr, on the Persian Gulf coast. The plant had been heavily damaged in repeated bombings by Saddam Hussein's air force in the course of the ongoing Iran–Iraq War. In January 1987, shortly before making his public declaration of Pakistan's nuclear status, Khan returned to Iran, this time, reportedly, to prepare a study for the Tehran regime on the feasibility of using the Bushehr reactor to produce plutonium for nuclear weapons.[20] According to an account by conservative journalist Kenneth Timmerman, Khan met with senior Iranian officials at an intelligence ministry guesthouse south of Tehran and, in a revival of his performance for Ali Bhutto more than ten years earlier, made the case for the superiority of enriched uranium as a path to the bomb.[21] It was at about this time that a member of Khan's supply network extended the offer to provide enrichment technology, plans, and bomb-making equipment to Iran.

The man who reportedly received the hand-printed order form in 1987 was Dr. Masud Naraghi, a US-trained laser and plasma physicist who was a senior project manager in the Atomic Energy Organization of Iran (AEOI). Naraghi, who now lives in the United States, says that it was Khan's European suppliers who "initiated" the sales. He insists he found out only later that the European "salesmen" were hawking Pakistani technology.[22]

Whatever the exact genesis of the offer, Iran followed through. But rather than ordering a full-blown, off-the-shelf uranium enrichment capability, the Iranians, by their own account, selected a more modest entry-level kit. This included parts from one or two disassembled first-generation Pakistani centrifuges along with technical specifications, designs, and instructions for the manufacture and assembly of additional machines.[23] Iranian officials have since said that they planned to follow Pakistan's model and develop an indigenous enrichment capability.[24] As former UN weapons inspector David Albright has noted, acquiring the plans and components would be "tremendously helpful" to the Iranians, allowing them to skip many of the difficult early research steps.[25]

Once the Iranians had placed their order with Khan's network, they received a summons to travel to Dubai later in 1987 to consummate the deal. The freewheeling Persian Gulf port city had long served as a transshipment point for much of the equipment Pakistan acquired for its own nuclear weapons program. Now it would increasingly become a hub for Khan's nuclear export business. It was in a dusty Dubai office that the first known transfer took place. Khan himself did not participate in the session. Instead his friend, a Dubai-based businessman named Sinawappu Seeni Mohamed Farook, represented the Pakistani side of the transaction. Farook, who was born in British India and lived in Sri Lanka before moving to Dubai, had a business that reportedly supplied goods to Khan's operation.[26] He would play a key role in the expansion of Khan's network. Also at the meeting was Farook's twenty-eight-year-old nephew, Bukhary Sayed Abu Tahir. Although Tahir, a Sri Lankan national, played only a minor role in the meeting with the Iranians, he would eventually supplant his uncle to become the central figure in Khan's nuclear smuggling enterprise.[27]

The Iranian side of the bargain was represented by a three-man delegation that included Masud Naraghi. Naraghi says that his role was primarily that of a technical expert, designated to ensure that the goods received from the Pakistanis checked out.[28] UN investigators believe that some of the Iranians at the meeting posed as employees of a front company to disguise their connections to Tehran's defense ministry.[29] Also at the meeting were three of Khan's European suppliers. One, a German named Heinz Mebus, was an old college friend of Kahn's from Berlin and had been one of the earliest suppliers to Pakistan's enrichment program.[30] At least one of the other suppliers at the meeting is believed to have been a German.[31]

The Iranians reportedly paid several million dollars for the items they had ordered.[32] Years later, after the deal was exposed, Iran told UN investigators that the Pakistanis had thrown in the equipment for casting uranium into bomb cores at no extra cost, presumably as an incentive to encourage future purchases.[33] The Iranians *would* buy more from Khan's network, but they would also shop elsewhere. Using the plans and specifications they had purchased from Khan and his cohorts as a kind of nuclear shopping list, Iran went bargain hunting, buying much of the equipment and technology they needed at lower prices from competitors in Europe, China, and Russia.[34] Iran quickly developed an extensive procurement network that would eventually rival and possibly exceed Pakistan's.[35]

But for all the purchases, Iran's centrifuge program made little progress initially. Scientists at the AEOI encountered serious technical difficulties and the enrichment effort foundered.[36] Within several years, Iran would return to the Khan network for additional help and equipment. When it did, B.S.A. Tahir would play a lead role in the deal.[37]

Nuclear "Cooperation"

At around the time of the Dubai deal, the heads of the Iranian and Pakistani atomic energy agencies entered into a secret nuclear cooperation agreement. Under the pact, Iranian engineers traveled to Pakistan to receive advanced training in nuclear technology.[38] These officially sanctioned but clandestine contacts may have helped obscure Khan's activities. Pakistani officials would later say that Khan's sales of nuclear plans and equipment to Iran began under the cooperation agreement to share ostensibly peaceful nuclear technology.[39] According to US intelligence sources, the CIA had picked up on the increased nuclear collaboration between the two countries but did not immediately detect the component and technical data transfers. Late the following year, US and allied intelligence services determined that Pakistan was helping Iran build a secret uranium enrichment plant outside of Tehran.[40] It was the US Defense Intelligence Agency (DIA) that first picked up message traffic among Pakistani military officers indicating Khan and his associates were engaged in a deal with Iran. According to a high-level DIA source, that evidence was turned over to the CIA, which launched a wider eavesdropping operation, tasking the National Security Agency (NSA) with targeting the top level of the Pakistani government, including intelligence and military officials. But for all the intelligence collected, Washington was unwilling to impose sanctions on its Cold War partner for its dangerous behavior.

Throughout 1987 and 1988, Pakistan's nuclear activity sped up dramatically as Islamabad scrambled to push its bomb program as far forward as possible before the clock struck midnight on its relationship with the United States.[41] As part of that process, Islamabad's nuclear procurement efforts in the United States and Europe also picked up markedly, according to intelligence sources. Increasingly though, those purchases would now feed into A.Q. Khan's lucrative new enrichment technology export business. All those involved in the Dubai deal had done very nicely for themselves. Now they would begin to formalize and expand the operation.

Meet the New Boss

On August 17, 1988, Pakistani President Mohammed Zia ul-Haq traveled to a military base near Bahawalpur, an hour's flight from Islamabad, to observe a target practice demonstration by a single American-made M-1 Abrams tank. At the time, Pakistan's military was considering making a major investment in the highly touted tank to help bolster its defenses in the wake of the Brasstacks scare. The demonstration turned out to be a flop, with the M-1 missing its target on each of ten attempts. At the conclusion of the embarrassing presentation, Zia was shuttled back to his waiting aircraft for the return flight to Islamabad. At the last minute, he invited US Ambassador Arnold Raphel to join him on the official presidential plane. Raphel, who had flown to Bahawalpur separately, accepted. Standing on the tarmac waving as the door to Pak One closed was Lt. Gen. Mirza Aslam Beg, the army vice chief of staff. Beg was the only general in Pakistan's chain of command who was not on the plane with Zia that day. He would fly back to Islamabad on a smaller aircraft, right behind Zia's.[42]

Zia's American-built Hercules C-130b transport plane took off on schedule at 3:56 P.M. But just minutes into the flight, air traffic controllers at Bahawalpur lost contact with the craft. Witnesses later said they saw the plane lurching up and down in the sky like an airborne roller coaster before it barreled into the earth. All thirty-two people on board were killed. In addition to Zia and Raphel, the dead included US Brig. Gen. Herbert Wasson, the head of the American military aid mission to Pakistan, and Gen. Akhtar Abdur Rehman, the chairman of Pakistan's Joint Chiefs of Staff, the second most powerful man in the country, after Zia. As head of the ISI for ten years, Rehman had been the main architect of the anti-Soviet campaign in Afghanistan.[43]

Zia's death, along with the deaths of most of Pakistan's military leadership, left an enormous power void. After eleven years of Zia's rigid rule, there was suddenly an entirely new political order in Islamabad. In accordance with Pakistan's constitution, the head of the Senate, Ghulam Ishaq Khan, became the country's acting president. The seventy-three-year old Ishaq Khan was a longtime advocate of Pakistan's nuclear program and a major supporter of A.Q. Khan's enrichment empire at Kahuta. Zia's successor as army chief of staff, a position he still held at the time of his death, was Gen. Alsam Beg. Beg, who had been one of a group of officers trained in guerrilla warfare tactics by the CIA during the 1950s in preparation for a possible Soviet invasion of Pakistan, was one of the principal tacticians behind the Afghan jihad. A committed Islamist with a deeply anti-Western worldview, Beg was also a fierce proponent of Pakistan's nuclear bomb program.[44]

Having come to power amid the trauma of Zia's death, Beg and Ishaq Khan chose to allow previously scheduled elections set for November 1988 to proceed. The contest pitted a wealthy businessman, Nawaz Sharif, who had the backing of Pakistan's military establishment, against the Western-educated Benazir Bhutto, daughter of Zulfikar Ali Bhutto, the father of the Islamic bomb and the man Zia had overthrown and hanged. Bhutto eked out a narrow victory in the election, and after considerable prodding by the United States, the military agreed to let her take office. As usual, though, there were conditions. Bhutto was not to interfere in military matters or issues related to the bomb. And while she would nominally serve as Pakistan's first democratically elected prime minister since her father, Beg and Ishaq Khan would maintain tight control over the nation's nuclear weapons program.[45] The CIA was convinced that Pakistan's military and nuclear establishments kept Bhutto in the dark about the full extent of the country's nuclear capabilities and the activities of Khan and his associates.[46]

Bhutto has stated that as prime minister she was repeatedly approached by Pakistani military officials and scientists seeking permission to export nuclear technology. She insists that she turned down all such requests.[47] Yet the former prime minister acknowledges that her efforts to "control the direction of nuclear policy" were "much resisted." She has also noted that she was barred from Khan's labs and had little influence over the Kahuta operation, which was protected by the military.[48]

At the top of the military heap was General Beg, a man with a grand vision for the bomb. Beg's Islamist views extended to sharing Pakistan's nuclear expertise with Iran. As army chief, Beg openly proposed selling or trading nuclear technology to Tehran. A.Q. Khan has reportedly claimed that Beg was aware of the transfers of centrifuge technology to Iran. In 1989, Iranian President Hashemi Rafsanjani reportedly told Bhutto that a group of Pakistani generals led by Beg had offered to provide Iran with nuclear weapons technology. Bhutto is said to have told both Rafsanjani and Beg that she objected to the scheme. But that was apparently not enough to curb Beg's enthusiasm. Following the 1991 Gulf War, Beg reportedly proposed that in order to avoid suffering the same fate as Iraq, Pakistan should sell nuclear technology to Iran as part of a "grand alliance" against the United States. Beg denies making such a proposal. Yet that same year, Beg told US Ambassador Robert Oakley, who had replaced Raphel, that he had reached an agreement with the head of the Iranian Revolutionary Guard to exchange Pakistan's bomb-making know-how for Iranian oil and conventional weapons. When the United States protested, the Pakistanis pledged to halt the deal.[49] But that did not put an end to Pakistan's export of nuclear technology.

The Pakistani nuclear black market scandal that would explode many years later is a direct outgrowth of the United States' failure to deal forcefully with Islamabad's bomb program when the opportunity existed. In 1988, with the Soviets not yet completely out of Afghanistan, the Reagan administration determined it necessary to continue the pretense that Pakistan did not possess a completed nuclear weapon. On November 18, President Reagan signed the final Pressler amendment certification of his term, officially declaring that Pakistan did not have a bomb.[50] US aid to Islamabad would continue, as would Pakistan's nuclear exports.

Nuclear Enrichment

In mid-December 1988, less than a month after Ronald Reagan certified that Pakistan did not have a bomb, Sinawappu Seeni Mohamed Farook, the man who reportedly presided over the transfer of centrifuge plans and equipment to Iran at the meeting in Dubai the year before, began the process of establishing a company in London that would become a spawning ground for A.Q. Khan's nuclear smuggling network. Farook, then forty-seven years old, had traveled to the UK for medical treatment. But he also had business on his mind and took along his nephew, B.S.A. Tahir, a bit player in the Dubai meeting who now served as a lieutenant to his domineering uncle in a collection of family-owned companies. Farook controlled a string of Dubai-based businesses known as SMB Group that had been founded in 1980 by Tahir's since-deceased father. Tahir managed the operations of the group, which included computer equipment and information technology firms as well as trading companies, warehouses, and an assortment of ill-defined enterprises. Now Farook sought to add a British wing to his Dubai empire.[51]

To help set up his new company, Farook turned to his London-based accountant, Abdul Mabood Siddiqui, a since self-professed longtime friend and associate of A.Q. Khan. Siddiqui, a British citizen originally from the Bengal region of what is now Bangladesh, was the senior partner in the West London accounting firm Reddy Siddiqui & Co. Siddiqui also had offices in Dubai, and it was there, his son later told investigators, that he first met Farook when his firm became the auditor for one of Farook's companies.[52]

Sitting in Siddiqui's London office, Farook and Tahir chose the name Foremost Trading Ltd. for their new British enterprise. Siddiqui would hold a minority stake in the firm and serve as an unpaid director but took no active role in the company's operations, according to corporate and court records. Farook and Tahir controlled the company entirely. For their banking services, they selected the main London branch of the Bank of Credit and Commerce International (BCCI).[53]

Foremost Trading began doing business. Farook and Tahir returned to

Dubai, leaving the day-to-day operations of their new company in the hands of Siddiqui's twenty-six-year-old son, Abu Bakr Siddiqui. Abu Siddiqui, who was born in East Pakistan, now Bangladesh, and holds dual British and Pakistani citizenship, ran the company out of a one-man office in Northwest London. For the next five years, Foremost dealt exclusively with a Dubai-based entity known as SMB Traders, which was controlled by Farook and Tahir. Farook, who Siddiqui later told investigators had "contacts" in the Pakistani government, was the company's director, but Tahir took the more active role. From Dubai, Tahir submitted orders to Foremost for electronic components, air compressors, industrial air conditioning equipment, and a variety of metals. Siddiqui contacted suppliers, obtained quotes, arranged purchases, and made shipments. Most of the orders were sent to SMB Traders in Dubai. Siddiqui would later recall a few orders going to the Civil Works Organization (CWO) in Karachi.[54] Although apparently unbeknownst to Siddiqui, CWO was the in-house construction company for Khan Research Laboratories (KRL).[55] A subsequent British Customs investigation would determine that most of the orders submitted to Foremost originated with Khan or a close associate in Pakistan, according to a pair of confidential Customs reports on the case. There is no evidence that Siddiqui knew at that time that Khan and his associate were behind the orders. Siddiqui, through his attorney, declined a request for an interview from the authors.[56]

Foremost's unusual business model proved reasonably successful, at least for its owners. In its first year of operation, Siddiqui later testified, the company made sales to SMB Traders of over £400,000 ($632,000). That figure jumped to £600,000 ($1 million) the following year before settling back to approximately £400,000 ($675,000) in 1991. But Foremost was charging only a 5 percent mark-up on the goods it sold to SMB Traders. (What SMB Traders charged Khan for the goods is not known.) Siddiqui initially received a salary of just £6,000 ($9,500), and there were growing tensions with Farook. Both Siddiqui and Tahir found the irascible businessman increasingly difficult to deal with. Siddiqui began considering other career options. While continuing to work for Foremost, he returned to school to study accounting and even worked part time for a restaurant owned by his family.[57]

Finally, Siddiqui and Tahir had had enough. In July 1994, they formed a new company, SMB Europe Ltd., that would replace Foremost and expand its operations. Both men served as directors, although Tahir held the majority stake in the firm. The original plan, by Siddiqui's account, was for the company to become a UK computer distributor, using one of the SMB Group's subsidiaries, SMB Computers, as a supplier. But as Siddiqui later told investigators, "there wasn't much work on that." Instead, Tahir and Siddiqui reverted to old habits, picking up where Foremost left off. In the early stages, SMB Europe received orders from Farook. But relations remained tense, and the two younger men soon broke off from Farook and formed their own supply chain. Tahir established a new company in Dubai, SMB Middle East, and began submitting orders to Siddiqui at SMB Europe in London. Once again, Siddiqui was sending components, air conditioning equipment, compressors, and other goods to Tahir in Dubai. In one or two instances, by his own account, Siddiqui sent shipments directly to the Peoples Steel Mills (PSM) in Karachi. PSM is now well known to investigators as a front for Pakistan's nuclear weapons program and is closely associated with A.Q. Khan. Siddiqui later testified that he learned sometime in 1995 that Tahir was forwarding the goods supplied from London on to PSM and an entity called the United Engineering and Trading Co. (UETC) in Rawalpindi. He also testified that he did not learn until 1996 that A.Q. Khan was behind UETC and only in late 1997 or early 1998 did he discover that Khan was involved in Pakistan's nuclear weapons program.[58] According to a source involved with the subsequent investigation, at the time both PSM and UETC appeared on British watch lists of companies suspected of WMD proliferation.

Atoms for Ayatollahs

For B.S.A. Tahir, the new, independent arrangement with Siddiqui was part of a broader development. By 1993, Tahir had begun to challenge Farook for control of the family business and was becoming an increasingly important figure in A.Q. Khan's nuclear smuggling network. A turning point was the network's second sale of centrifuge technology to Iran. Having made limited headway in their efforts to enrich uranium based on the designs and equipment purchased from Khan's team in 1987, the Iranians were poised to try again, this time on an expanded scale. The opportunity, according to Iranian officials, came when Tahir, presumably acting on Khan's instructions, approached an employee of an Iranian state agency that was purchasing computer equipment from an SMB Group company and offered to sell Tehran complete centrifuges and technical documentation. The offer reportedly went all the way up the line to Iranian President Rafsanjani. Tehran's leadership proved interested and dispatched two representatives to meet with Tahir and Farook. Discussions took place in Dubai in August and December of 1993. The Iranians ultimately chose to purchase five hundred first-generation centrifuges, known as P1s and, for reasons that are unclear, a set of plans similar to those purchased from Khan's network in 1987. They also bought designs for more advanced P2 centrifuges.[59]

Iran took delivery of the first batch of P1 components in January 1994. Tahir arranged additional shipments from Dubai in March and May of that year. After inspecting the goods, the Iranians made payment to Tahir in October. An Iranian representative carried two briefcases stuffed with about $3 million in cash to an apartment used by Khan on his frequent visits to Dubai. Tahir then distributed the money throughout the network. A final shipment of components was sent to Iran the following year.[60]

Contact between the Iranians and Khan's network would continue. Between 1996 and 1999, there were as many as ten meetings. Iran claims these were to discuss technical problems they encountered with the P1s. The Iranians repeatedly complained about the poor quality of the components they received, many of them castoffs from Kahuta, which had by then upgraded to P2s. Iran insists that at no time did it discuss the P2 designs with the Pakistanis. In fact, the Iranians maintain that because of the difficulties with the P1s they did not even begin work on the more sophisticated P2s until 2002. UN investigators appear skeptical of these claims, and Tahir has asserted that Iran was sent three P2s in the mid-1990s. If true, Iran could have used them as templates to build more.[61]

Whether or not Iran received actual P2s in the 1990s in addition to the designs, it is clear that Khan and his gang had once again given a major boost to Tehran's enrichment program. The five hundred P1s and ongoing assistance from Pakistan allowed Iran to move from basic research and testing to development of a small but operational enrichment facility. Despite the technical challenges, Iranian scientists gained valuable knowledge and expertise much faster than would otherwise have been possible and soon became self-sufficient in building centrifuges. The P2 designs, and possibly even the machines themselves, laid the basis for a much more sophisticated program. During the 1990s, Iran secretly built a highly secure, partially underground facility in Natanz to house the fifty thousand centrifuges that, with Khan's help, it planned to manufacture.[62]

US intelligence, which had long monitored Khan's activities closely, kept a sharp eye on his network's continued dealings with Iran. By the mid-1990s, the United States had developed extensive intelligence showing that Pakistan had sold centrifuge designs and equipment to Iran. Intelligence reports revealed that Iran was actively pursuing a nuclear weapons capability based on the plans obtained from Pakistan and was dealing with many of the same companies through which Pakistan procured components for its bomb program.[63] In early 1997, Israeli intelligence delivered what one former White House nonproliferation official calls a "devastating briefing" to US intelligence and nonproliferation officials, presenting clear proof that Iran was pursing an "aggressive missile program" in parallel with a uranium enrichment research and development effort that was being aided by Pakistan. The United States also picked up signs that Iran was buying equipment from Khan. But despite clear indications of the Khan network's bolstering of Iran's enrichment program, the Clinton administration did not move to halt or expose the activity. While the administration worked hard to curtail Russian and Chinese assistance to Iran's nuclear program, it chose to continue monitoring Khan's contributions without taking action.

Several factors lay behind the adoption of that strategy. Gary Samore, a top nonproliferation official in the Clinton administration, has said that the United States held off going public in hopes of learning who Khan's other customers were and catching Iran with more than "a hole in the ground."[64] But in doing so, the United States was also allowing Khan's sales to continue and Iran's nuclear program to advance. According to other Clinton administration nonproliferation officials, there was also a concern within the intelligence community that going public with the information about Khan's sales would expose intelligence sources and methods. To the extent those considerations played a role in shaping US policy, it suggests that protecting the intelligence operation took precedence over stopping the spread of nuclear weapons.

Another factor, according to former Defense Department nonproliferation specialists, was the CIA's working relationship with Pakistani intelligence. Although as a result of Pakistan's ongoing bomb-building activity the first Bush administration had reluctantly cut off aid to Pakistan in 1990, following the Soviet withdrawal from Afghanistan, the CIA maintained the close ties with the Inter-Services Intelligence (ISI) that had grown out of their joint efforts in the anti-Soviet Afghan jihad. Throughout the 1990s, the agency relied heavily on the Pakistani intelligence service in a variety of operations, including supplying covert assistance to Muslim forces during the Bosnian war and, ironically, keeping tabs on Afghanistan's Taliban regime, a creation in large part of the ISI. To protect the cooperative relationship with ISI, according to former Pentagon nonproliferation experts, some US intelligence officials resisted cracking down on Khan's network, which enjoyed the support and protection of Pakistan's intelligence service.

An interest in seeing Pakistan deal with the proliferation problem may also have contributed to the administration's failure to take action. In 1993, Benazir Bhutto, who had been ousted from office in 1990 by Pakistan's military-backed president Ghulam Ishaq Khan, won reelection as the country's prime minister. Bhutto's new government quickly vowed to halt Pakistan's nuclear cooperation with Iran. The Clinton administration publicly accepted that pledge at face value, despite Bhutto's well-understood inability to corral Pakistan's nuclear establishment in general and A.Q. Khan's operation in particular.[65] Thus, while embracing Bhutto and her attempts to cut off Pakistan's nuclear dealings with Iran, the United States effectively stood by while Khan's nuclear peddling continued.

"One-Stop Shop"

After closing their first nuclear deal with Iran in 1987, Khan and his associates began looking for new customers. Iraq was apparently a target of one of their early pitches. In October 1990, the month that President George H.W. Bush invoked the Pressler amendment, cutting off aid to Pakistan, and just two months after Iraq invaded Kuwait, one of Khan's intermediaries reportedly approached Iraqi intelligence officials with an offer to sell Saddam Hussein's regime a package of uranium enrichment technology, similar to the deal extended to Iran. The Iraqis seem to have rejected the offer, at least in part out of concern that it could be a CIA sting operation.[66] Even though the Iraqi deal did not materialize, Khan and his network did not give up.

By the mid-1990s, Khan had developed a widespread reputation as the go-to guy for nuclear know-how and technology. It was an image he actively cultivated. For years, he had talked openly about his country's nuclear weapons program, generally exaggerating his own role in that effort. He had traveled widely through Europe, Asia, Africa, and the Middle East, constantly on the prowl for anyone who might lead him to a potential customer for his illegal wares. And he had published ostensibly scholarly articles on various aspects of uranium enrichment that were little more than thinly veiled advertisements for his goods. Khan's lab would eventually begin hawking its goods at trade shows, handing out glossy brochures bearing the KRL seal and the words "Government of Pakistan" that advertised everything from conventional weapons, such as multibarreled rocket launchers and armor-piercing antitank ammunition, to centrifuge components, such as inverters, vacuum-pump technology, and even complete centrifuges. KRL representatives assured interested parties that all of the advertised goods had been approved for export by the Pakistani government. KRL also held conferences on such subjects as vacuum technology and "Vibrations in Rotating Machinery." And in case anybody missed the point, Khan's friend, publisher Zahid Malik, produced a fawning biography sponsored by Khan himself titled *Dr. A.Q. Khan and the Islamic Bomb.* Not the Pakistani bomb. The *Islamic* bomb. The book first appeared in Urdu in 1989, followed by an English-language version in 1992.[67] But despite these overt promotions, US officials would later express surprise at learning that Khan had been selling nuclear technology.

It was not until mid-1997 that the US Department of Commerce officially listed KRL as being engaged in "weapons proliferation." The designation, which was also applied to the Pakistan Atomic Energy Commission (PAEC) and Pakistan's small-scale reprocessing facility, required US companies to inquire whether an export license was required for goods sent to these locations. Although the United States had known about the activities of these facilities for years, their official listing as weapons proliferators had been held up by State Department concerns that the designation would harm US foreign relations and by intelligence agencies' claims that the move would compromise sources.[68] Even when the Pakistani facilities were finally listed, it had little impact on their activities, as they had long since turned to procuring what they needed through front companies, none of which were designated.

By the time the Clinton administration officially named KRL a nuclear weapons proliferator, Khan's network had already lured a major new customer. It was in 1997, according to Tahir, that Libya approached Khan, asking for help in developing a uranium enrichment capability. Libyan leader Muammar el-Qaddafi had been one of the early backers of Ali Bhutto's Islamic bomb effort in the 1970s. While that partnership had not worked out to Qaddafi's satisfaction, he had never abandoned his quest for the bomb. During the 1980s, Libya had tried fitfully to build a nuclear program, buying equipment from Germany and elsewhere and bringing in foreign engineers to help. But without the kind of leg up that Khan's pilfering from Urenco had provided or that Khan himself had given to Iran, Libya had little to show for its efforts. Finally, in 1995, Qaddafi decided to make another push for the bomb. Two years later he turned to Khan for help.[69]

The first contact, by Tahir's account, took place in an Istanbul café in mid-1997. There Khan, accompanied by Tahir, met with a man known only as Karim and Libya's deputy prime minister, Matuq Mohamad Matuq, the head of Qaddafi's nuclear program. The Libyans said they wanted to buy centrifuges for their enrichment program. Khan indicated he would be glad to oblige. Later that year, Libya received the first batch of equipment, twenty complete P1 centrifuges, along with the components for another two hundred units.[70]

Additional meetings among Khan, Tahir, and the Libyans took place over the next several years. One would be in the city of Casablanca, in Morocco. Others took place in Dubai. The Libyans told Khan they weren't sure how to proceed. The centrifuges they'd received were a start, but were not enough for a full-fledged enrichment program, and Libya didn't have the ability to build a complete plant. The Libyans made clear they wanted to "speed things up" by using Khan's network as a "one-stop shop." They wanted to buy a complete uranium enrichment facility, using Khan's gang as their exclusive supplier.[71] It was the opportunity Khan had been waiting for.

Under the deal that was eventually worked out, Khan and his network would deliver all the plans and equipment necessary to build a complete P2 centrifuge enrichment plant capable of producing enough weapons-grade uranium to build ten bombs a year. Also included would be training of Libyan engineers and ongoing technical assistance; in other words, full customer service. The price was $100 million. As a bonus, Khan threw in copies of the blueprints for the fully tested nuclear weapon he had obtained from the Chinese years earlier.[72]

This would be a massive undertaking, far larger than anything Khan and his network had engaged in previously, other than the building of the Kahuta facility. But this would be a more or less private enterprise, involving dozens of middlemen, engineers, manufacturing facilities, suppliers, and shippers all spread out over at least four continents. It would also involve large sums of money and careful coordination. And running the whole operation would be Khan's right-hand man, B.S.A Tahir.

"Dr. Sahib"

As Tahir's role in Khan's increasingly audacious ventures expanded, his other business partner, Abu Siddiqui, quietly toiled away in London, filling orders that rolled in from Pakistan via Dubai. Now, however, those orders became evermore exotic.

It was while Tahir was arranging the shipment of centrifuges to Iran in the spring of 1994 that he and Siddiqui had set up SMB Europe in London. The new company did quite well, according to Siddiqui, sending £800,000–1 million ($1.3–1.6 millon) worth of industrial equipment and electronic goods to Tahir in Dubai in the course of a year. But Siddiqui was only earning a salary of only £12–15,000 ($19–24,000) per year. And in 1995, he learned that Tahir was forwarding the goods he bought from SMB Europe on to Pakistan, presumably at a profit. Siddiqui decided he wanted a bigger piece of the action.[73]

With Tahir's consent, in April 1995, Siddiqui established his own trading company, Orland Europe Ltd. Like its predecessors, Orland would sell primarily to Tahir's front companies in Dubai. But it would also deal increasingly with Khan's conduits in Pakistan and with Khan himself. The types of goods being ordered also changed. In addition to the computer equipment, electrical components, air conditioning units, and metals of the past, Siddiqui was now sending heavy industrial cranes, high-heat furnaces, vehicles, highly specialized measuring devices, calibrators, and high-strength aluminum. Occasionally shipments would be detained by British Customs authorities because they lacked the necessary export licenses or because of concerns about the designated end user, such as Peoples Steel Mills. But licenses were applied for and obtained, end users were switched to other Khan-related entities, and shipments were rerouted through Dubai. One way or another, things got through.[74]

As before, Khan initially submitted orders to Siddiqui through Tahir. When questions or problems arose, correspondence would also be channeled through Tahir by fax. But eventually, Khan, who Siddiqui says he originally understood to be a high official with Peoples Steel Mills, began faxing orders to Orland directly. Siddiqui later suggested to British investigators that Khan may have been trying to establish a smoother operation. The direct orders from Khan were, for the most part, for the kinds of goods and materials Siddiqui had been supplying right along. On occasion, however, Khan would slip in requests for personal items such as gardening equipment or cars. And in some instances, Khan, Tahir, or even Farook would ask Siddiqui to provide assistance to Khan's daughter, Dina, who was then attending graduate school in the UK.[75]

By 1997, Siddiqui was communicating with Khan on a more or less weekly basis. When difficulties with orders cropped up, Khan addressed them himself. In November 1997, for example, Khan sent a request to Orland for a price quote on a coordinate measuring machine, a sophisticated device used in manufacturing high-precision parts such as centrifuge components. Khan included specific instructions on where Siddiqui should say the machine was being sent. "Dear Mr. Abu," Khan wrote. "Please get quotation with detailed literature for Dubai. They may not allow to send it to us. Regards, DS." DS was short for "Dr. Sahib," a term of respect that Siddiqui and others used in addressing the eminent scientist. The following month, Siddiqui received a fax from one of Khan's close associates, Mohammad Farooq (not to be confused with Tahir's uncle), indicating that the machine would be sent to Peoples Steel Mills. Siddiqui later told investigators that he believed at the time that Farooq was an official at PSM and also connected to United Engineering and Trading Co. In fact, Farooq was a centrifuge expert and the head of the Foreign Procurement Division of KRL. Not long after receiving the fax from Farooq, Siddiqui wrote to Khan with a problem. The manufacturer of the measuring machine, Baty International, had indicated that an export license would be required and wanted to know the intended use of the equipment. "I told [them] the intended end user is PSM and nature of business is steel mill," Siddiqui wrote to Khan. "Can we say the intended use is steel forging?" Khan responded in a handwritten note, scrawled on Siddiqui's fax. "My Dear Abu," he wrote, "The machine is fine measuring machine to measure the dimensions of precision component parts. You can say it is meant for the quality control department of PSM. Best regards, DS."[76] Prosecutors would later assert that the device, along with other items Siddiqui exported, was destined for KRL.[77]

It was at around the time of these exchanges with Khan that Siddiqui, by his own account, learned from his brother-in-law in Karachi, who happened to work in the food-irradiation division of the PAEC, that the man he'd been dealing with at PSM was the head of a well-known laboratory in Kahuta and engaged in "testing nuclear things." Siddiqui later testified that he was not initially concerned by this revelation, since he dealt with Khan in other capacities and believed he was supplying goods to PSM and UETC. On that basis, he proceeded with the coordinate measuring machine order. An export license was obtained from British trade authorities in March and the device was shipped to PSM on May 2, 1998.[78]

Shortly after the measuring machine left the dock, Siddiqui gained new insight into his Pakistani client's activities. On May 11, 1998, India detonated three nuclear devices at an underground site in Pokhran. It was New Delhi's first nuclear test since the "peaceful" Smiling Buddha explosion twenty-four years earlier. Two days later, India tested two more atomic devices. In response to India's action, Pakistan carried out its first-ever nuclear test, detonating five devices on May 28. Then, in an audacious display of one-upmanship, Islamabad exploded a sixth device two days later. For most Pakistanis, the tests were cause for wild celebration. No longer would they live in fear of an Indian invasion. Their country had proved itself the equal not only of its bitter rival but also of far more developed Western states. Pakistan had become only the eighth nation on earth to develop a nuclear capability and the first Islamic state to have the bomb. A.Q. Khan, already well known and widely admired in Pakistan, was now, deservedly or not, a certified national hero. Washington's response did little to alter that perception. The Indian and Pakistani nuclear displays automatically triggered severe economic sanctions under a 1994 law intended to deter atomic testing. Congress, however, prompted in part by concerns about the impact on agricultural exports from the United States, quickly granted President Clinton authority to waive the sanctions. Clinton made full use of the waiver, with no accompanying requirement that Pakistan rein in its nuclear activities.[79]

On the day of Pakistan's final test, Khan gave what was for him a relatively restrained interview to a Pakistani newspaper. In it he declared that Pakistan had made the decision to test because it had "no other option" in the face of India's action. He also took the opportunity to explain that his labs had been producing enriched uranium for more than eighteen years, and that Pakistan had possessed the capability to build a bomb since the end of 1984. On the subject of the briefly imposed US sanctions, Khan stated, "Sanctions do not affect our programme. We are totally independent, self-reliant and make everything here."[80] He did not mention the role the United States had played in helping Pakistan become "self-reliant" or the assistance he was then providing other countries in their quest to achieve the same status.

In London, Abu Siddiqui watched the BBC television coverage of the Pakistani tests. The report showed images of the granite mountain above the test site casting off a cloud of chalky dust that settled over the gray peaks, turning them white. It also featured an interview with Khan in which he discussed the great event. Just days later, on June 2, Siddiqui sent a fax to Khan congratulating him on his achievement. He would later tell investigators that it was not his view that what Khan had done was okay. He noted that he had family ties to Bangladesh, India, and Pakistan and that he would never knowingly have supplied a nuclear weapons program.[81] It is clear, however, that he had, presumably unknowingly, done just that.

Oh Bhai, Where Art Thou?[82]

While Siddiqui was only belatedly getting to know Khan, he and Tahir had become quite close. For Siddiqui, Tahir was like an elder brother. In fact, he referred to Tahir as "Bhai," a South Asian term meaning "brother." And Siddiqui clearly looked up to Tahir, who was, by all appearances, a successful businessman with close connections to influential people. Although Tahir, for his part, was clearly using Siddiqui, he treated him reasonably well in some respects. He had, for example, gone along with Siddiqui's wish to start his own company, thereby allowing him to keep more of the profits from their trade with Khan. And Tahir, or possibly someone behind Tahir, had provided Siddiqui with an interest-free loan to buy a house.[83]

Tahir had also developed a relationship with Siddiqui's father. That bond grew tighter during a series of mysterious trips to Africa, which the elder Siddiqui embarked on with Tahir and Khan. The destination on each of the four excursions was Timbuktu, a remote desert outpost in Mali. The trips, which took place between 1998 and 2002, included side visits to such locales as Sudan, Nigeria, and Niger, each, like Mali, a country rich in uranium deposits. Joining Khan on these adventures, along with Abdul Siddiqui and Tahir, was, among others, Army Brig. Gen. Sajawal Khan Malik. Brigadier General Malik was among a group of Khan's top aides and associates detained and questioned by Pakistani authorities after the nuclear smuggling scandal broke in 2003. All were eventually released. Another regular on the African trips was Khan's old college friend from Holland, Dutch businessman and engineer Henk Slebos, long alleged to be one of Khan's major suppliers. In the mid-1980s, Slebos was convicted in the Netherlands of attempting to sell nuclear-related equipment to Pakistan. In 2005, a Dutch court convicted him on charges of sending equipment used for uranium enrichment to Pakistan between 1999 and 2002.[84]

The first of Khan's African jaunts began in the latter part of February 1998. Khan and his traveling companions assembled in Dubai on February 20 and left for Casablanca two days later. The group spent two days in the Moroccan port city. By one account, it was during that visit or the one-day stopover on the return leg of the journey that Khan and Tahir held their second meeting with the Libyans to work out details of the uranium-enrichment plant deal.[85]

From Casablanca, Khan and his fellow travelers flew to Mali's capital, Bamako, where they spent a day touring the city. They then took a $4,000 chartered flight from Bamako to Timbuktu, arriving on February 25. Oddly, after all that effort and expense, the group spent only a few hours in the impoverished city. They traveled home along the same route by which they had come. They would return to Timbuktu in 1999, 2000, and 2002, each time following a different course. Abdul Siddiqui, who wrote a memoir about the trips, offers no explanation for Khan's purpose in making the journeys, other than to travel and sightsee.[86] There is no suggestion that Abdul Siddiqui was involved in any of Khan's nuclear deals.

In one of many odd insights, Abdul Siddiqui's memoir reveals that throughout the 1998 trip, Khan persistently urged Tahir to find a wife and settle down. The pressure from Khan, who was something of a father figure to Tahir, apparently worked. Just days after returning from the Timbuktu adventure, Tahir called the elder Siddiqui from Dubai asking if Abu could "accompany him on a trip to Singapore, Sri Lanka and Malaysia to find a prospective bride." The answer, it seems, was yes. Adbul Siddiqui's book offers no details about his son's wife-hunting mission with Tahir except to say that "the final choice fell on a girl named Nazimah in Malaysia." The winner of the bride lottery was, in fact, Nazimah Binti Syed Majid, the daughter of a former Malaysian diplomat. Tahir had become an increasingly well-known figure in Malaysia over the past several years, investing in several up-scale businesses in Kuala Lumpur and hobnobbing with the rich and famous. Among his newfound Malaysian associates was

Kamaluddin Abdullah, the son of the country's then-foreign minister and soon-to-be prime minister. Kamaluddin and Nazimah would, through an investment vehicle of which Tahir was a director, become major stakeholders in a Malaysian company that would play a key role in the Khan network's supply of centrifuge equipment to Libya.[87]

Tahir's marriage to Nazimah was a high-society affair with the cream of Malaysia's business and political communities attending. It also drew celebrants from around the world, including Abdul and Abu Siddiqui, A.Q. Khan, several of Khan's longtime suppliers, and the cast and crew of the recent Timbuktu adventure. The ceremony took place in Kuala Lumpur on June 28, 1998,[88] one month after Pakistan's first nuclear test, an event that some of the guests had helped make possible. For Khan and the members of his network in attendance, there were multiple reasons to celebrate. In addition to Tahir's marriage and Pakistan's successful bomb test, there was also the fact that Khan and his network had embarked on a major nuclear deal with Libya that held the promise of vast rewards.

Following Tahir's wedding, the Siddiquis returned to London, Abdul to his accounting practice, Abu to his export business. For Abu, it may have been something of a letdown. He had now met the famous Dr. Khan and his associates and had a glimpse of Tahir's glamorous new lifestyle. It was a far cry from his own far more modest existence. And to make matters worse, there was little evidence his situation would improve. Although Siddiqui was, by his own account, then selling about £1 million ($1.6 million) worth of goods and equipment to Khan and Tahir each year, he was still earning just a 5 percent commission. And since his return from Tahir's wedding, sales had started to drop off. By fall, there were almost no orders coming in. Things got so bad that Siddiqui made a feeble attempt to scrounge up new customers by placing an ad on the Internet. He even considered closing Orland altogether.[89] It seemed his Bhai and Dr. Sahib had moved on to bigger things and no longer needed his services.

Then in November, a large order came in from Tahir. It was for bars and sheets of high-strength aluminum alloy. There were eighteen items of various dimensions requested. The order had originated with Mohammad Farooq, the man Siddiqui would later learn was the director of foreign procurement at KRL. At the time, Siddiqui was simply glad for the business. He had supplied aluminum before and saw no reason not to proceed with the order. "I had been exporting aluminum for seven to eight years and [had] no problem," he later testified. Siddiqui found a British supplier that could fill the order, obtaining the aluminum from American manufacturers. In mid-December, Siddiqui confirmed the order with the supplier and arranged delivery details. He then left on vacation.[90]

Shortly after Farooq's order arrived at Orland, British Customs authorities received a tip from MI6. The British intelligence service, which, like the CIA, had been monitoring Khan's operation for years, informed Customs that Siddiqui was in the process of arranging a shipment of metals that included licensable aluminum to KRL through one of Tahir's front companies in Dubai. Customs, which had authority for controlling such exports, began to investigate. On January 11, 1999, shortly after Siddiqui returned to London, Customs agent Maxine Crook, accompanied by another officer, stopped by Orland for an "educational visit." The stated purpose of the visit was to explain export-licensing procedures to Siddiqui. But the officers were also hoping to gain some insight into his operation. Crook later testified that she told Siddiqui that some goods, including certain metals, required export licenses. She explained that if he was uncertain about a particular shipment, he should contact the British Department of Trade and Industry for clarification. She then asked Siddiqui whether he dealt in electrical components or metals. According to her testimony, he told her there had been some metals orders in the past but that he had none at present. Crook explained that some shipments might require a license because of where they were going. Siddiqui, Crook testified, said that he had done some business with Pakistan. Crook advised him that if he had any suspicion that an item was going to be used in a WMD program he had an obligation to contact DTI. She also noted that he had done business with UETC, CWO, and PSM and explained that he should consult DTI about any orders from them. Crook added that Dubai was a well known "diversionary point" for goods going to "countries of concern," including Iran, Iraq, China, and Pakistan. She suggested that Siddiqui inquire whether any goods shipped to Dubai would be diverted elsewhere.[91]

Crook testified that she then asked Siddiqui whether he had any current orders. It was at that point, she testified, that he produced the aluminum order from Farooq. Crook asked whether Siddiqui had checked with DTI about the order. Siddiqui, according to Crook, said he had not. She advised him that he should. She then asked for a copy of the order. Siddiqui provided her with one.[92] With that, the "education" was complete.

Two days after Crook's visit, Siddiqui received a price quote by fax from the company supplying the aluminum. Siddiqui replied, accepting the terms. On February 1, Agent Crook sent Siddiqui a letter, reviewing the main points of her visit. Siddiqui wrote back on February 23, indicating he had taken note of the letter and would be "careful."[93]

By early May, Siddiqui's supplier had obtained the aluminum from various companies in the United States. Siddiqui made payment of more than £65,000, according to court records. Shortly afterward, he sent an invoice to Tahir for the same amount. On May 7, Siddiqui arranged for the aluminum, twelve metric tons in all, to be shipped from Thamesport, on the Isle of Grain, east of London. On the export declaration he listed the consignee as Sama Machinery and Equipment in Dubai, a company Siddiqui later testified he understood to be controlled by Tahir. Siddiqui also testified that he realized the shipment was probably destined for Pakistan.[94]

Three days later, Customs agents, who had been monitoring the order for months, seized the shipment at the dock. Four of the aluminum bars in the order, because of their specific dimensions, required an export license that Siddiqui had never obtained. In conjunction with the seizure, Customs agents conducted a search of Siddiqui's home and office, taking all of the corporate records and faxes detailing millions of dollars worth of sales of electrical components, metals, machinery, and equipment that had ended up at Khan's lab over the course of a decade. They also raided his parents' house. There, according to court records and sources familiar with the investigation, they discovered, among other things, a KRL brochure describing the uranium enrichment process and a photograph of Abdul Siddiqui standing beside A.Q. Khan in Casablanca. They also unearthed a Pakistani accounting magazine dated May–June 1998. The cover story focused on Pakistan's nuclear weapons program. It reported how Khan had set out to restore the balance of power in South Asia following India's 1974 nuclear test. Khan had determined, according to the magazine, that the only way to achieve that goal was to "buy whatever we can from the international market" to build a nuclear weapons program. The operation, the magazine noted, had been shielded by a high degree of secrecy.[95]

Operation Akin

Following the raid on Orland, Customs agents began the arduous process of unraveling Siddiqui's long history of dealings with Tahir and Khan. As the agents pored through the evidence, it soon became clear that what had started as a simple export violations case had developed into something much bigger. Leading the investigation, code-named Operation Akin, was a thirty-year-old agent named Atif Amin. A British-born Muslim of Pakistani descent, Amin was an investigating officer on a special counterproliferation team that had been set up within the Customs services' Commercial Fraud Branch. He had joined the team three years earlier and had been the lead investigator on two politically charged, high-profile arms-smuggling cases.[96] While those experiences made Amin an ideal candidate to lead the Siddiqui investigation, nothing could have prepared him for the interference he would run into as he zeroed in on A.Q. Khan's nuclear smuggling ring.

Amin and his team quickly set to work sifting through the records seized from Siddiqui's office and tracing the circuitous trail. On July 19, they brought Siddiqui in to Customs' headquarters for a lengthy formal interrogation in the presence of his lawyer.[97] The agents also made arrangements to arrest Tahir if he returned to the UK, according to Western law enforcement sources.

By late summer, Customs had developed a relatively complete picture of the London end of Siddiqui's activities. But to build a case against him on anything more than simple export license violations they would have to prove where the goods he shipped ended up. That evidence lay in Dubai. In August, Customs submitted a formal request to Dubai's Ministry of Justice, seeking authorization to carry out an investigation in the emirate. As part of the request, Customs provided a list of individuals and entities they planned to look into, as well as phone numbers, e-mail addresses, and bank accounts they wanted to trace. Although Dubai routinely cooperated with British inquiries into such things as cigarette and drug smuggling, it had consistently rejected requests to conduct investigations into WMD proliferation. True to form, when Customs' request to carry out an inquiry into the Siddiqui case arrived, Dubai balked. Only after months of stalling and repeated appeals from British authorities did the Dubai authorities finally relent. In March 2000, the Justice Ministry granted approval for the investigation. Arrangements were made for Amin to travel to Dubai the following month.

Before his departure, Amin and his fellow investigators brought Siddiqui in for another extended round of questioning.[98] During more than seven hours of interrogation, they obtained valuable new information about Khan's operation. Then, shortly before Amin was to leave for Dubai, he received a curious warning. According to a source familiar with the investigation, MI6 sent a message through liaison channels informing Amin that he should steer clear of a Dubai company called Desert Electrical Equipment Factory, even if he came across indications of something going on there. Although Desert Electrical had not come up in the Siddiqui investigation, its owner was reportedly one of Tahir's partners in SMB Computers. Libyan officials have since reportedly told investigators that Desert Electrical's facilities were being used at the time to manufacture centrifuge components and to train Libyan scientists.[99]

Duly warned, Amin set off for Dubai in early April. Although British Customs officials refused requests for an interview with Amin for this book, it has been possible to piece together a picture of his investigation. Shortly after arriving in Dubai, Amin checked in with Malcolm Nesbit, the Customs liaison officer at the British embassy. Nesbit, who had been posted to Dubai from his usual duties as the head of the Customs investigation unit in his native Newcastle, welcomed his fellow officer. The two men discussed how Amin's investigation should proceed. Nesbit, a man of medium build in his midforties, explained that it would be necessary to get the cooperation of the Dubai police and offered to make arrangements for Amin through his contacts in the department. Nesbit also suggested that, because of the sensitive nature of the case, Amin should meet with the MI6 chief of station. Amin agreed and Nesbit set up a briefing.

The meeting with the station chief took place in the highly secure section of the embassy building reserved for MI6. According to a source familiar with the exchange, Amin briefed the station chief on the Siddiqui case and outlined the investigation he planned to conduct in Dubai over the coming weeks. He also provided the station chief with the list of companies he intended to visit. The station chief, a rather thin, short man in his early forties, asked that Amin keep him informed of anything interesting that turned up. He expressed no concerns about the investigation, except to suggest that it might be best if Amin did not get too close to one of the institutions on the list, Habib Bank. Habib was the bank through which Tahir sent payments to Siddiqui.[100]

As arranged by Nesbit, Amin next checked in at police headquarters, a sprawling, white building off a main road in Dubai's Al Tawar district. There he met with senior police officials and explained the scope of his inquiry. The police provided Amin with office space in a small, secure interview room where he could work and keep documents. They also assigned an officer to assist Amin in his investigation. The designated officer was a young lieutenant named Alwari Essam, who served in the Interpol section of the department's criminal investigations division. Lt. Essam, a tall, athletic young man with a closely cropped beard, had been on the force only a few years but had participated in several smuggling investigations involving drugs and other contraband. He had not, however, been involved in a nuclear smuggling case. It was simply not something Dubai investigated. When pressed, local officials typically expressed skepticism and suggested nothing was happening.

Essam shared that attitude initially. He was dubious of the investigation and laughed at the idea that Dubai was a hub of nuclear black marketeering. But he would gradually become convinced.

From the start, Essam got along quite well with Amin. The two men were roughly the same age, and the fact that Amin was a Muslim and of South Asian descent made him seem less foreign and intimidating than the British investigators Essam had encountered in the past. Amin did not tell Essam everything he knew about the case in the beginning, discussing it only in general terms. But as the investigation progressed, Amin revealed more and more. The idea was to track Siddiqui's exports to the front companies in Dubai and from there, if possible, to the ultimate recipients in Pakistan. They would also trace phone numbers and e-mail addresses found in Siddiqui's records, as well as the source of payments made to Siddiqui through Dubai banks. Amin also wanted to develop a clearer picture of Tahir's role in the larger smuggling network.

For more than two weeks, Amin and Essam crisscrossed Dubai, visiting the companies that had turned up in Siddiqui's records. They went to shippers, traders, manufacturers, banks, corporate registrars, telephone companies, and Internet service providers asking questions, looking through records, and tracking down leads. Early on they discovered connections between some of the front companies and high-ranking Dubai officials. As Essam looked at the evidence the investigation uncovered, listened to Amin's conversations with owners and managers of companies, and saw the fear the visits to those companies generated, his skepticism gradually faded. And as his understanding of the case grew, he became increasingly helpful — and enthusiastic. "It's like a James Bond movie, isn't it?" he asked Amin.

While Essam was becoming a convert, the real James Bonds seemed, at best, disinterested in Amin's investigation. As agreed, Amin periodically reported to MI6 on the progress of the inquiry, including new front companies he had discovered and apartments that Tahir had arranged for Khan's use during trips to Dubai. At one point the station chief remarked, presumably based on intercepted communications, that Khan and his network were aware of the investigation but did not believe Amin would turn up much. But MI6 offered Amin no useful information or assistance. The general sense, according to a source familiar with the briefings, was one of displeasure that the inquiry was taking place at all. But the spies, and Khan's gang, would soon show a much greater interest.

Khan "Comes Here All the Time"

On a Wednesday morning, two and a half weeks into the investigation, Amin and Essam paid a visit to a company that had not been on the original list. The company, Green Crest Industries (M.E.) Ltd., was a Pakistani-owned plastic-bag manufacturer based in Dubai's Jebel Ali Free Zone.[101] Its offices were located in an industrial facility at the end of a dusty road in the expansive free-trade area. Speaking with Green Crest's manager, Amin explained that he was conducting an investigation into a British exports violation case. He then told the manager that an entry in the suspect's phone book listed a Dubai phone number for Dr. A.Q. Khan that was registered to Green Crest. Amin added that apartments and post office boxes rented for Khan by Tahir also appeared to be associated with Green Crest. The manager and several other employees of the company, all Pakistanis, denied any knowledge of the matter and said they did not know A.Q. Khan. At that point, another employee who was wandering past said in Punjabi, "Sure we do. He has a flat and he comes here all the time." Amin, who speaks Punjabi, overheard the comment. He also understood the manger's stern Punjabi rebuke of the loose-lipped employee. The atmosphere grew hostile, and Amin and Essam, realizing they could get no further, left.

Green Crest's chairman was a prominent Pakistani businessman by the name of Shaik Muhammad Farooq (not to be confused with Tahir's uncle or the procurement director at KRL).[102] In a 2006 telephone interview, Farooq, who passed away in 2007, stated that Green Crest had "absolutely no relationship" with Tahir. The one exception, he said, is that Green Crest had an apartment in the same building as Tahir on Al Maktoum Street in Dubai and in 1994 they had traded units. Tahir, he said, had taken Green Crest's larger unit on the ninth floor of the building while Green Crest had taken Tahir's smaller apartment on the seventh floor. Beyond that, he said, there was no connection. Asked in a subsequent interview whether Khan had ever visited Green Crest, Farooq offered a confusing reply. "He never visited our factories," Farooq said. "He never visited our office. He never visited. Except sometimes he is there and he is inviting a lot of people including other businessmen for dinner or so otherwise no. Absolutely baseless.... I'm 100 percent sure that he never visited us."[103]

Whether or not Khan ever visited Green Crest, there is no question that Farooq and his companies have long been associated with Dr. Sahib and Pakistan's bomb program. According to the book *The Islamic Bomb*, in 1979, Pakistan's procurement guru, S.A. Butt, used one of Farooq's companies, Asiatic Chemical Industries Ltd., as "a conduit for the Pakistan Atomic Energy Commission."[104] During the early 1980s, according to a secret US State Department cable, another of Farooq's companies, Arshad Amjad & Abid (Pvt.) Ltd., purchased a coordinate measuring machine from a Japanese manufacturer and resold it to the Pakistan Chemical Corp., a company identified in A.Q. Khan's biography as a front for KRL.[105] Arshad Amjad & Abid also appears on a secret British Security Service (MI5) list of companies of "proliferation concern" for their procurement of goods for "weapons of mass destruction programmes."[106] And Khan's biography gives special praise to Farooq and Arshad Amjad & Abid for playing "a very commendable and daring role" in obtaining goods and equipment for Khan's Kahuta facility from a long list of countries including the United States, Britain, Canada, Switzerland, Germany, Italy, Holland, France, and Japan.[107]

Farooq himself seems to have been well acquainted with Khan. He served as vice chairman of the A.Q. Khan Institute of Biotechnology and Genetic Engineering, of which Khan was chairman, and was on the board of other institutions with Khan. And in 2001, Farooq hosted a reception in Khan's honor at which the good doctor informed the audience that Pakistan's bomb program was "indebted" to America's "wholehearted" support for the Afghan war. And in 2002, Farooq reportedly accompanied Khan during a bizarre incident in which Khan and a group of associates allegedly attempted to take over administrative control of a mental health hospital by force.[108]

The Libyan Connection

Following the episode at Green Crest, Amin and Essam headed for a company called Deepsea Freight Services, a shipping agency through which, Siddiqui's records showed, some of the goods he exported had been cleared for delivery to Islamabad. The manager of Deepsea is K. Hafeez Uddin, a South Asian in his midforties. Sitting in Hafeez's cramped office, Amin explained that he was conducting an investigation into shipments that had come from the UK. He asked if he could see the records for two of the Pakistani companies to which shipments had been forwarded, United Engineering and Trading Co. and Allied Engineering. Hafeez was visibly nervous. Essam explained that it was a British investigation and that he was not suspected of any wrongdoing. That seemed to put Hafeez at ease. Hafeez reached around to a bookcase behind his desk and pulled down a large file. Amin began looking through it. In the front section he found records of shipments going from Siddiqui to Tahir's companies in Dubai. Behind those he found paperwork for those same shipments going onward to UETC and Allied; precisely the information the Customs investigation sought. But in back of those was something completely unexpected: air waybills for shipments going from UETC to Tahir's front companies in Dubai. Amin started looking through them. Hafeez got fidgety. Amin kept looking. Then he spotted something remarkable: records of shipments going from UETC to Dubai that were marked for transshipment to Libya. Some of the shipments were identified as ring magnets, key components in centrifuges. And the consignee on some of the ring magnet shipments was Desert Electrical, the company Amin had been told to avoid.*

Although Amin had no authority to compel Hafeez's cooperation, he asked if he could take the file. According to a source involved with the investigation, Hafeez flatly refused and would not allow Amin to make copies. In telephone interviews with one of the authors of this book in 2006, Hafeez repeatedly stated that British and Dubai authorities had taken the records from him at this office.[109] He later told a different story. "The Dubai police call us and we went there," he said. "They took all our

*Desert Electrical's phone and fax numbers now belong to another company. Desert Electrical did not respond to a letter sent to the fax number seeking comment. For more details, see chapter ten.

files."[110]* The source involved with the investigation says that Amin and Essam left Deepsea without the files.

In the 2006 telephone interviews, Hafeez acknowledged that Deepsea had cleared two or three shipments from the UK through customs in Dubai. He denied, however, that Deepsea sent any shipments to Pakistan. But when pressed, he again gave a different account. "Two or three shipments have come from [the] U.K. and gone on to Pakistan," he said. "Not Pakistan. We don't even know that is where it's gone. . . . It was cleared."[111]

Hafeez also denied that Deepsea had handled shipments coming from Pakistan. "We have not done any shipment from Pakistan to Dubai," he said. But upon further questioning, Hafeez was more equivocal. "I don't remember but I don't think we are receiving any shipments from Pakistan," he said. "I don't know." Hafeez said that a representative of one of Tahir's companies sometimes came to Deepsea to pick up records of shipments sent to Dubai. He added, however, that he did not know Tahir. "We don't know much about Tahir [or] what Tahir is doing," he said. "And this company [Deepsea] has nothing to do with Tahir."[112]

A source involved with the British Customs investigation says there is no question Deepsea handled Pakistani cargo. "The fact is that Deepsea received multiple shipments from Siddiqui and forwarded them on to Pakistan," the source said. "It also received multiple shipments from KRL-related companies destined for Tahir's front companies in Dubai." When one of the authors of this book called Hafeez for further comment in 2007, Hafeez hung up.[113] Neither Hafeez nor Deepsea is accused of any wrongdoing.

Revealing A.Q. Khan in Dubai

After leaving Deepsea, Amin and Essam drove back to police headquarters, discussing the morning's events. When they arrived at headquarters there was widespread commotion in the building. Amin and Essam were confronted by a group of officers. Amin was ushered into the

*Hafeez said he was told only that the records concerned shipments of "contraband" goods. He said the files were not returned to him. He also said Deepsea did not know what the shipments contained. (Author's interviews with K. Hafeez Uddin, February 27, March 8, and March 28, 2006.)

windowless briefing room he had been using as an office. Essam was escorted into another room. After about forty minutes, Essam entered Amin's office. He explained that he had been told by his superiors that what he and Amin had been doing had to stop. They would no longer be allowed to conduct interviews in the field. From then on, witnesses and suspects would be invited to police headquarters. Amin would be allowed to submit questions, but the police would ask them on his behalf. If witness or suspects refused to come they would not be compelled. And if Amin needed records or other materials, the police would get them for him. And, Essam said, Amin would have to turn over all the records of his investigation to the police. That meant everything: documents obtained from the companies Amin had visited, witness statements, notes, photographs. All of it. Even the material Amin brought over from England.

Essam told Amin that he had been severely reprimanded by his superiors and members of Dubai's internal security service. They had accused him of helping Amin to "reveal A.Q. Khan in Dubai." Essam said that the security service wanted to know why Amin had been trying to find out where Khan stayed in the city. They had suggested that Amin did not really work for British Customs but was, in fact, an agent of MI6 and that there was a plot to take Khan out. Amin told Essam that the security service officers had been watching too many movies.

Amin was angry about the new restrictions. They would make it impossible to conduct a meaningful inquiry. But there was little he could do about it immediately. He told Essam that they would press ahead as best they could while trying to get the restrictions lifted. In the meantime, Amin wanted to see the Deepsea file. He put in the required request with the police.

Later that afternoon the records from Deepsea arrived and were delivered to Amin's office. The folder was noticeably thinner than it had been when Amin and Essam saw it earlier in the day. As Amin began looking through the file it was clear that documents were missing. When he got to the back of the folder, he discovered that the air waybills for the shipments designated for Libya were gone. "Where the hell are they?" he asked Essam. They could only guess.

Amin said the situation was intolerable. Evidence had disappeared, and the new interview procedures would make getting useful information virtually impossible. Witnesses might not show up, and even if they did they would not have their records with them. Amin wanted to interview Tahir and his associates next. And because Tahir was a suspect, the interviews would have to be conducted under British rules of law. That meant interviewing him directly and advising him of his rights. Amin mentioned going to see Tahir right away. But by then it was late Wednesday afternoon, the end of the workweek in Dubai at the time.* Amin and Essam agreed that they would go see Tahir in his office at SMB Computers on Saturday, the start of the workweek.

"Traitor"

After leaving police headquarters Wednesday evening, Amin called Malcolm Nesbit and told him what had happened that day. Nesbit suggested that MI6 be informed. He said that what Amin had run into fit into something the station had been working on. Amin asked what he meant. Nesbit said that several months earlier MI6 had asked him to monitor shipping containers going to Libya. MI6 had wanted to know when the containers arrived in Dubai, how long they stayed, when they left port, and what they were carrying. Nesbit said MI6 hadn't told him what they were looking for, but he now realized it probably had something to do with what Amin was working on. Nesbit indicated he would let MI6 know what Amin had discovered. Amin spoke with the station chief by phone later that evening. The station chief asked Amin to meet him at the embassy the following morning.

Amin was staying at the Sheraton Hotel, almost directly across Dubai Creek from the embassy. On Thursday morning, he rose early, ate breakfast, and then took a water taxi, known as an *abra*, across the creek. Because it was the weekend, the embassy was closed. Amin presented his identification to the security officer at the gate of the compound. Once inside the embassy, the station chief came down to meet him. The two men went upstairs to the secure offices of MI6. Only one other agent was working in the station that day.

Amin told the station chief what had happened the day before. This time, the station chief expressed interest. He didn't mention anything about Libya but said they needed to file a report. He asked Amin to go

*Dubai has since switched to a Friday-Saturday weekend. See: "Have Your Say: UAE Weekend," Gulfnews.com, May 17, 2006.

through everything that had happened in detail. Amin narrated. The station chief wrote. Amin corrected the station chief several times when he noticed him embellishing some of the facts. When they had finished the report, known as a CX or intelligence report, the station chief sent it to London. That report has since been made available to other Western intelligence services.

On Thursday evening, Amin went out to dinner with a friend, a young woman he had met through Nesbit's secretary. He returned to his hotel late and went to bed.

At about 2 A.M., the phone rang in Amin's room. It was Nesbit. "Are you awake?" Nesbit asked. "I am now," Amin said. "We need to speak with you," Nesbit said. "Meet us downstairs." There was no need to explain who the other person would be.

Amin dressed and went down to the lobby. The hotel seemed deserted. Amin sat in one of the lounge chairs near the center of the cavernous room and waited. At about 2:30 A.M., the MI6 station chief came through the hotel's main entrance. Nesbit entered a few moments later, looking back over his shoulder. Amin rose, and the three men greeted each other. They moved to a table on one of the raised platforms off to the side of the dimly lit lobby. A waiter came by, and they ordered coffee and tea.

When the waiter was gone, the station chief leaned over to Amin and whispered, "You're at risk here." Amin gave him a puzzled look. The station chief leaned in again and whispered, "You're in *danger!*" Amin seemed to hold back a laugh. He looked at Nesbit, then back at the station chief. "What are you talking about?" Amin asked. "You can't stay here," the station chief said earnestly. "You can't keep doing what you're doing. You have to get out." "Where is this coming from?" Amin asked. The station chief said he had received a telex from London. He didn't have all of the details, but he knew that Khan and his associates in Dubai were angry about Amin's investigation. They had been discussing how they could stop him from finding out more. They had mentioned possible physical reprisals. Khan had even talked about Amin personally, saying that as a Muslim he was betraying the "cause" and calling him a "traitor." Although the station chief did not say so, he made it clear that MI6 was monitoring Khan's phone calls.

"These people are dangerous," the station chief said. "They have assets in the local mafia that they use for smuggling. They won't hesitate to kill people." Amin said nothing.

"You might not be safe in the hotel," the station chief continued. "Maybe you should stay with Malcolm." Nesbit suggested Amin might be safer at the embassy.

"I'm fine here," Amin said. "You're overreacting. I'm fine."

Unable to convince Amin of the danger, the station chief gave up. Amin and Nesbit agreed to talk the next day. Amin headed back to his room.

On Friday afternoon, Amin and Nesbit spoke by phone. Amin pressed Nesbit to help get the investigation back on track. Nesbit said he would set up a meeting with senior police officials to try to smooth things out. The two men would try to work out an agreement that would allow Amin to interview suspects and witnesses in the field. Most important, they would try to get Amin's files back. Amin and Nesbit agreed they would go to police headquarters first thing Saturday morning.

In the course of the conversation, Amin mentioned that he felt MI6 was playing up the threat from Khan's network. Nesbit agreed and suggested they were doing so because his investigation was compromising their monitoring of Khan's dealings with Libya, the very thing, unbeknownst to Nesbit, MI6 had tried to warn Amin from doing. It seemed both Khan and MI6 shared an interest in shutting down Amin's inquiry.

On Friday evening, Amin went to one of the local street markets, known as souks, with another British Customs agent who was conducting a cigarette smuggling investigation in Dubai. Although they both served in Customs' investigative service, they had met for the first time only a few days earlier in Malcolm Nesbit's office. Wandering through the crowded marketplace, looking for gifts to take back home, Amin told his colleague about his investigation and what he had encountered. Amin made clear that despite the threats and the interference from the police he was going to press ahead with his inquiry. In fact, he said, he was going to confront the ringleader of Khan's smuggling network the very next day.

As Amin was explaining all this, his mobile phone rang. It was his boss, Mickey Bispham, the acting head of his unit. Bispham, who was known for his abrupt manner, didn't mince words. "Get your ass on the next flight to London," Bispham said. Amin protested. "That's just a knee-jerk reaction, Mickey," Amin said. "We need to discuss this calmly. I'm here on the ground. You don't have a feel for what's going on here. I can sort out my own security. I can stay at the embassy." Amin also told Bispham that he and Nesbit were going to police headquarters the next day to get his documents back and straighten things out. His investigation wasn't over yet. He still had to see Tahir. But Bispham wasn't interested. Orders had come from above. There was nothing to discuss. Amin was to return to London immediately.

Amin hung up the phone disgustedly. He looked at his colleague for a moment, then said he had to leave. He'd been ordered to return to London. He had no choice. He asked his colleague to find Lieutenant Essam and let him know what had happened. Amin then turned and walked away.

"Finding Out Too Much"

Back in London, Amin and his team set about trying to salvage what they could of the Siddiqui case in the wake of the aborted Dubai investigation. Without the records and witness statements Amin had collected, they were going to have to rely exclusively on the evidence obtained in the raid on Siddiqui's office to make their case. Customs' legal division submitted requests though the Foreign Office for return of the material that had been seized from Amin in Dubai, but nobody held his breath.

Amin later wrote a briefing report on the Dubai investigation and the problems that had been encountered. The two-page, single-spaced report noted that documents had been removed from company files after Amin's initial visits, that the Dubai security service had instructed the police to halt Amin's company visits and alter the format of the investigation to prevent the direct questioning of witnesses, that the security service had reprimanded the police for helping to reveal A.Q. Khan's presence in Dubai, that all evidence and witness statements had been taken from Amin, and that Amin's investigation had been cut short.

The other problem Amin had encountered in Dubai was the threat from Khan's network. Given that, Customs security officers met with Amin several times to discuss improving security around his home. They also looked into any associates Khan might have in England who could pose a threat to Amin. Though they judged the threat to him to be minimal while he was on British soil, MI6 received indications that he remained in danger if he returned to Khan's turf. Investigators later learned that Khan's henchmen had tried to locate members of Amin's extended family in Pakistan. The investigators also intercepted communications in which Khan had vowed to take action against Amin if he traveled to Pakistan.

Shortly after Amin retuned from Dubai, he spoke with a senior Customs official named Euan Stewart, then the Assistant Chief Investigation Officer for the Commercial Fraud Branch at Customs headquarters. Stewart welcomed Amin back to London and congratulated him on the work he had done in Dubai. "You're a damn good investigator," Stewart said. "You did a great job." Stewart indicated that he was well aware of the circumstances under which Amin was pulled out of Dubai. At the time, Stewart had been in Washington attending a meeting of high-level US and British officials, including representatives of MI6. During the meeting he had spoken with a senior MI6 official who complimented him on the Dubai investigation. "Your man's turned over far more stones over there than we've managed in the last few years and he's found lots of insects crawling around underneath," Stewart reported the MI6 officer saying. But the officer had then added an ominous note, telling Stewart, "If I was you, I'd get my man out of there." As Stewart explained to Amin, "You were finding out too much." The question was too much for whom — Khan and his gang, or MI6? The answer may have been both.

Stewart made clear that whatever else MI6 knew about Khan's dealings with Tripoli, which turns out to have been a lot, it had not been aware of the ring magnets going though Dubai. For all of its electronic wizardry and monitoring of shipments, it was only through Amin's old-fashioned detective work that the spy agency learned of those transfers. But while Amin had turned up valuable intelligence, he had also created what MI6 and the policy makers who control it perceived to be a quandary: Should they act on the intelligence, disrupt Khan's network, and expose Libya's nuclear program, or should they continue their monitoring operation? They chose the latter option. In fact, it would be another three years before MI6 and its American counterpart finally deemed the time right to take action — a move that would be accompanied by great fanfare and self-congratulation. In the meantime, Khan's network had been allowed to continue peddling its dangerous goods.

"Exceptional Circumstances"

In early August 2001, Abu Siddiqui went on trial in London on seven counts of illegally exporting goods and equipment that could be used in Pakistan's nuclear weapons program. Crown Prosecutor Mukul Chawla described Siddiqui to the jury as a "procurement agent" for A.Q. Khan's labs. He argued that Siddiqui had exported the goods in full knowledge that they were destined for Pakistan's bomb program. B.S.A. Tahir's name figured prominently in the proceedings. He was described in testimony as the middleman in Khan's nuclear deals.[114] Many of the details of Khan's Dubai smuggling operation were revealed in open court. But because Dubai refused to return the records seized from Atif Amin, that material was unavailable for use as evidence.

On August 29, 2001, the jury returned guilty verdicts against Siddiqui on three counts of violating British export regulations.[115] Judge George Bathurst-Norman set sentencing for October. In between Siddiqui's conviction and his sentencing, the terrorist attacks of September 11 occurred. Washington and London now deemed it essential to secure Islamabad's assistance in the war on terror. Suddenly, Pakistan was once again a vital British and American ally. And, as in the past, it became imperative that Islamabad not be embarrassed over its nuclear program for fear of losing its cooperation, this time in combating the Islamic militants spawned by the anti-Soviet Afghan jihad.

At the sentencing hearing on October 8, 2001, Judge Bathurst-Norman acknowledged that the charges on which Siddiqui had been convicted would normally carry a "very substantial" prison term. But in this case, Bathurst-Norman said, there were "exceptional circumstances." Bathurst-Norman said he was satisfied that Siddiqui lacked "political or religious motives" and had been unaware that the goods he exported were bound for Khan's lab. Overall, Bathurst-Norman found that Siddiqui was "someone perhaps who is too inclined to take people on trust" and had been "blinded" to facts that were "absolutely staring [him] in the face." On that basis, he sentenced Siddiqui to a twelve-month suspended sentence and a £6,000 ($10,000) fine.[116] As an individual familiar with the case remarked, "The sentence was generally perceived to be remarkably lenient." In a scenario eerily reminiscent of earlier nuclear smuggling cases in the United States and Canada, Siddiqui walked out of court essentially a free man.*

In December 2001, just weeks after Siddiqui's sentencing, a new company was set up in Malaysia to manufacture what turned out to be centrifuge components destined for Libya under a contract with Siddiqui's former business partner, B.S.A. Tahir. The Malaysian plant replaced the

---

*Judge Bathhurst Norman did not respond to a letter from the authors seeking comment.

manufacturing functions formerly carried out at Desert Electrical in Dubai, which had ceased production operations around the time of Atif Amin's investigation.[117] Tahir, Khan, and the rest of the network would continue their smuggling of nuclear technology for another two years while Western intelligence looked on.

"They Could Have Blown the Whistle on This Long Ago"

In the aftermath of the Siddiqui case, Atif Amin went about his job, but his short-circuited Dubai investigation and the knowledge of what might have been loomed in the background. In early 2004, when Khan's nuclear smuggling operation was finally exposed, Amin could only have shaken his head in astonishment as he listened to British and American officials take credit for their great "success."[118] But for Amin, the most galling moment came in October 2004, when, during the presidential debates, President Bush declared that Khan and his associates had been "brought to justice."[119] Amin knew, as few others could, that the president's assertion was nonsense.

Amin and his colleagues knew how close their investigation had come to tracing out Khan's network and revealing its customers years earlier. But that opportunity had been squashed. The question was why. Was Khan's nuclear smuggling allowed to continue to keep Pakistan happy? Or did MI6 and the CIA think they could contain the damage and reel in other, bigger fish by continuing to monitor Khan's enterprise? Had a deal been cut with Libya or one of Khan's other customers? Amin wasn't a political strategist. He was, in essence, a cop, and all he knew was that nuclear technology had been allowed to spread unnecessarily for more than three years and that the country he had dedicated his life to protecting was in greater danger as a result. That was the upshot of his curtailed investigation. And for that reason, Amin was sickened by the notion of Bush trying to get political mileage out of the affair. It was, he later told colleagues, utter hypocrisy. "They knew exactly what was going on all the time," Amin said. "If they'd wanted to they could have blown the whistle on this long ago."

10
End Game

THE BRITISH CUSTOMS investigation in the Abu Siddiqui case had publicly exposed major elements of A.Q. Khan's nuclear smuggling network. Key figures, including B.S.A. Tahir, had been named, front companies revealed, and details of the network's mode of operation laid bare, all in open court. Yet there was little coverage of the case and not a single news organization bothered to follow the extensive trail of evidence that had been laid out at trial. An important opportunity to learn more about Khan's dealings had been squandered.

But of far greater consequence was the failure of Western intelligence agencies to act on the information Atif Amin had turned up in Dubai. In addition to tracking down previously unknown elements of Khan's network, Amin had uncovered clear indications that Khan and his cohorts were transferring nuclear technology to Libya. In doing so, however, he ran up against two powerful forces. One was the Khan network itself. Faced with exposure, the network shifted into fight *and* flight modes, lashing out with threats against Amin while simultaneously shutting down compromised operations and transferring them overseas. Khan's associates were also able to use their influence with the authorities in Dubai to derail Amin's investigation.

The other force that Amin ran into was politics. The United States and Great Britain had long been aware of Pakistan's nuclear smuggling. Yet since the late 1970s, successive US and British governments had — for political reasons — allowed the activity to continue. The decisions were made at the presidential and ministerial level. But once the policy was in place, the CIA, MI6, and their sister services in France, Germany, and elsewhere enforced it with ruthless efficiency. If that meant standing by and watching as Khan's operatives carried out their business, that is what the intelligence services did. And if it meant shutting down investigations that threatened that policy, they did that too.

The CIA and MI6 had kept close tabs on Khan's activities for years and watched his network transform from a procurement operation to a proliferation enterprise. By the mid-1990s, the CIA had developed detailed information about Khan's dealings with Iran and North Korea and determined that his nuclear smuggling network was expanding. Together with MI6, the agency began an aggressive intelligence operation to target Khan's growing black market nuclear venture. The joint Anglo–American effort focused on identifying the far-flung elements of the network and gathering extensive intelligence on each. In addition to closely monitoring Khan's movements and communications, the spy services tagged procurement agents, front companies, engineers, and manufacturers, and tracked the movement of money and equipment. They also engaged in old-fashioned spycraft, recruiting informants from within the network and conducting "covert entries" to gain information.[1] The National Security Agency and its British counterpart, the Government Communications Headquarters, also participated in the effort, expanding their long-running monitoring of Khan and his labs to track the communications of the wider network and listen in on conversations with its customers.

By 1999, the British and American spy services had determined that Khan and his associates were engaged in an effort to provide nuclear technology to Libya. Key players and companies involved in the Libyan operation had been identified, and orders and shipment were being tracked. Some details, however, remained elusive. By April 2000, according to a watered-down British review of the Anglo–American intelligence effort, MI6 and the CIA had developed an "evolving" but "incomplete" picture of the Khan network's supply of enrichment equipment to Libya.[2] That, of course, was precisely when Atif Amin's investigation turned up solid evidence that centrifuge components were being directed to Libya through Khan's front companies. But rather than encouraging Amin's investigation and helping to recover the records seized by the Dubai authorities, MI6 persuaded British Customs to pull Amin out of Dubai on hyped-up claims of a threat from Khan's gang. From the perspective of MI6, allowing Amin to keep digging risked exposing sources planted deep inside Khan's network, compromising the ongoing monitoring operation. While the political decision had been made not to act against Khan's smuggling ring, it was nevertheless essential that the flow of intelligence about its activities continue. Although, as MI6 had acknowledged, Amin's straightforward investigative techniques had uncovered valuable new information that had gone undetected despite years of covert surveillance, the decision was nonetheless made to shut down the inquiry. Amin's successful criminal investigation and the potential for learning even more about Khan's network were sacrificed for what MI6 and its political masters deemed a higher purpose.

Despite all that was known about the Khan network's nuclear dealings, British and American policy remained unchanged. In Washington, the CIA argued that moving too soon would harm chances of rolling up the network completely. But that was merely political cover. The instructions were to watch and wait. The CIA and MI6 fell in line. As the British review of the joint operation later reported, even after overwhelming evidence that the Khan network was supplying a uranium enrichment plant to Libya had been obtained, "action to close down the network" was "deferred to allow the intelligence agencies to continue their operations to gather further information on the full extent of the network."[3] Given that logic, Atif Amin's investigation simply could not be allowed to continue. Rather than using the information that Customs had turned up as an opportunity to shut down large portions of a dangerous nuclear smuggling ring, round up the perpetrators, and expose the nuclear programs of Libya and Iran before they had a chance to take root, the United States and Britain chose to stand back and observe. It was a policy that would continue to guide British and American action for another three years, even as the threat of nuclear weapons falling into the hands of terrorists became a paramount concern. Only when the political conditions were right would Washington and London choose to act. But even then, Pakistan would be protected.

Watching and Waiting

By the late 1990s, the British and American intelligence agencies had developed a clear understanding of the broad dimensions of Khan's operation. They also picked up indications that a major deal was in the works. The volume of goods ordered through Khan's procurement agents picked up dramatically, reaching levels that far exceeded the needs of the Khan Research Laboratories (KRL). MI6 began to observe gatherings of known members of Khan's network in Dubai, and Khan himself made frequent trips to the emirate. Evidence eventually emerged that the network was mass-producing centrifuge components. And in Libya there were signs that Qaddafi had begun to reconstitute his nuclear program.[4]

In 1998, indications emerged of a new dimension to Khan's dealings. In April, when Pakistan tested its new medium-range, KRL-built Ghauri missile, US intelligence analysts quickly determined that it was essentially a carbon copy of North Korea's Nodong missile. Pakistan had maintained close political and military ties to North Korea since the early 1970s, when Ali Bhutto first established diplomatic relations with the secretive communist state. In the early 1980s, as Islamabad began seeking a delivery system for its soon-to-be-completed nuclear bomb, North Korea provided technical assistance to Pakistan's nascent missile development program. That aid culminated in a 1993 deal, long sought by the Pakistani military, in which Pyongyang provided Prime Minister Benazir Bhutto with the designs for the Nodong missile. Bhutto and other Pakistani officials have since claimed that Islamabad planned to pay the North Koreans from an "invisible account" set aside for covert activities. But whatever the original terms of the arrangement, circumstances soon changed. Just months after Pakistan received the Nodong plans, North Korea reached a deal with the United States to close its nuclear reactor and suspected plutonium reprocessing facilities in exchange for guarantees of fuel oil and future delivery of safeguarded nuclear power reactors. The North Koreans were quickly cheating on the accord and blocking full-scale UN inspections of the facilities. And while the Clinton administration remained focused on Pyongyang's plutonium program, Kim Jong Il began seeking an alternative route to the bomb. His quest soon led him to Khan's lab.[5] US intelligence began tracking military cargo planes transiting between North Korea and Pakistan, according to former Clinton administration nonproliferation officials. The CIA reported that A.Q. Khan was making repeated trips to Pyongyang. And when Pakistan conducted its nuclear tests in May 1998, just a month after the first launch of the North Korean–designed, KRL-built Ghauri missile, North Koreans were in attendance.[6] Something was clearly going on.

US officials repeatedly expressed their concerns to Islamabad about the North Korean–Pakistani link. President Clinton personally took up the matter with Prime Minister Nawaz Sharif twice and, following Sharif's ouster, with Pakistan's new military ruler, Gen. Pervez Musharraf. Both Pakistani leaders simply denied a deal existed, and the nuclear barter continued. Khan has reportedly since told investigators that the transfers to North Korea were carried out with the knowledge of senior Pakistani military officials, including Musharraf.[7] And, clearly, Pakistani commanders must have had some idea about how North Korean missile technology ended up in Pakistan's possession, and what had been carried aboard the military flights between Islamabad and Pyongyang. But whatever the exact state of official Pakistani knowledge about the deal, the Clinton administration chose not to publicly expose the deadly trade or to discontinue the waiver of economic sanctions as a means of forcing Islamabad to curb Khan's activity. As a result, the Pakistanis simply ignored the American "concerns," and Khan's nuclear dealings went on unabated.

While the Clinton administration was lecturing Islamabad about the nuclear barter arrangement with North Korea, US and British intelligence services were alerting other countries to the Khan network's wider activities. According to a former US intelligence operative, by no later than April 1999 the CIA and MI6 had informed Malaysian intelligence that B.S.A. Tahir was scouting Malaysia as a possible manufacturing site for "restricted components." The warnings, which came in the midst of the British Customs investigation — based on a tip from MI6 — of Abu Siddiqui, became more specific in the months that followed. In June, the intelligence agencies told the Malaysians that Tahir's activities were part of a deal involving A.Q. Khan, according to the former intelligence operative. And by fall the Malaysians were being informed that Tahir was attempting to purchase controlled goods through various agents, including Siddiqui, and that he had approached a major Malaysian conglomerate about booking shipping-container space. Late in the year, the CIA told the Malaysians that the deal in which Tahir and Khan were involved was with Libya. At around the same time, according to a former Israeli government minister and cabinet member, George Tenet informed Israeli officials that Tripoli had embarked on an aggressive new nuclear project with Pakistani help.

George Tenet kept President Clinton closely apprised of the mounting intelligence on Khan's network.[8] By 2000, the evidence of the deal with Libya had become apparent enough to convince the administration that some action was necessary. Yet, as the British Customs investigation demonstrated, the policy of watch and wait remained firmly in place. Unwilling to shut down the smuggling network, the United States settled on a plan to present Pakistan with a limited package of information, sufficient to demonstrate Khan's trafficking in centrifuge technology without compromising sources and methods. The scheme yielded predictably unsatisfactory results. When the Americans confronted Musharraf with photographic evidence of Khan's nuclear dealings, the general feigned surprise and denied any knowledge of the matter.[9] Yet it was clear the Americans would have to be appeased. Musharraf ordered Inter-Services Intelligence (ISI) to investigate, a move unaccompanied by any significant risk of embarrassment. In the later part of the year, the intelligence service began informing Musharraf that Khan was making unauthorized trips to Dubai for meetings with "dubious characters." When questioned about his activities, Khan gave evasive answers and made clear his refusal to abide by any restrictions. Musharraf, knowing he had to be seen as doing something about Khan's activities, removed the scientist as head of KRL and banished him from the facility. But to avoid popular backlash for taking action against the man Musharraf himself referred to as a national hero, the general couched the move as a retirement and gave Khan a cabinet position as a special presidential advisor.[10] No further penalties were imposed, and Khan remained free to travel — and to continue his nuclear wheeling and dealing.

When the Bush administration took office in early 2001, George Tenet, who stayed on as CIA director, immediately began educating senior officials about Khan's activities and the depth of the intelligence agency's penetration of the smuggling network.[11] The warnings proved alarming enough to prompt the new administration to put diplomatic pressure on Pakistan to reign in Khan's operation.[12] The administration also tried a new tactic. In late May 2001, Deputy Secretary of State Richard Armitage announced to London's *Financial Times* that the United States had "concerns" about Pakistani proliferation, centering on "people who were employed by the nuclear agency and have retired."[13] While the public diplomacy initiative was novel, it fell far short of actually taking action against Khan's network. It would, in any case, prove a short-lived experiment.

Just a little more than three months after Armitage's diplomatic warning shot, al Qaeda terrorists crashed passenger jets into the Pentagon, the World Trade Center, and a field in Pennsylvania, killing nearly three thousand people. Once again, the United States determined it needed Pakistan's assistance. US "concerns" about proliferation could be set aside. Now only one issue mattered. On September 12, Armitage confronted the Pakistanis with a stark choice: Side with the United States or with the terrorists. Siding with the United States meant cutting off all ties to the Taliban and cooperating with American reprisals for the terrorist attacks.[14] Reining in A.Q. Khan was not on the list of demands. The perceived need to ensure Islamabad's support in the fight against al Qaeda and the Taliban, groups that had emerged from US support for Pakistan and the mujahedin in one of the closing battles of the Cold War, led the Bush administration to lift the remaining ban on military aid to Pakistan, put in place because of its nuclear activity.[15] The public (and private) warnings about Khan's nuclear trafficking would cease. Islamabad's collaboration in the war on terror could not be jeopardized.

Just weeks after the 9/11 attacks, British Prime Minister Tony Blair traveled to Islamabad to help make sure Pakistan chose the right side in the us-versus-the-terrorists contest. Although MI6 had kept Blair fully briefed on the Khan network's clandestine dealings with Libya and Iran, the prime minister delicately sidestepped the issue of nuclear proliferation in his discussions with Pakistani officials. Instead, Blair focused on the rewards available to Pakistan in exchange for its cooperation in the antiterror campaign.[16]

A stark indication of the new reality came in early December, when George Tenet flew to Pakistan for an emergency meeting with Musharraf. Newly acquired intelligence indicated that a pair of Pakistani nuclear scientists with ties to the Taliban and other radical Islamic groups had met with Osama bin Laden and his top aide to discuss the development of nuclear weapons. Tenet went to Islamabad to demand that Musharraf mount a thorough investigation. "The President needs to be assured that you will not let your country's scientists act on behalf of bin Laden," Tenet told Musharraf. The general gave his assurance and the resulting investigation, in which the United States participated, confirmed the scientists' contacts with al Qaeda. The scientists and their cohorts were quickly rounded up and the main suspect, Sultan Bashiruddin Mahmood — the man who had preceded A.Q. Khan as the first head of Pakistan's uranium enrichment program — was placed under permanent house arrest.[17]

While US concern about the incident was clearly warranted, the reaction is notable for the contrast to Washington's approach to dealing with the Khan network. US policy makers chose not to confront Musharraf about Khan's nuclear trade with Libya, Iran, and North Korea, ostensibly for fear of exposing the joint CIA–MI6 operation that had penetrated Khan's network.[18] In fact, this was simply a variant on the cover story for the longstanding watch-and-wait policy, combined with the new requirement that the US avoid jeopardizing Islamabad's cooperation in the war on terror. While the United States had justifiably concluded that the threat of terrorists acquiring a nuclear capability required an urgent response, it was content to allow the transfer of nuclear technology to state sponsors of terror to continue, under the watchful eye of Western intelligence. It would be another year before the United States and Britain determined that the political climate was right for putting the intelligence they had collected on Khan's network to use.

Operation Aquarium

Note: What follows in the remainder of this chapter, relating to the activities of Peter Griffin, is a summary of information drawn primarily from two confidential British Customs reports and a publicly released Malaysian police report. It has been supplemented by information from media accounts as well as by comments from some of the participants in the events discussed. The authors have attempted to engage Mr. Griffin in a discussion of these events. In an e-mail exchange with one of the authors, Mr. Griffin indicated he might at some point be willing to participate in an interview. He then stopped responding to the authors' messages. The authors also sent letters to Mr. Griffin and to his attorney, offering Mr. Griffin an opportunity to comment on the information obtained by British Customs investigators that is summarized below. Mr. Griffin's attorney, Nigel Tait, responded that his instructions were that Mr. Griffin "does not wish to avail himself of the opportunity to comment."[19] *The authors then sent Mr. Tait a letter listing the nature of the information contained in the Customs reports and requesting that he advise as to whether he believed any of the information was inaccurate or incomplete. Mr. Tait did not respond to the letter. The authors also sent a letter seeking comment on these matters to Mr. Griffin's son, Paul, who is briefly discussed below.*[20] *The authors have not received a response to the letter. Finally, the authors offered to show Peter Griffin the portions of this book that refer to him and invited him to comment. The authors have received no response to that offer.*

Although the United States and Britain remained unwilling to take action against A.Q. Khan's nuclear smuggling operation, MI6 and the CIA continued their aggressive surveillance of his network. According to the

British review of the joint operation, in March 2002 the agencies "pulled together the strands of intelligence" they had collected. "The conclusions showed the wide spread of Khan's network and that he had moved his base outside Pakistan and was now controlling it through his associates in Dubai," the review stated. "At the same time, intelligence showed that he had now established his own production facilities, in Malaysia. He was being helped in his activities by a network of associates and suppliers, including B.S.A. Tahir."[21] In fact, the intelligence services knew a great deal more, particularly about the network's deal with Libya. As George Tenet noted in his 2007 memoir, "we knew virtually everything there was to know."[22]

According to a confidential British Customs report, in December 2002 MI6 provided Customs with a trove of materials concerning a Dubai-based company called Gulf Technical Industries (GTI). The records, according to the report, concerned GTI's suspected involvement in supplying machinery and equipment to Libya's weapons of mass destruction program. Customs' involvement in the case stemmed from the fact that one of GTI's directors was British national Peter Griffin, a man well known to UK investigators and nonproliferation experts. Griffin's British company, Weargate Ltd., had been at the center of the inverter uproar in the late 1970s, and his subsequent exports to Pakistan had, over the years, drawn the interest of investigators.[23] An air of mystery has surrounded his activities. As the German magazine *Der Spiegel* reported, "For years, Griffin has been seen as a man with good connections to the intelligence community, especially with the British."[24] While there has long been speculation about Griffin's possible involvement in supplying Pakistan's bomb-building program, and more recently in perhaps assisting Libya's, he has never been charged with a crime.

Since the late 1970s, Griffin's name has appeared in scores of news reports and at least eight books in connection with his possible involvement in nuclear trafficking. He has consistently denied any wrongdoing. In 2004, he reportedly acknowledged having been a supplier to Pakistan for two decades. He insisted, however, that Britain's Department of Trade and Industry had approved all of the sales.[25] He has also said that none of his Libyan-related transactions involved nuclear technology and has denied assisting Tripoli's weapons program in any way.[26] In 2005, Griffin accepted substantial libel settlements from two major British news outlets over reports of his alleged nuclear dealings. *The Guardian* newspaper agreed to pay Griffin £50,000 ($96,000) plus legal expenses after reporting that a 2004 Malaysian police report, based largely on statements given by B.S.A. Tahir, had concluded that Griffin knowingly helped Libya establish a facility for the production of centrifuge components for its nuclear weapons program.[27] The BBC subsequently agreed to pay Griffin £30,000 ($53,000) plus legal fees for a radio broadcast that made similar allegations based on the same Malaysian police report.[28]

While Griffin has vigorously challenged all suggestions that he or his companies were involved in nuclear trafficking, British Customs investigators have presented a very different picture.

Griffin, a trim, white-haired Welshman, had established Dubai-based GTI in June 2000 in conjunction with a local sponsor, Ahmad Hassan Rashid Amhad al Abbar. In August 2001, he moved to a small village in the Provence region of southeastern France, and his son, Paul, eventually took over day-to-day management of the company.[29]* But a British Customs report states that the elder Griffin "continued to play a key role in organizing GTI business from his home in France."

The British Customs team assigned to the Griffin investigation, code-named Operation Aquarium, was the same unit that had carried out the investigation in the Abu Siddiqui case. Although the team's one-time leader, Atif Amin, had since moved on, having been seconded to the British Home Office to work on communications surveillance policy, his former colleagues would build on the knowledge and experience they had accumulated in the earlier investigation as they began looking into Griffin's affairs.[30] After processing the material received from MI6, the Customs agents of Operation Aquarium worked with French authorities to arrange a search of Griffin's home. The agents also made arrangements with Spain's joint military and civilian police force, the Guardia Civil, to carry out a simultaneous raid on a Spanish machine tool supplier, Comercial Nork S.L., with which GTI had done business, according to a July 2004 Customs report of the investigation.

The raids took place on June 20, 2003.[31] Griffin was not in his well-protected million-dollar villa in the town of Figanières when the authorities arrived, and he did not return during the search. The Customs report indicates that he was in Dubai on business. Law enforcement and intelligence sources suggest he may have been tipped off. French

*Paul Griffin did not respond to the authors' interview request, sent through his attorney.

investigators, accompanied by a British Customs officer, seized two personal computers from Griffin's home, along with a large cache of documents and computer disks, according to the Customs report. At Comercial Nork, the report says, the Guardia Civil confiscated four computers and a variety of documents. French and Spanish authorities made the seized material available to British Customs. Much of that information has since been distributed through a secret US-run intranet database shared by the intelligence services of the United States, Britain, Canada, Australia, New Zealand, and several Western European nations. Just more than a year after the raids, on July 1, 2004, British Customs issued a confidential, twelve-page interim report on Operation Aquarium. In the latter part of 2005, Customs produced a fifty-one-page follow-up. Both reports have been placed in the Western intelligence database. The authors have gained access to the reports.

On Griffin's computers, investigators found tens of thousands of files, including text documents, spreadsheets, e-mail messages, letters, digital images, and reports. The 2005 Customs report states that a "large amount" of that material related to the manufacture of components for uranium-enrichment centrifuges. This included detailed engineering and assembly instructions and long lists of parts for the machines. Investigators also discovered many records concerning the design and supply of two machine shops that the Customs reports indicate were intended for the production of centrifuge parts. One of these, according to the reports, was to be set up in Libya, the other in Malaysia.

Customs submitted the material from Griffin's computers for analysis by a high-level technical expert at the UK branch of Urenco, the uranium enrichment consortium from which A.Q. Khan had stolen the centrifuge designs on which Pakistan built its enrichment program. The 2005 Customs report states that the Urenco expert concluded that the material from Griffin's computers "relates to centrifuge components of early Urenco design." All of the part descriptions found on Griffin's computers matched Urenco's component designations, the report says. Some of the parts listed in the documents on Griffin's computers were identified by German names. Urenco confirmed that those names were the correct designations for the components of the German-designed G2 centrifuge used by Urenco. It was the pilfered G2 designs that A.Q. Khan used as the basis for Pakistan's P2 centrifuge. The Customs report states that Urenco concluded that some of the parts identified in the material from Griffin's computers "would not have any other application" than as components for a G2 centrifuge.

Urenco also analyzed records from Griffin's hard drives concerning the production of "P1 components," which the Customs report indicates is a reference to parts for the first-generation Pakistani centrifuges, which were based on Urenco designs and developed at A.Q. Khan's labs. The Urenco analysis determined that some of the records on Griffin's computer called for the use of slightly different grades of materials for P1 parts than those specified in the original Urenco designs, according to the Customs report. The report gives no indication as to who modified the plans. A source familiar with the Urenco analysis believes the modifications represent an effort to substitute materials that were available when the superior materials called for in the original designs could not be acquired due to export restrictions.

The 2005 Customs report indicates that among the mass of material on Griffin's computers investigators found records showing "substantial" payments to "DS" (which the report indicates is a reference to "Dr. Sahib"), B.S.A. Tahir, and several individuals who have since been arrested or face charges in connection with their alleged involvement in Khan's nuclear smuggling ring. At Griffin's home, investigators discovered a 1993 letter from Griffin addressed to "Hennie and Qadir." The 2004 Customs report states that the tone of the letter suggests "a close personal relationship" between Griffin and A.Q. Khan. Inside Griffin's home, investigators also recovered a commemorative box from Tahir's 1998 wedding in Malaysia.

One of the items investigators found on Griffin's hard drive was a document titled "Facts of situation" concerning the British Customs investigation in the Abu Siddiqui case. The Customs reports indicate that the document "passes negative comment" on the Siddiqui investigation and expresses worries about the bad publicity the case could generate in Germany, Holland, and the UK, the three partner countries in Urenco. The document has since been shared with Western intelligence agencies through the intranet database operated by the United States, according to a US intelligence source.

Another document on Griffin's computer was a chronology he produced, according the 2005 Customs report, of the events surrounding the aluminum order in the Siddiqui case. The report states that Griffin later told British investigators that the chronology was produced with Tahir and Siddiqui's lawyer, who had visited Dubai.

The British Customs reports indicate that much of the material found on Griffin's computers concerned the machine shop intended for Libya, known as "Project 1001" or "P1001." That material, the reports say, included detailed studies produced by Griffin listing machinery and equipment that would be required for the project. The 2005 Customs report states that, "Early feasibility studies produced by PG [Peter Griffin] show that the Project was for a specific purpose to manufacture component parts for centrifuges." There are references throughout the records, according to the report, to the manufacturing of parts for both "P1" and "P2" machines. The report indicates that Griffin also produced a carefully designed floor plan for the machine shop that appeared to be based on an early layout used by Pakistan for its uranium enrichment program. The 2004 Malaysian police report states that Tahir "said that the plans for Machine Shop 1001 was [sic] prepared by Peter Griffin."[32] In 2005, Griffin's attorney issued a statement saying, "The Malaysian police, although they referred in their report to allegations against Mr. Griffin made by Mr. Tahir, made no findings of wrongdoing against Mr. Griffin."[33] The 2005 British Customs report states that, based on an analysis of the material on Griffin's computers, "it is clear that he designed a precision machine shop 'Project 1001,' specifying all the equipment needed for the production of P1 and P2 centrifuge parts."

The Malaysian police report states that Tahir identified Griffin as the "middleman" in Project 1001.[34] According to the British Customs reports, records from Griffin's computers show that Griffin procured the equipment for P1001 through GTI. But those records do not suggest that GTI sent the machinery to Libya. Rather, the Customs reports indicate that GTI imported the equipment to Dubai and turned it over to Tahir. The reports say that Tahir is believed to have arranged the transshipment of the goods to Libya, evading export controls. The 2005 report notes that UN inspectors later found much of the equipment at the "Project 1001" machine shop being set up in Janzour, Libya, for the manufacture of centrifuge components. The machines were still in their shipping crates. The report states that some bore labels reading "Gulf Technical Industries LLC" and "GTI Dubai." Testifying in 2006 at the German trial of one of the alleged participants in the scheme to supply the Libyan program, Griffin reportedly stated that the equipment he obtained could be used for civilian purposes and that he learned about centrifuge technology only in February 2004 with the release of the Malaysian police report in which he had been incriminated. "I have only then started to search on the Internet to see what Tahir had dragged me into," Griffin testified.[35]

One of the documents found on Griffin's computer appears to lend insight into some of the arrangements for the Libyan machine shop project. It is a letter from Griffin and the local sponsor for GTI, Ahmad al Abbar, addressed to the manager of a Dubai bank. Portions of the letter are reproduced in the 2004 Customs report. Dated July 20, 2000, just a month after Griffin and al Abbar set up GTI, the letter states that Griffin had been working on "several large projects for Libya which were 'on hold' due to Western embargoes" that were then in place. Meanwhile, according to the letter, excerpts of which also appear on the Western intelligence intranet database, according to a US intelligence source, Griffin was "still doing business with Pakistan in the main." That business, the letter states, was "transferred to Dubai," where al Abbar became Griffin's sponsor, setting up a company that Griffin ran called Al Abbar Medical and Scientific Supplies, Engineering Division. The 2005 Customs report indicates that other records found on Griffin's computers show that most of the business done by Al Abbar Medical and GTI was with Pakistani companies associated with KRL. These included Peoples Steel Mills, the Civil Works Organization, and Allied Engineering. The report states that some orders showed "KRL" as the client. The report describes GTI and Al Abbar Medical as "nothing more than KRL procurement companies with a small amount of legitimate business." Records from Griffin's computers for the period between 1997 and 1999 show Allied Engineering and another Pakistani company identified in the Customs report as being associated with KRL, Industrial Engineering, as the clients for P1001. The report indicates that it was Tahir who forwarded these orders to Pakistan. In a letter to the authors of this book, Al Abbar Medical stated that Griffin was an employee at Al Abbar Medical and GTI and that "any actions that he might have taken with regards to the transactions in question were an abuse of the trust placed in him by both companies." The letter also stated that Al Abbar Medical "does not under any circumstances conduct any questionable business dealings."[36]

In their letter to the Dubai bank manager, Griffin and Ahmad Al Abbar explained that Griffin ran Al Abbar Medical with the intention of eventually "exploiting" the two Libyan opportunities. One of those, the letter states, involved the supply of £48 million ($72 million) worth of "laboratory equipment" for nine hundred Libyan secondary schools over a period of five years. The other was a deal to "supply and equip a precision machine shop with an initial budget of USD 10 million." When it "became obvious early in 2000 that the sanctions on Libya were lifting," the letter states, Griffin and al Abbar formed GTI to "exploit the new opportunities." Soon after GTI was formed, according to the letter, there were two positive developments. One was that "a prepayment/working capital injection of USD 2 million was received from our colleague in Libya who will be participating in the benefits of the business."* The other was that "the machine shop has been re-validated and the first order of two lathes are [sic] due within the next three weeks." The letter states that an advance payment for the lathes equivalent to 56,565,460 Spanish Pesetas ($337,000) had been "transferred to the Spanish Principals" on July 13, 2000. The 2004 British Customs report indicates that a purchase record found on Griffin's computer shows a payment to "Nork PL" on July 13, 2000, in the amount of "56.565.460." The GTI reference number for the purchase was "1001.01.04."

The lathes GTI purchased from Spain are highly specialized machines used to shape metal cylinders to a high degree of precision. Known as flow-forming lathes, they can, when properly equipped, be used to produce such things as the critically important rotors and high-tolerance casings for ultra-centrifuges. In those circumstances they are subject to strict international export controls. But the lathes GTI purchased were reportedly not outfitted with the computer control systems necessary to produce centrifuge components.[37] According to Comercial Nork, no export licenses were required.[38] Computer controls for the lathes were reportedly purchased separately by unnamed persons and added to the machines after they were shipped from Dubai.[39] Griffin has stated that he handed the lathes over to Tahir with the understanding that they were intended for a machine shop being set up in Dubai by the Libyan National Oil Company.[40] In late 2000, Tahir sent one of the lathes, valued at $257,680, to a South African company alleged to be involved in Khan's network, according to an indictment filed in South Africa. The South African company returned the lathe to Dubai a year later. That lathe eventually ended up in Libya.[41] There are conflicting reports as to whether the other lathe wound up in Libya or Malaysia.

---

*When asked in a 2006 e-mail message from one of the authors who the Libyan was, Griffin responded, "There wasn't one." (E-mail, Griffin to Armstrong, June 18, 2006.)

Another document from Griffin's hard drives seems to reveal the genesis of his dealings with Comercial Nork. According to the 2005 British Customs report, a document labeled C1001.200 and dated May 1, 1999, indicates that Griffin attended a machine tool trade show in Bilbao, Spain, in March 1998. In the document, the report states, Griffin commented on Spain's "relaxed" export controls. He also identified two Spanish companies through which he could obtain the equipment he needed, and ultimately settled on Comerical Nork.

Other records found on Griffin's computers show that in addition to the two flow-forming lathes, he purchased much of the equipment for P1001 from Comercial Nork, according to the British Customs reports. A representative of Comercial Nork confirmed that the company sent approximately fifty machines to GTI over a period of several years. Harbil Echaniz, a former co-owner and director of Comercial Nork, said none of the equipment her firm sold to GTI was dual-use or required special export licenses. Echaniz said her company dealt directly with Peter Griffin and understood that the equipment was intended for a machine shop that was to be set up in Dubai to produce spare parts for the oil industry. All of the equipment, she said, was shipped to Dubai. She said Comercial Nork was told it would not need to send anyone to install the machines because workers would be sent to Spain to learn how to install and operate the equipment. Echaniz said that she and other representatives of Comercial Nork traveled to Dubai twice and met with Peter Griffin. When they asked whether the machine shop was ready, they were told it was not and that the equipment was being kept in storage. Echaniz said she learned from media reports that appeared after the Pakistani nuclear smuggling scandal broke that the machines had been sent elsewhere.[42] Comercial Nork is not accused of any wrongdoing.*

---

*In what may be a remarkable coincidence, the Spanish newspaper *El Pais* reported that in November 2001, Spain's intelligence service, the National Intelligence Center, warned the government agency responsible for regulating military-related exports that machinery being exported to Dubai by an unnamed Spanish company could be used to build uranium enrichment centrifuges. The Dubai client for the equipment, according to *El Pais,* was Peter Griffin's company GTI. In February 2002, the paper reports, Spain's secretary of state for commerce asked the unnamed company for technical specifications on the Dubai shipments. According to *El Pais,* "the case seemed resolved" until June 19, 2002, when Spain received a request from British Customs for an "urgent registry" on the Spanish firm. This resulted in Spain's Guardia Civil launching an investigation code-named Operation Aquarius. (Miguel Gonzalez, "Libya Connection," *El Pais,* March 1, 2004.)

The 2004 Malaysian police report states that it is alleged (by an unidentified source) that on two occasions "Griffin arranged to send 7 to 8 Libyan technicians to Spain" to attend training courses on operating flow-forming lathes.[43] The British Customs reports state that Griffin was "instrumental" in arranging for Libyan engineers to travel to Spain for training on equipment for P1001 and "arranged for their visa applications." The 2005 Customs report notes that Griffin also attended some of the training sessions. Echaniz says that Griffin arranged for "teams" of Libyans to be trained at Comercial Nork on four occasions. Some of the Libyans attended more than one training session, she says. According to Echaniz, the training was simple and routine, the same as was offered to other customers purchasing equipment. Echaniz said that when the first group of Libyans arrived their visas did not allow them to remain in Spain long enough to complete their training. As a result, she said, Comercial Nork took the Libyans to the local police and arranged to have their visas extended.[44]

According to the British Customs reports, records found on Griffin's computers show that between 1997 and 2002 he procured a wide range of goods and equipment for P1001. While much of the equipment came from Spain, the reports say the records show that he also used suppliers in Italy, Australia, Switzerland, and Turkey. Among the goods GTI obtained for P1001, the reports state, were a computer-controlled surface grinder, a universal cylindrical grinder, a radial drilling machine, a vacuum horizontal furnace, a cyanide treatment tank, a universal hardness tester, an ultrasonic cleaning machine, a large quantity of aluminum, and a variety of measuring devices. The 2005 Custom report states that UN inspectors later discovered a vacuum horizontal furnace in Libya, still packed in its crate. The report notes that a crude attempt had been made to black out the words "GTI Dubai" on the side of the crate.

Another piece of equipment acquired by GTI, according to the British Customs reports, was a high-precision vertical boring machine from Switzerland. The highly specialized machine is export-controlled because of its dual-use capability. The 2005 report states that Griffin produced an end-user certificate for the machine, stating that it was going to a "private end-user" in Dubai and listing his office address. The report says that Griffin later told British investigators that after importing the machine to Dubai, there had been a change in plans and that he gave the machine to Tahir. The report adds that Griffin "eventually conceded that he had did

[sic] this to 'get around' export controls." UN investigators later found the machine in Libya, still in its crate. A picture included in the Customs report shows that the crate was marked "GTI Dubai."

In all, the 2005 Customs report states, Griffin facilitated the supply of roughly $10 million worth of machinery, equipment, and materials for P1001. "From the evidence available," the reports says, "it can be shown clearly that Peter Griffin participated in providing Libya with a machine shop specifically designed to produce gas centrifuge components." Such a machine shop, the report adds, "would clearly help Libya to develop their capability to produce a nuclear weapon."

Project 1027

On Griffin's hard drives investigators also found documents concerning the design of a Malaysian machine shop that the 2005 British Customs report describes as being "similar to P1001." One of those documents was a preliminary study produced by Griffin in February 2001 titled "KL Project 1027 Malaysia." The study, according to that report, "specifies the machinery needed to establish a workshop capable of large scale manufacturing of P2 components." It has since been shared with Western intelligence services through the American-run intranet database, according to a US intelligence source.

The same month that the Project 1027 study was produced, Griffin and Tahir reportedly met with officials of a politically connected Malaysian oil and gas services conglomerate to discuss a possible deal to manufacture finely tooled aluminum machine parts. Officials of the Malaysian company, Scomi Group Berhad, have said they believed the parts were intended for use in the oil industry.[45] Tahir was a director of the holding company that was the principal shareholder of Scomi. Two of the main investors in the holding company, called Kaspadu, were Tahir's wife, Nazimah, and Kamaluddin Abdullah, the son of Maylaysia's then-foreign minister and soon-to-be prime minister, Abdullah Ahmad Badawi.[46]

Griffin has reportedly acknowledged that he and Tahir met with Scomi officials but says the discussion concerned the export of tanker trucks. He says the deal never went through.[47]

The 2005 British Customs report states that Griffin was "heavily involved in the planning of P1027." It notes that there were more than one hundred documents on Griffin's computer related to the project, including schematics, designs, cost analyses, and "production schedules for the P2 centrifuge component parts." The report also states that records from Griffin's computers show that he procured at least seven of the machines listed in his original study for the project. The Malaysian police report states that "Griffin presented a feasibility study" for the project recommending "the type of machinery needed." One such piece of machinery, according to the report, was a "Cincinnati 150 Machining Centre." The report adds that the same machine was "purchased and installed by Urs Tinner," an engineer brought in as a full-time technical consultant on the project in April 2002 on the recommendation of Tahir "after Peter Griffin was found unsuitable for the job."[48] Tinner, the son of Swiss engineer Friedrich Tinner, a longtime associate and alleged supplier to A.Q. Khan, was reportedly a CIA mole recruited to spy on the Malaysian operation. He was arrested in Germany in 2004 and subsequently extradited to Switzerland. As of late 2006, Tinner remained in custody in Switzerland, where he, his father, and his brother, Marco, are under investigation on suspicion of assisting Libya's nuclear weapons program.[49]

Scomi has said that it signed a contract with GTI in December 2001 to manufacture a variety of machined aluminum parts. GTI was reportedly represented by Tahir. The contract, worth about $3.5 million, reportedly made GTI Scomi's biggest customer. To fill the order, Scomi set up a new manufacturing unit known as Scomi Precision Engineering (SCOPE) at an industrial park outside of Kuala Lumpur, Malaysia's capital. SCOPE purchased materials and equipment from around the world and began producing what turned out to be an array of centrifuge components, including casings, top ends, crash rings, and flanges.[50] It sent most of the finished parts to a Dubai outfit called Aryash Trading Company between December 2002 and August 2003, according to the Malaysian police report. But the police report includes a copy of a "delivery note/packing slip" dated August 1, 2002, showing a consignment from SCOPE of 832 high-strength aluminum tubes addressed to Gulf Technical Industries in Dubai. The report also includes a copy of a "proforma invoice" from SCOPE for the same number of tubes of identical dimensions addressed to Desert Electrical Equipment Factory in Dubai, the company Atif Amin had discovered listed as the recipient for centrifuge components designated to be sent to Libya from a KRL front company more than two years earlier. The police report concludes that "the consignment was directed to Desert Electrical" on the "instructions of Urs Tinner"[51] Desert Electrical is reportedly defunct, and its phone and fax numbers now belong to another company.[52]

Griffin has denied being involved in the SCOPE deal. He has reportedly said that Tahir falsified documents to make it appear that SCOPE was doing business with GTI. Griffin has also said that he forbid Tahir from importing items through GTI, but that Tahir ignored the order.[53]

Take Down

In August 2003, just weeks after the raids on Peter Griffin's home and Comercial Nork, five giant shipping containers packed with "quality aluminum" parts were loaded onto a cargo ship in Malaysia. Spy satellites tracked the vessel as it made its way across the Indian Ocean and through the Strait of Hormuz. When it docked in Dubai, the containers were transferred to a German-registered ship known as the *BBC China,* which the US had been monitoring for nearly a year. After the *BBC China* left Dubai, Urs Tinner reportedly informed his CIA handlers that the cargo was destined for Libya. When the ship passed through the Suez Canal, Washington and London finally pulled the trigger. The orders went out to seize the vessel. On October 4, at the request of the United States and Britain, German and Italian authorities interdicted the ship and diverted it to a port in southern Italy. There, agents boarded the vessel and secured the five shipping containers marked with the SCOPE logo, none of which had been listed on the ship's manifest. Inside the containers, investigators found thousands of carefully packed centrifuge components.[54] The 2005 British Customs report indicates that of the nine types of components discovered on the *BBC China,* eight appear on a March 2001 list, found on Peter Griffin's computers, of parts to be produced in Malaysia.

The seizure of the *BBC China* began the unraveling of A.Q. Khan's nuclear smuggling network. Confronted with the evidence seized from the ship, Muammar el-Qaddafi agreed — in what was the culmination of a years-long ritual dance among the United States, Britain, and Libya — to give up his nuclear program.[55] Under pressure from the US, Pervez Musharraf ordered the investigation that resulted in Khan's public mea culpa, followed by Musharraf's pardon of the man he called his "hero."[56] Khan is currently under house arrest. The Pakistanis also detained eleven of Khan's top aides as part of the investigation, including KRL procurement director Mohammad Farooq. All were subsequently released, and Islamabad declared the investigation "closed."[57]

In May 2004, Malaysian authorities detained B.S.A. Tahir under the country's Internal Security Act.[58] Although Tahir was found not to have violated Malaysian law, information provided by British Customs that the 2005 Customs report states had been found on Peter Griffin's computers allowed the Malaysians to hold Tahir under their security law for up to two years without trial. Tahir cooperated with Malaysian and Western investigators, revealing new details about the Khan network.[59] He was released in the early fall of 2006 and immediately "disappeared," according to US and Malaysian sources.

On September 1, 2004, South African authorities, acting on information provided them by British Customs that, according to the 2005 report, had been obtained from Griffin's computers, raided two engineering firms. Hidden behind a false wall at one of the companies, investigators discovered eleven shipping containers filled with what prosecutors allege was a complete piping system for feeding uranium gas into centrifuges. A flow-forming lathe purchased by GTI and sent to South Africa by Tahir had already been returned to Dubai, according to an indictment in the case.[60] Two of the principals of the companies are awaiting trial in South Africa as of this writing.[61]

In the spring of 2006, Peter Griffin testified at the trial of Gotthard Lerch, the German engineer whose first dealings with Pakistan date to the 1970s, now accused of overseeing procurement in the South African wing of Khan's network. Griffin reportedly shed little light on Lerch's alleged role in the Libyan deal and denied any witting involvement in the scheme. Griffin devoted much of his testimony to arguing that he had been victimized by Khan and Tahir. A mistrial was declared amid charges prosecutors had withheld evidence. Lerch's attorneys had complained they were denied access to Western intelligence files and argued their client was the victim of a Western intelligence plot. Lerch is scheduled to be retried.[62]

On June 12, 2005, British authorities detained Peter Griffin at Luton Airport outside London, according to the 2005 British Customs report. The authorities questioned Griffin for two days, the report says, before

*The report notes that Griffin was "bailed to return on 23 January 2006."

releasing him without charges.* The Customs report states that, "From material available, Peter Griffin appears to be a key member of the A.Q. Khan network working closely with B.S.A. Tahir and others. He has a long history of personal and business dealings with Dr. A.Q. Khan and there is evidence to show that for many years, Peter Griffin has been engaged in the proliferation of technology and equipment to Pakistan's WMD programme." In addition, the report says, Griffin "played a hugely significant role in assisting the Libyans in their quest to develop a nuclear weapon."

British Customs declined to comment on the status of its investigation of Griffin. A source familiar with the investigation says there is "no appetite" to bring charges against him. The source says that "other agencies" have an interest in seeing that Griffin is not prosecuted because of concerns about "political and other sensitive ramifications." As a result, the source believes the case will be "swept under the carpet."

Today Griffin lives comfortably in an eighteenth-century farmhouse on a five-acre estate near the French town of Cognac.[63] Some US and European intelligence officials have suggested that Griffin, like others who have had dealings with A.Q. Khan, may have been cooperating with Western authorities, perhaps for a very long time. Asked by one of the authors in a June 2006 e-mail exchange whether he had provided assistance to any Western intelligence service, Griffin offered a one word reply: "Later."[64]But later never came. Griffin, through his attorney, declined repeated requests for an intrview for this book.

Epilogue

FOLLOWING THE PUBLIC exposure of A.Q. Khan's nuclear smuggling ring in early 2004, the Bush administration proudly declared that the "criminal enterprise" had been put "out of business," the network "dismantled," and the culprits "brought to justice."[1] Yet more than three years later, few of those responsible for one of the worst acts of proliferation in history face any criminal charges, most remain free, and there is evidence that Pakistan's nuclear smuggling continues.[2] And while Libya has given up its effort to build weapons of mass destruction, the status of North Korea's centrifuge program remains unknown as of this writing, and Iran is now on the verge of large-scale uranium enrichment, a development that has the Bush administration making increasingly warlike noises and many Middle Eastern nations scrambling to establish their own nuclear programs.[3] Meanwhile, none of the engineers or technicians trained by Khan's gang to supply the Libyan operation are accounted for, leaving a potential pool of skilled workers available to nations with nuclear ambitions or to terrorist groups bent on acquiring an atomic device. Yet despite all this, George Tenet and other former US intelligence officials responsible for implementing the watch-and-wait strategy that led to this state of affairs insist the outcome was a "major victory."[4] A greater victory would have been to shut down Khan's trade in atomic technology years earlier, along with the nuclear programs of Libya, Iran, and North Korea. The upshot of America's failure to do so is the stuff of a 3 A.M. nightmare.

The horrifying prospect we face today of terrorists armed with nuclear weapons is a direct outgrowth of shortsighted policies implemented decades ago. In order to inflict a major Cold War defeat on the Soviet Union, American policymakers forged an alliance with Islam's most extreme elements and gave them the means to wage war against a common enemy. In the process the United States helped foster the Islamic fundamentalist movement that would give rise to the likes of the Taliban and al Qaeda. Pakistan provided vital assistance, helping support the anti-Soviet jihadis, and in exchange Washington ignored, and in some cases actively covered up, Islamabad's development of nuclear weapons.

The fateful compromise accepted by Jimmy Carter in the late 1970s and perpetuated by Ronald Reagan and George H.W. Bush set the stage for a world in which stateless terrorist networks can acquire the makings of an atomic bomb.

Throughout the 1990s and into the new millennium, Democratic and Republican administrations alike held fast to the misguided strategy they had inherited. Although official US–Pakistani relations grew strained in the post–Cold War period, Islamabad maintained back-channel influence in Washington, quietly aiding US intelligence efforts and helping the CIA keep tabs on the Taliban regime that Pakistan itself had helped install and continued to support. And as the CIA accumulated mounting evidence of Pakistani trafficking in nuclear technology, the Clinton administration chose not to act, even as despotic regimes such as Libya, Iran, and North Korea were assembling the infrastructure of a nuclear weapons program. Meanwhile, the militant Islamic forces Washington and Islamabad had helped unleash gathered strength and prepared for their next jihad.

After the attacks of 9/11, Pakistan once again became an indispensable US ally, this time in helping to combat the terrorists Washington and Islamabad had jointly helped nurse into being. Determined to maintain Pakistan's support in that effort, the Bush administration continued to watch and wait as A.Q. Khan and his cohorts carried out their deadly business. By the time the United States finally took action against Khan's operation, the situation had long since spun out of control.

Some analysts, including former UN weapons inspector David Albright, have suggested that the United States' failure to move against Khan's network sooner was the result of an intelligence failure. They argue, in essence, that the United States lacked the information it needed to take action earlier.[5] But that is not true. The CIA and other Western intelligence agencies had been watching Khan's smuggling operation closely for decades and had highly detailed knowledge of its activities. As one Defense Intelligence Agency source put it, "We could have intervened years earlier, based on what we knew." High-level policy makers in Washington, London, and elsewhere were fully aware of that intelligence and yet, for political reasons, chose not to act.

In the end, as former CIA nonproliferation analyst Richard Barlow notes, the long delay in clamping down on Khan's sales of nuclear technology was the result of a policy failure rather than a lack of actionable intelligence. Barlow says that when he was monitoring Pakistan's nuclear program in the mid- to late 1980s there was already more than enough evidence of malfeasance to justify taking action. "I refuse to believe that the intelligence dried up," he says. He believes the Khan network's subsequent nuclear trafficking was "totally and utterly" avoidable. "We clearly could have shut it down," Barlow says. "What we have here is the US government waiting for fifteen years and watching while nuclear weapons were spread to our enemies and now it's too late to do anything about it."[6] That, in Washington, is what passes for "success."

Even after the United States belatedly outed Khan's operation, it failed to take appropriate steps to ensure that the smuggling network had been wrapped up. Although Khan reportedly told Pakistani investigators that senior military commanders, including then–army chief of staff (now also president) Gen. Pervez Musharraf, were aware of and approved his transfers of nuclear technology, a Pakistani military panel concluded that Khan had acted in a strictly "private" capacity.[7] No Pakistani military figure, nor any of Khan's Pakistani cohorts, have been charged in the affair. And Khan himself was granted a pardon by President Musharraf. Former prime minister Benazir Bhutto believes that in exchange for the pardon Khan was asked to "fall on his sword in order to save other more powerful people."[8] But despite the obvious inadequacies of the Pakistani "investigation," the Bush administration, fearful of losing Islamabad's cooperation in the war on terror, blithely accepted Pakistan's assertions that Khan's nuclear trade had been conducted without Islamabad's knowledge and that his network had been eliminated. US ambassador to Pakistan Ryan Crocker expressed Washington's satisfaction with Islamabad's inquiry into the matter.[9] So satisfied was the United States, in fact, that rather than imposing sanctions on Pakistan as might have been expected, Washington saw fit to declare its partner in the antiterrorism campaign a "major non-NATO ally" and arranged a massive new military assistance package, complete with the sale of another consignment of F-16 fighter jets.[10]

While busy lavishing rewards on Pakistan in the wake of the proliferation scandal, the Bush administration chose not to press for access to A.Q. Khan, whom Islamabad declared off limits to American and international investigators.[11] As a result, important questions about the activities of Khan's network remain unanswered.

One is whether any country other than Libya received copies of the bomb design. American and international investigators strongly believe, but cannot prove, that Khan passed the plans to Iran. If he did, then the US failure to halt Khan's smuggling earlier further increased the odds that Iran could eventually produce the nuclear arsenal that the West now fears. In addition, some intelligence experts also suspect that Khan gave the bomb designs to North Korea, although hard evidence is lacking.[12]

An equally disturbing question is whether all of Khan's customers have been identified. US officials and intelligence experts have suggested that Khan had several still-undisclosed clients. Among the countries considered possible customers are Syria, Malaysia, Indonesia, Myanmar, Sudan, Algeria, and Abu Dhabi. Two close American allies, Egypt and Saudi Arabia, are also on the list of suspects. The Saudis were early backers of Pakistan's bomb program and later reportedly helped fund Islamabad's nuclear and missile program purchases from China. According to documents revealed by a dissident Saudi diplomat, between 1985 and 1990 Riyadh poured an estimated $5 billion into an Iraqi effort to build a bomb as part of a pact with Saddam Hussein's regime to provide the kingdom with access to any weapons that were produced. Since the late 1990s there have been persistent reports of Saudi–Pakistani nuclear collaboration. In 1999, the Saudi defense minister, Prince Sultan bin Abdul Aziz al Saud, toured Khan Research Laboratories with Pakistani Prime Minister Nawaz Sharif and army chief of staff Pervez Musharraf. Prince Sultan also received a briefing on the KRL operation from A.Q. Khan. The following year, Khan led a Pakistani delegation on a visit to Saudi Arabia. Prince Sultan hosted the group, and Khan delivered a speech in which he thanked Riyadh for contributing to the success of Pakistan's 1998 nuclear test. Khan and his top aides are believed to have made multiple trips to the kingdom. In the fall of 2003, just as Khan's empire was crashing down, reports surfaced that Pakistan had entered into a secret pact to provide Saudi Arabia with nuclear weapons technology in exchange for low-priced oil guarantees. By one account Islamabad also agreed to provide Riyadh with missile delivery systems for the weapons that, while remaining under Pakistani command, would be based on Saudi soil. In 2006 the Saudis began leading a drive by Middle Eastern states to develop ostensibly peaceful nuclear programs in the face of a rising threat from a nuclear-armed Iran.[13] Thus to counter the risk posed by an Iranian nuclear program that was aided by a Pakistani effort which the Saudis helped back, Riyadh now proposes taking the entire region nuclear.

Another ongoing concern is the possibility that elements of Khan's nuclear smuggling network are still operating. Of the more than fifty people believed to have been actively involved in the black market venture, only a handful are in custody or face charges. Some are unaccounted for entirely. And it is far from clear whether all of the companies used by the network have been identified or that all of its workshops have been shut down. Even as the Bush administration crowed that Khan's enterprise had been eliminated, evidence emerged that at least some portions continued to function. In March 2004, Swiss police foiled an attempt to ship sixty tons of high-strength aluminum tubes of the type used for centrifuges from a Russian manufacturer through intermediaries in Western Europe and Dubai to entities associated with Pakistan's nuclear weapons program. Shipping records indicated that one of Khan's longtime associates had a hand in the deal. When the plot was interrupted, a second attempt was made to send the tubes to Pakistan, this time through an intermediary in the United Kingdom who was not previously known to be involved in Khan's operation. That shipment was seized by authorities in Dubai. While the tubes were apparently destined for Pakistan, proliferation expert Andrew Koch notes that — as had been the practice in the past — they could subsequently have been rerouted to one of the Khan network's customers. As Koch told a congressional panel in 2006, "This incident is just one illustration of what many inspectors working on the Khan affair believe — that parts of his former network remain intact, operating either on their own or with other similar networks built by national governments such as Iran."[14]

Since that incident, additional concerns about Pakistani proliferation have emerged. In January 2006, the British newspaper *The Guardian* reported that a European intelligence assessment had determined that even after Khan went on Pakistani television to "confess" in February 2004, Islamabad's clandestine nuclear purchases continued. "Since the beginning of 2004 extensive procurement efforts for the Pakistani nuclear sector have been registered," the intelligence report stated, according to *The Guardian.* The report, representing the combined knowledge of the intelligence services of Britain, France, Germany, and Belgium, and prepared on behalf of the European Union, noted that the range of materials and equipment being purchased by the Pakistanis "clearly exceeds" the requirements for Islamabad's bomb program, suggesting that the surplus goods are being traded on the black market.[15] Among the items on Pakistan's nuclear shopping list, according to another account of the intelligence report, are high-strength aluminum tubes, ring magnets, machine tools, and chemicals, as well as equipment used to produce liquid- and solid-fueled missiles. The intelligence report is also said to identify twenty Pakistani government offices, laboratories, private companies, and trading organizations believed to be involved in the nuclear purchasing effort.[16] Khan's former base of operation, KRL, remains a key part of that effort, continuing, according to the report, to obtain equipment from the West though a maze of front companies.[17]

The US response to Pakistan's continued nuclear smuggling has been disturbingly familiar. The Bush administration has failed to invoke applicable sanctions against Pakistan and, in at least one case, capitulated when Islamabad blocked an official US investigation from going forward in Pakistan. On New Year's Day 2004, US Customs agents in Denver arrested a South African electronics salesman with Israeli citizenship on charges of illegally sending restricted technology, including American-made switches that can be used to detonate nuclear weapons, to a suspected front for Pakistan's bomb program. The South African, a former Israeli army major named Asher Karni, was accused of violating American export laws by arranging the shipment of two hundred triggered spark gaps, small electronic devices that can be used to set off atomic bombs, as well as specialized oscilloscopes that can be used to test nuclear weapons, from the United States through South Africa to a businessman in Islamabad with close ties to the Pakistani military. The Pakistani businessman, Humayun Khan, heads a family firm that has done business with Pakistan's Ministry of Defense for more than forty years and reportedly served as a purchasing agent for Islamabad's nuclear program as far back as 1975. US officials reportedly suspect that the Pakistani government was the actual customer behind Khan's purchase of the triggered spark gaps. The oscilloscopes eventually turned up at a Pakistani company which the United States has barred from purchasing equipment that can be used in the manufacture of nuclear weapons.[18]

Following Karni's arrest, investigators from the Departments of Commerce and Homeland Security planned to travel to Islamabad to interrogate Humayun Khan. But when the investigators asked the State Department to clear the trip, they were denied permission. Senior US officials told the *Los Angeles Times* that Washington had asked Islamabad to cooperate in the case but that the requests had not been forceful or public. The paper reported that the case had become part of a "tug-of-war" between "federal agencies that enforce US nonproliferation laws and policy makers who consider Pakistan too important to embarrass" because of its assistance in the war on terror.[19] As in the past, halting the spread of nuclear weapons fell victim to foreign policy objectives deemed more pressing.

Asher Karni eventually pleaded guilty to violating American export control laws and, in exchange for his cooperation in the case, was sentenced to three years in prison. Based in part on information provided by Karni, in April 2005 a federal grand jury indicted Humayun Khan on related charges. He has denied any wrongdoing. Despite a standing extradition treaty between Islamabad and Washington, he has not been sent to the United States to stand trial.[20]

Nonproliferation expert Leonard Weiss has argued that the Karni affair represents a violation of the Solarz amendment, the 1985 law passed in the wake of the Nazir Vaid case that bans aid to any country that attempts to smuggle nuclear-related technology out of the United States. Weiss, who served on Sen. John Glenn's staff during the 1970s and 1980s and was the chief architect of the Nuclear Nonproliferation Act of 1978, told a congressional panel in 2006 that he believes the Bush administration has "ignored" the violation "because it wants Musharraf's help in the war on terror."[21] As Humayun Khan himself explained to the *Los Angeles Times,* "It's all about politics. . . . You [the American government] close one eye and open the other at particular times to these things that have been going on."[22]

The Karni case also underscores concerns about the possibility of terrorists getting hold of nuclear weapons. American investigators were reportedly alarmed by Humayun Khan's involvement in the Karni deal because of his suspected ties to Islamic militant groups, including the All Jammu and Kashmir Muslim Conference (AJKMC), a Pakistani party that is believed to support combatants in the disputed region of Kashmir.[23] Though there is no evidence that Khan intended to make restricted technologies available to militants, the threat of ideologically motivated or profit-hungry individuals sharing atomic technology or material with terrorists is clear.

There are also fears that terrorists could receive technical assistance in building a nuclear or radiological device from sympathetic individuals. During the 1990s, al Qaeda launched an aggressive campaign to recruit Muslim students and scientists to assist in its quest for chemical, biological, and nuclear weapons. Incidents such as Osama bin Laden's pre-9/11 meeting with two Pakistani nuclear scientists suggests how perilously close the terrorist organization has come to succeeding in that effort. And the prospects remain high for attracting new recruits from within Pakistan's nuclear establishment, which includes many scientists with such extreme fundamentalist views that, as one Pakistani source puts it, "their beards sweep the floor."

An even more frightening scenario involves the prospect of Muslim extremists gaining control of Pakistan's nuclear arsenal. As of this writing, mounting unrest in Pakistan raises the possibility that President Musharraf's regime could be pushed aside in a military coup or overthrown by militant Islamists angry about Islamabad's cooperation with the West. As in Pakistan's nuclear establishment, there are many within the country's military and intelligence services who are sympathetic to the goals of the Taliban and al Qaeda. While military experts believe it is unlikely that unauthorized personnel could gain access to Pakistan's nuclear warheads, Pentagon planners have nonetheless war-gamed just such a situation. Their unsettling conclusion is that the United States does not have the capability to quickly "lock down" all of Pakistan's nuclear facilities in the event of a coup or a terrorist attack. Should such a situation arise, therefore, the United States would face some very stark choices, including the consideration of nuclear strikes to destroy Pakistan's arsenal rather than allowing it to fall into hostile hands.[24]

While Pakistan remains a source of ongoing concern, fears also exist that the countries where the Khan network sowed its nuclear seeds could now become shopping grounds for al Qaeda and or other like-minded groups. The US watch-and-wait policy gave both Iran and North Korea time to advance the nuclear programs they built with Khan's assistance. As early as 2004, US military commanders warned Congress that North Korea could provide nuclear weapons to al Qaeda or other terrorist organizations.[25] Pyongyang's 2006 nuclear test, while small by atomic bomb standards, only adds to concerns that the desperately poor communist state could try to sell its nuclear know-how and technology to a terrorist organization. Iran, a charter member in President Bush's "axis of evil" and a US-designated "state sponsor of terror," has long been viewed as a potential source of nuclear material for terrorists. Its development of a uranium enrichment capability, with the help of Khan's network and time provided by US policy, has significantly increased those fears.

It is also feared that Iran could provide assistance to other states with nuclear aspirations. Tehran has established an extensive network for procuring restricted goods and equipment for its nuclear and missile programs, primarily from Europe, Russia, and North Korea. Some of the Khan network's middlemen are said to be involved in the process.[26] As Iran gains proficiency in uranium enrichment, the possibility exists that it could, as Pakistan did, use its network to begin selling its technology and expertise to other countries seeking to develop nuclear weapons.

Beyond worries about the possible direct sale or transfer of know-how or equipment, Iran's nuclear program has already become a spur for the proliferation of atomic technology in the Middle East. As concerns about Iran's nuclear ambitions have grown in recent years, at least eighteen states in the region have expressed interest in establishing their own nuclear programs. Saudi Arabia has reportedly already begun experimenting in the field and has joined with five Arab Gulf states in launching an atomic energy program set to begin in 2009. Roughly a dozen countries in the region, including those that have joined with Saudi Arabia, have recently called on the UN's International Atomic Energy Agency (IAEA) for help in starting their nuclear efforts. Like Iran, all insist that they seek the technology only for peaceful purposes. But as has been well understood since the early days of the Atomic Age, presumably peaceful civilian nuclear programs can readily be diverted into efforts to develop atomic weapons. At an Arab summit meeting in March 2007, officials of twenty-one governments in and around the Middle East warned that the rush to counter a nuclear Iran could result in "a grave and destructive nuclear arms race in the region."[27] And as the ranks of nuclear states expand, the risks of technology leakages and spiraling proliferation grow.

To address mounting worries about proliferation not only in the Middle East, but also in Asia, Latin America, and beyond, some have proposed reviving a concept nearly as old as atomic weapons themselves: the creation of a nuclear-fuel "bank." As in the ill-fated Atoms for Peace program, such a bank would store enriched uranium and distribute it to member states for use in nuclear energy programs. Russia took the early initiative, first offering to enrich uranium for Iran as a means defusing tensions over Tehran's nuclear program and then proposing, in early 2006, to set up a series of nuclear-fuel centers under IAEA supervision. In September 2006, billionaire philanthropist Warren Buffett pledged $50 million to establish an IAEA-run fuel bank that would serve as a supplier of "last resort" to any nation that agreed not to produce its own fuel. The Bush administration has expressed support for similar proposals. In April 2007, the two US senators from Indiana, Republican Richard Lugar and Democrat Evan Bayh, introduced legislation calling for the creation of a repository in which nations could store their excess nuclear fuel.[28] Each of these plans seeks to prevent nations from building nuclear weapons by placing the necessary fuel under international safeguards. But they also share the fatal flaw inherent in the original Atoms for Peace program, namely that in promoting nuclear energy they would spread know-how and technology that can be used to produce fuel for atomic bombs. As an expert panel appointed by President Truman warned in 1946, any effort to halt the development of nuclear weapons based solely on a system of international safeguards and inspections is doomed to fail and provides "no prospect of security against atomic warfare."[29] Yet President Eisenhower proceeded with just such a plan. Of the nine states now believed to possess nuclear weapons, all but the original three (the United States, the Soviet Union, and Great Britain) built their bombs by piggy-backing off their ostensibly "peaceful" nuclear programs. Nevertheless, the world now seems poised to repeat this dangerous and failed experiment.

While backing new variations on the old Atoms for Peace scheme, the Bush administration has also entered into a nuclear cooperation agreement with India that reverses three decades of US nonproliferation policy and threatens to set off a new arms race in South Asia. The agreement, signed by President Bush in December 2006, creates an exception to the US Atomic Energy Act, which bars trade with nations, such as India, that are not signatories to the nuclear Non-Proliferation Treaty (NPT), the main international accord governing nuclear weapons. Under the exception, US companies will be allowed to invest in and build civilian nuclear facilities in India. They will also be allowed to sell India nuclear fuel for those facilities. In exchange, India has agreed to allow international inspections of its civilian nuclear plants. But under terms of the agreement, only fourteen of India's twenty-two reactors are designated as civilian. The remaining eight are considered military and are therefore exempt from inspection. Those military reactors can be used to produce fuel for bombs. The agreement has promoted loud complaints of unequal treatment from Pakistan, which has not been granted a similar exception. Many fear that the US–India nuclear alliance could provoke Pakistan into expanding its own nuclear arsenal, triggering a renewed regional arms race. Despite such concerns, the Bush administration has pushed forward with the agreement, driven by a desire to enlist India's support in the war in Iraq and as a hedge against Chinese ascendance in Asia. Once again, short-term foreign policy considerations have won out over nonproliferation interests.

While effectively ignoring proliferation concerns in the case of India, the Bush administration has proved schizophrenic on the subject in other instances. In dealing with Iraq, which did not have nuclear weapons, the administration chose the military option in order to prevent Saddam Hussein's regime from developing weapons of mass destruction. Yet in taking on North Korea, which does have nuclear weapons, the administration has employed a combination of hard-nosed diplomacy and lucrative rewards in an effort to persuade Pyongyang to give up its arsenal. The resulting message for Iran and other states with similar ambitions is that they can enhance their security by developing nuclear weapons.

And yet the Bush administration has persisted with this approach. While paying lip service to diplomacy, the administration has relied primarily on saber rattling in its confrontation with Tehran, warning that Iran will suffer "meaningful consequences" if it does not mend its ways.[30] That threat has been reinforced by news reports that the United States has prepared contingency plans for an assault on Iran that would include major air strikes, possibly with the use of earth-penetrating "bunker-buster" nuclear weapons to take out Tehran's underground WMD facilities.[31] Iran has responded with predictable defiance, digging in its heels and pushing ahead with its uranium enrichment program. The Bush administration has further undermined efforts to convince Iran to halt its nuclear program by promoting plans to add new weapons to America's nuclear arsenal.[32] Such a stance is both hypocritical and counterproductive, helping to convince Tehran and others that they need nuclear weapons for their own protection, and allowing them to argue that the United States reserves the right to possess such weapons for itself and its allies while denying it to others.

The difficulties the Bush administration now faces in dealing with Iran and North Korea are largely a result of US failure to confront these problems years earlier. Had the United States exposed Khan's dealings with Tehran and Pyongyang at an early stage, it might have convinced both countries to abandon their nuclear ambitions or, at a minimum, dealt a severe blow to their efforts. Yet for political reasons, the United States chose not to do so.

While allowing Pakistan's nuclear smuggling to continue, the United States supplemented its watch-and-wait strategy with attempts to sabotage the enrichment programs of Khan's clients. "The idea was to get targeted countries to spend resources on faulty designs," according to a source who participated in these efforts for US intelligence. Yet as the CIA discovered, intention and outcome can be two very different things. According to a book by *New York Times* reporter James Risen, in 2000 the CIA provided Iran with a set of top-secret bomb plans containing a hidden design flaw that intelligence officials hoped would render any weapon made from them inoperable. But the scheme backfired when a Russian defector used by the CIA to deliver the plans reportedly alerted Iran to the ruse. As a result, the Iranians may have gained valuable bomb-building information from the tainted designs.[33] While similar CIA efforts may have yielded less calamitous results, a high-level French intelligence source is dismissive of the cloak-and-dagger tactics. He argues that covert operations should not have been treated as a substitute for taking action to halt Pakistan's nuclear smuggling. "It is likely this effort was an afterthought to help justify why they allowed the Khan network to procure and then proliferate," he said.

Whatever operations the CIA mounted in hopes of stalling the nuclear programs of A.Q. Khan's customers, the fact remains that the United States allowed dangerous weapons technology and know-how to spread unnecessarily for decades. In the early 1960s, both as a presidential candidate and as president, John F. Kennedy warned repeatedly of the dangers of allowing new nations to develop nuclear weapons. In a debate with then–vice president and Republican presidential hopeful Richard Nixon in October 1960, Kennedy argued that it would be "a great mistake" not to make "every effort" to "provide for some control over these weapons."[34] Others echoed that warning. In early 1965, a blue-ribbon panel on proliferation headed by former Deputy Secretary of Defense Roswell Gilpatric delivered a report to President Johnson declaring that "preventing the further spread of nuclear weapons is clearly in the national interest despite the difficult decisions that will be required." The panel concluded that, as a "matter of great urgency," the United States should "substantially increase" its efforts to rein in atomic weapons.[35] Yet despite a clear understanding of the danger, the United States for years promoted the development of technology that could be used to build nuclear bombs. It did so as part of a Cold War effort to counter the influence of the Soviet Union in the developing world. But in placing short-term foreign policy objectives ahead of a commitment to curbing nuclear weapons, the United States opened the door to one the gravest problems the world faces today.

In the decades that followed, the United States maintained its myopic view, shoving aside concerns about the spread of the world's most dangerous weapons in pursuit of presumably greater goals. In the process, it unbridled the twin horrors of widely available nuclear technology and a militant fundamentalist movement seeking the means to destroy its enemies. Those who helped chart this reckless course remained stubbornly unwilling to acknowledge its inherent dangers. In early 1998, more than a year after Osama bin Laden issued a fatwa effectively declaring war on the United States, a French reporter asked former national security advisor Zbigniew Brzezinski, the man who persuaded President Carter to overlook Pakistan's development of nuclear weapons in order to secure Islamabad's help in backing the anti-Soviet Muslim insurgents in Afghanistan, whether he regretted having "given arms and advice to future terrorists." Brzezinski was unrepentant. "What is most important to the history of the world?" Brzezinski responded. "The Taliban or the collapse of the Soviet empire? Some stirred-up Moslems or the liberation of Central Europe and the end of the Cold War?"[36]

Less than eight months after Brzezinski uttered those words, al Qaeda–linked terrorists set off simultaneous car-bomb explosions at the US embassies in Kenya and Tanzania, killing 224 people, including twelve Americans, injuring some five thousand more, and bringing Osama bin Laden and his terrorist gang to worldwide attention as a major threat. The following year, bin Laden declared efforts to acquire weapons of mass destruction a "duty."[37] Today bin Laden and his followers remain committed to fulfilling that duty, and the opportunities for them to do so are greater than ever, thanks to the failed American policy that Brzezinski helped implement.

The sad reality is that the United States and its dubious ally, Pakistan, have helped create a post–Cold War environment far more dangerous than the one we faced in the days of the Soviet Union. As a legacy of policies initiated long ago, we now live in a world in which rogue regimes and borderless religious extremists can acquire nuclear weapons. That is the new balance of terror.

NOTES

Introduction

1. Dexter Filkins, "N. Korea Aid to Pakistan Raises Nuclear Fears," *Los Angeles Times*, August 23, 1999.
2. GlobalSecurity.org, "Pakistan's 'Nuclear Bombshell' Evokes Proliferation 'Nightmare,'" February 6, 2004, http://www.globalsecurity.org/wmd/library/news/pakistan/2004/wwwh40107.htm.
3. See, for example: David Sanger, "Confronting the Nuclear Threat America Didn't Want to be True," *New York Times*, February 8, 2004; John Lancaster and Kamran Khan, "Pakistani Scientist is Pardoned," *Washington Post*, February 6, 2004; and David Rhode and Talat Hussain, "Delicate Dance for Musharraf in Nuclear Case," *New York Times*, February 8, 2004.
4. Gollust, David. "US Pakistan React," GlobalSecurity.org, February 5, 2004, http://www.globalsecurity.org/wmd/library/news/pakistan/2004/pakistan-040205-33e7b063.htm.
5. "Iraq and Weapons of Mass Destruction," Remarks of CIA Director George Tenet at Georgetown University, February 5, 2004, available at www.cia.gov.
6. "President Bush Discusses Progress in the War on Terror," White House Press Release, July 12, 2004, available at http://www.whitehouse.gov/news/releases/2004/07/print/20040712-5.html.
7. Sadaqat Jan, "Pakistan says A.Q. Khan nuclear proliferation probe is over," Associated Press, May 2, 2006.
8. Bryan Bender, "Nuclear Threat Expected to Pose a Major Challenge," *Boston Globe*, October 2, 2004, p. A4; and "Presidential Candidate Cites Nuclear Terror Threat," *Global Security Newswire*, March 29, 2007.

Chapter 1

1. See http://www.wunderground.com for week of September 27–October 1, 2004.
2. Tim Spicer, *An Unorthodox Soldier: Peace and War and the Sandline Affair* (Edinburgh, Scotland: Mainstream Publishing, 1999), pp. 208–9.
3. BBC News, October 1, 2004.
4. See, for example: "The Final Judgement," *The Independent* (London), October 7, 2004.
5. "Transcript of the Candidates' First Debate in the Presidential Campaign," *New York Times*, October 1, 2004.
6. James Chace, "After Hiroshima: Sharing the Atom Bomb," *Foreign Affairs* 75, no. 1 (January/February 1996).
7. Ibid.
8. Harry Truman, Special Message to the Congress, October 3, 1945, available at http://www.presidency.ucsb.edu/ws/index.php?pid=12327. Regarding discussions with the Soviet Union, see Center for Global Security Research, "Atoms for Peace After 50 Years," December 2003, appendix D, p. D-2.
9. Chace, "After Hiroshima."
10. Ibid.
11. "A Report on the International Control of Atomic Energy," (The Acheson-Lilienthal Report), available at http://honors.umd.edu/HONR269J/archive/AchesonLilienthal.html.
12. Ibid.
13. Chace, "After Hiroshima."
14. Ibid.
15. The Baruch Plan, presented to the UN Atomic Energy Commission on June 14, 1946, available at http://www.atomicarchive.com/Docs/Deterrence/BaruchPlan.shtml; Leonard Weiss, "Atoms for Peace," *Bulletin of the Atomic Scientists* (November/December 2003). *Note:* Estimates of the number of the number of nuclear weapons in the US stockpile

circa 1946 vary. A 1994 analysis in the *Bulletin of the Atomic Scientists* places the number at nine. See Robert S. Norris and William M. Arkin, "Estimated US and Soviet/Russian nuclear stockpiles, 1945–1994," *Bulletin of the Atomic Scientists* (November/December 1994).

16. Chace, "After Hiroshima."
17. The Baruch Plan.
18. Weiss, "Atoms for Peace."
19. Chace, "After Hiroshima."
20. Atomic Energy Act of 1946, Pub. L., no. 79-585, Section 10, subsection B, paragraph 2A; and Weiss, "Atoms for Peace."
21. Center for Global Security Research, "Atoms for Peace After 50 Years," pp. D-5–D-6; and Weiss, "Atoms for Peace."
22. Ibid.
23. Ibid.
24. Center for Global Security Research, "Atoms for Peace After 50 Years," pp. D-5–D-6.
25. Ibid, pp. D-6–D-7.
26. Ibid, p. D-7.
27. Ibid, pp. D-7–D-8.
28. Ibid, p. D-8.
29. President Dwight Eisenhower, "Atoms For Peace," December 8, 1953, available at http://www.eisenhower.archives.gov/atoms.htm.
30. Center for Global Security Research, "Atoms for Peace After 50 Years," p. D-9; and Weiss, "Atoms for Peace."
31. Weiss, "Atoms for Peace."
32. Atomic Energy Act of 1954 (see http://www.nrc.gov/about-nrc/governing-laws.html, accessed May 16, 2007); and Weiss, "Atoms for Peace."
33. Ibid.
34. Weiss, "Atoms for Peace."
35. Ibid.
36. Ibid.
37. A.Q. Khan, "Uranium Enrichment at Kahuta, a Decade of Pakistani Experience," *Dawn*, August 1, 1986. Reprinted as "A Decade of Uranium Enrichment at Kahuta" in Zahid Malik's *Dr. A.Q. Khan and the Islamic Bomb*, (Islamabad: Hurmat Publications, 1992), chapter 10, pp. 90–99.
38. Center for Global Security Research, "Atoms for Peace After 50 Years," p. D-12

Chapter 2

1. Agreement for Cooperation Between the Government of the United States of America and the Government of Pakistan Concerning Civil Uses of Atomic Energy, *TIAS* 3315, 6 *UST*, August 11, 1955; Charles Bevans, *US Treaties and Other International Agreements* (Washington, DC: US Government Printing Office, 1955), pp. 2665–70.
2. George Perkovich, *India's Nuclear Bomb* (Berkeley: University of California Press, 1999), p. 48; and Shahid-Ur-Rehman, *Long Road to Chagai* (Islamabad: Print Wise Publication, 1999), p. 22.
3. Perkovich, *India's Nuclear Bomb*, p. 23.
4. Perkovich, *India's Nuclear Bomb*, p. 23; Dennis Kux, *The United States and Pakistan 1947–2000* (Baltimore: Johns Hopkins University Press, 2001), pp. 49–50; and Jamshed Nazar, "A History of US–Pakistan Relations," *Chowk*, December 12, 2003.
5. Perkovich, *India's Nuclear Bomb*, p. 23.
6. Dennis Kux, *The United States and Pakistan 1947-2000*, Baltimore: Johns Hopkins University Press, 2001, pp. 66-74.
7. Perkovich, *India's Nuclear Bomb*, p. 48; and Carey Sublette, "Pakistan's Nuclear Weapons Program: The Beginning," January 2, 2002, available at http://nuclearweaponarchive.org/Pakistan/PakOrigin.html.

8. Shahid-Ur-Rehman, *Long Road to Chagai*, pp. 22–23.
9. Nuclear Threat Initiative, "India Profile: Nuclear Chronology 1955–1959," http://www.nti.org/e_research/profiles/India/Nuclear/2296_2345.html.
10. Shahid-Ur-Rehman, *Long Road to Chagai*, pp. 22–23.
11. Perkovich, *India's Nuclear Bomb*, pp. 48–50.
12. Perkovich, *India's Nuclear Bomb*, p. 50; and Weiss, "Atoms for Peace," (November/December 2003).
13. Seymour Hersh, "On the Nuclear Edge," *New Yorker*, March 29, 1993.
14. Perkovich, *India's Nuclear Bomb*, p. 48
15. Memorandum from W. Park Armstrong to Fisher Howe, et al., "National Intelligence Estimate on the 'Fourth Country Problem,'" May 3, 1957, enclosing letter to Director of Central Intelligence Allen W. Dulles, April 25, 1957; available at http://www.gwu.edu/~nsarchiv/NSAEBB/NSAEBB155/, Document 1.
16. National Intelligence Estimate 100-6-57, "Nuclear Weapons Production in Fourth Countries: Likelihood and Consequences," June 18, 1957; available at http://www.gwu.edu/~nsarchiv/NSAEBB/NSAEBB155/, Document 2.
17. National Intelligence Estimate 100-2-58, "Development of Nuclear Capabilities by Fourth Countries: Likelihood and Consequences," July 1, 1958; available at http://www.gwu.edu/~nsarchiv/NSAEBB/NSAEBB155/, Document 3a.
18. Zeba Khan, "Abdul Qadeer Khan: The Man Behind the Myth," YesPakistan.com, 2001, http://www.yespakistan.com/people/abdul_qadeer.asp. Also see Zahid Malik, *Dr. A.Q. Khan and the Islamic Bomb* (Islamabad: Hurmat Publications, 1992), pp. 44–47.
19. Steve Weissman and Herbert Krosney, *The Islamic Bomb* (New York: Times Books, 1981), pp. 40–41; Pakistan People's Party, "Zulfikar Ali Bhutto," www.ppp.org.pk/zab.html.
20. "Zulfikar Ali Bhutto," www.ppp.org.pk/zab.html.
21. Zulfikar Ali Bhutto, *If I Am Assassinated* (Sahibabad, India: Bell Books, 1979), p. 117.
22. Salmaan Taseer, *Bhutto: A Political Biography* (London: Ithaca Press, 1979), p. 154.
23. Bhutto, *If I Am Assassinated*, p. 117. Also see Perkovich, *India's Nuclear Bomb*, p. 48.
24. Bhutto, *If I Am Assassinated*, p. 117.
25. Perkovich, *India's Nuclear Bomb*, pp. 65, 108.
26. Cable No. 462, US Embassy (Karachi) to Secretary of State (Rusk), November 18, 1964. Available at William Burr, "Documents on the US Atomic Energy Detection System," National Security Archive Electronic Briefing Book No. 7, National Security Archive, Washington, DC, http://www.gwu.edu/~nsarchiv/NSAEBB/NSAEBB7/nsaebb7.htm, Document 10.
27. Perkovich, *India's Nuclear Bomb*, p. 108.
28. Khushwant Singh, "Pakistan, India and The Bomb," *New York Times*, July 1, 1979. For an alternate version of this comment, see Patrick Keatley, "The Brown Bomb," *The Guardian* (Manchester), March 11, 1965, p. 10.
29. Dennis Kux, *The United States and Pakistan 1947–2000*, (Baltimore: Johns Hopkins University Press, 2001), pp. 158–159.
30. George Perkovich, "Could Anything Be Done To Stop Them?: Lessons from Pakistan," July 26, 2006, Nonproliferation Policy Education Center, http://www.npec-web.org/Essays/20060726-Perkovich-CouldAnythingBeDone.pdf. Also see Perkovich, *India's Nuclear Bomb*, p. 108.
31. Perkovich, *India's Nuclear Bomb*, pp. 108–10.
32. Weissman and Krosney, *The Islamic Bomb* (New York: Times Books, 1981), p. 50.
33. John F. Kennedy, Nixon/Kennedy debates, "The Presidential Countdown: Mr. Lodge: A Profile," broadcast on CBS Television Network and CBS Radio Network on October 11, 1960. Available at John F. Kennedy Presidential Library, Boston, Mass.
34. Perkovich, *India's Nuclear Bomb*, p. 52.
35. McGhee to Rusk, "Anticipatory Action Pending Chinese Communist Demonstration of Nuclear Capability," September 13, 1961, available at (http://www.gwu.edu/~nsarchiv/nsa/DOCUMENT/950428.htm.

36. Secretary of Defense (McNamara) to the President, memorandum, "The Diffusion of Nuclear Weapons with and without a Test Ban Agreement," February 12, 1963, available from the National Security Archive, Washington, DC, http://nsarchive.chadwyck.com/nsa/documents/NP/00941/all.pdf.
37. Nuclear Threat Initiative, "India Profile: Nuclear Chronology, 1960–1964," http://www.nti.org/e_research/profiles/India/Nuclear/2296_2346.html; and Secretary of State (Rusk) to US Embassies, airgram, circular CG-769, March 3, 1961, in Virginia Foran, "US Nuclear Non-Proliferation Policy, 1945–1991," no. 00755, available from the National Security Archive, Washington, DC, http://nsarchive.chadwyck.com/nsa/documents/NP/00755/all.pdf.
38. Cable No. 462, US Embassy (Karachi) to Secretary of State (Rusk).
39. Department of State, "Addendum, INR Contribution of NIE 4-65, Likelihood of Further Nuclear Proliferation," November 4, 1965. National Security Archive, http://www.gwu.edu/~nsarchiv/NSAEBB/NSAEBB155/, Document 11b.
40. Weissman and Krosney, *The Islamic Bomb*, p. 41.
41. Ibid., pp. 50–51. Also see Dennis Kux, *The United States and Pakistan 1947–2000*, pp. 161–62; and Malik, *Dr. A.Q. Khan and the Islamic Bomb*, p. 184.
42. "Atom Bomb Pledge By Pakistan," *The Times* (London), November 22, 1965.
43. Kux, *United States and Pakistan*, p. 212.
44. Weissman and Krosney, *The Islamic Bomb*, p. 41; Kux, *United States and Pakistan*, p. 175; and Pakistan People's Party, "Zulfikar Ali Bhutto," www.ppp.org.pk/zab.html.
45. Zulfikar Ali Bhutto, *The Myth of Independence* (London: Oxford University Press, 1969), pp. 152–55.
46. Weissman and Krosney, *The Islamic Bomb*, p. 41; and Kux, *United States and Pakistan*, p. 178.
47. Kux, *United States and Pakistan*, pp. 184–87, 204; and Weissman and Krosney, *The Islamic Bomb*, pp. 40-42.
48. Kux, *United States and Pakistan*, pp. 199–205
49. Bhutto, *If I Am Assassinated*, p. 122.
50. Kux, *United States and Pakistan*, p. 207.
51. Ibid.
52. James B. Simpson, ed. *Simpson's Contemporary Quotations* (Boston: Houghton Mifflin,1988), accessed at http://www.bartleby.com/63/19/19.html.
53. Bhutto, *If I Am Assassinated*, p. 118.
54. Weissman and Krosney, *The Islamic Bomb*, pp. 43–45.
55. Ibid., p. 45; and Shahid-Ur-Rehman, *Long Road to Chagai* (Islamabad: Print Wise Publication, 1999), p. 16.
56. Weissman and Krosney, *The Islamic Bomb*, p. 45.
57. Shahid-Ur-Rehman, *Long Road to Chagai*, p. 18.
58. Weissman and Krosney, *The Islamic Bomb*, pp. 45–46.
59. Shahid-Ur-Rehman, *Long Road to Chagai*, p. 18.
60. Weissman and Krosney, *The Islamic Bomb*, pp. 53–54. Regarding Bhutto's visit to Jordan, see "Bhutto Says He Won't Sever Tie with Soviet on Bangladesh Issue," *New York Times*, January 26, 1972. For another account of the trip, see Sultan M. Khan, *Memories & Reflections of a Pakistani Diplomat* (London: The London Centre for Pakistan Studies, 1997), pp. 414–19.
61. Weissman and Krosney, *The Islamic Bomb*, pp. 54–55.
62. Ibid., pp. 56–57.
63. Ibid., pp. 52, 55.
64. Ibid., pp. 59–61. Also see: Marie Colvin, "How an Insider Lifted the Veil on Saudi Plot for an 'Islamic Bomb,'" *Sunday Times* (London), July 24, 1994.
65. Weissman and Krosney, *The Islamic Bomb*, pp. 62–63. Although the Canadian reactor went into operation in 1972, the agreement to purchase it had been completed in 1965.

Bhutto himself, then serving as foreign minister, had negotiated the deal (Bhutto, *If I Am Assassinated*, p. 117; and Malik, *Dr. A.Q. Khan and the Islamic Bomb*, p. 125)

66. The Chronicle of Pakistan, "1974," http://pakistanspace.tripod.com/74.htm.
67. Weissman and Krosney, *The Islamic Bomb*, pp. 62–64.
68. Ibid., pp. 53, 64, 60. Regarding Libya sending cash to Pakistan, also see John K. Cooley, "Qaddafi's Great Aim for Libya Is a Nuclear Capability of Its Own," *Christian Science Monitor*, November 12, 1980.
69. Weissman and Krosney, *The Islamic Bomb*, p. 64. Faisal's successor, King Khalid (who reigned from 1975 through 1982) also reportedly took an interest in Pakistan's bomb program. Zahid Malik's less than entirely reliable biography of A.Q. Khan reports, "According to a former Secretary for Mr. Bhutto, the latter [Bhutto] had requested five hundred million dollars aid from King Khalid of Saudi Arabia during his visit to Pakistan. He was told that Saudi Arabia would provide one billion dollars, provided Mr. Bhutto submitted a complete outlay of the nuclear programme.... Due to elections [in 1977] and the subsequent agitation, Mr. Bhutto could not pursue the matter." (*Dr. A.Q. Khan and the Islamic Bomb*, p. 86.)
70. Thomas Woodrow, "The Sino-Saudi Connection," *China Brief* 2, no 21 (October 24, 2002).
71. Colvin, "How an Insider Lifted the Veil."
72. Weissman and Krosney, *The Islamic Bomb*, pp. 64–65, 162.
73. Ibid., pp. 65, 162.
74. Michael Evans and Richard Ford, "BCCI Accounts Used to Pay for Nuclear Arms," *The Times* (London), July 22, 1991; Dan Atkinson and John Willcock, "BCCI Aided Nuclear Project," *Guardian* (London), July 26, 1991; Jonathan Beaty and S.C. Gwynne, "Not Just a Bank," *Time*, September 2, 1991.
75. Weissman and Krosney, *The Islamic Bomb*, p. 65.
76. Department of State, "Pakistan and the Non-Proliferation Issue," January 22, 1975. National Security Archive, http://www.gwu.edu/~nsarchiv/NSAEBB/NSAEBB6/index.html, Document 20.
77. National Security Decision Memorandum 289, March 24, 1975, Gerald R. Ford Library, http://www.fordlibrarymuseum.gov/library/document/nsdmnssm/nsdm289a.htm.
78. Department of State, Swartz (Eur/Rpe) to Hartman (Eur), "Demarche to Pakistan on Nuclear Fuel Reprocessing," January 30, 1976. National Security Archive, http://www.gwu.edu/~nsarchiv/NSAEBB/NSAEBB6/ipn21_1.htm.
79. Department of State, memorandum of conversation, "Proposed Cable to Tehran on Pakistani Nuclear Reprocessing," May 12, 1976. National Security Archive, http://www.gwu.edu/~nsarchiv/NSAEBB/NSAEBB114/index.htm, Document 3. Pakistani officials later confirmed that Qaddafi had offered to finance the French reprocessing plant "in return for access to the plutonium it produced." The officials claimed, however, that Pakistan rejected the offer (Don Oberdorfer, "Pakistan: The Quest for Atomic Bomb," *Washington Post*, August 27, 1979).

Chapter 3

1. Steve Weissman and Herbert Krosney, *The Islamic Bomb* (New York: Times Books, 1981), p. 128.
2. George Perkovich, *India's Nuclear Bomb* (Berkeley: University of California Press, 1999), p. 179. Also see: Bernard Weinraub, "Atom Test Buoys Indians' Morale," *New York Times*, May 20, 1974, p. 65.
3. Quoted in Weissman and Krosney, *The Islamic Bomb*, p. 129.
4. Perkovich, *India's Nuclear Bomb*, p. 179.
5. Shahid-Ur-Rehman, *Long Road to Chagai* (Islamabad: Print Wise Publication, 1999), p. 43.
6. "Pakistan Expresses Concern," *New York Times*, May 19, 1974, p. 19.
7. Kathleen Teltsch, "Pakistan Disputes India on A-Arms," *New York Times*, June 8, 1974, p. 13.
8. Zahid Malik, *Dr. A.Q. Khan and the Islamic Bomb* (Islamabad: Hurmat Publications, 1992) p. 59.

9. Shahid-Ur-Rehman, *Long Road to Chagai*, pp. 43–45. Regarding Prime Minister House, see Lewis Simons "Bhutto Wants US Aid in Rebuilding Pakistan Military," *Washington Post*, July 6, 1973, p. A16.
10. Shahid-Ur-Rehman, *Long Road to Chagai*, pp. 43–45.
11. Perkovich, *India's Nuclear Bomb*, p. 28; and Federation of American Scientists, "Plutonium Production," http://www.fas.org/nuke/intro/nuke/plutonium.htm.
12. Shahid-Ur-Rehman, *Long Road to Chagai*, pp. 29–30, 45.
13. Shahid-Ur-Rehman, *Long Road to Chagai*, pp. 27–37. Regarding early negotiations with the French, see Weissman and Krosney, *The Islamic Bomb*, p. 75.
14. Shahid-Ur-Rehman, *Long Road to Chagai*, p. 30.
15. Weissman and Krosney, *The Islamic Bomb*, pp. 74–75.
16. Shahid-Ur-Rehman, *Long Road to Chagai*, pp. 30–37.
17. Ibid., pp. 38–39.
18. Ibid, p. 38; e-mail, Riazuddin to author, November 3, 2006; and Riazuddin's Pakistan Academy of Sciences biography, http://www.paspk.org/fellow8.htm.
19. Shahid-Ur-Rehman, *Long Road to Chagai*, pp. 17, 38; and Riazuddin's biography, on the Pakistan Academy of Sciences' Web site: http://www.paspk.org/fellow8.htm.
20. Shahid-Ur-Rehman, *Long Road to Chagai*, pp. 38–39; and http://www.paspk.org/fellow8.htm.
21. Shahid-Ur-Rehman, *Long Road to Chagai*, pp. 40, 45. Shahid-Ur-Rehman writes that shortly after the June 1974 meeting, Riazuddin and two members of the team charged with developing the explosive charge for the bomb traveled to the US for a prearranged visit to a high explosives laboratory at a university in the state of Washington. The lab was reportedly headed by a scientist who had played a key role in the Manhattan Project's development of the explosive used for the first atomic bombs. Upon arrival in Washington, however, Riazuddin and his colleagues were informed that the lab visit had been canceled due to Riazuddin's affiliation with the PAEC. Despite the setback, the Pakistani delegation managed to visit "a small lab on the borders [sic] of New Mexico" before leaving the United States and, on their way home, toured "a commercial lab for explosives in Germany." (*Long Road to Chagai*, p. 46.)
22. Perkovich, *India's Nuclear Bomb*, p. 186. More specifically, Pakistan announced it would build eight new reactors by 1990 and another sixteen by 2000. See Zahid Malik, *Dr. A.Q. Khan, From Kahuta to Hills of Chaghi* (Islamabad: A Pakistan Observer Publication, 1998), p. 130.
23. Department of State, "Pakistan and the Non-Proliferation Issue," January 22, 1975, National Security Archive, http://www.gwu.edu/~nsarchiv/NSAEBB/NSAEBB6/index.html, Document 20.
24. National Security Decision Memorandum 289, March 24, 1975, Gerald R. Ford Library, http://www.fordlibrarymuseum.gov/library/document/nsdmnssm/nsdm289a.htm.
25. Pakistan Academy of Sciences, http://www.paspk.org/indexa.htm; click on "Current Fellows," then on "page 6" (accessed June 20, 2007).
26. Malik, *Dr. A.Q. Khan and the Islamic Bomb*, p. 58.
27. Ibid. Elsewhere, Malik reports that after witnessing the surrender ceremony on television, Khan was "submerged in a sea of sorrow." With "tears rolling down his checks," Malik writes, Khan climbed to the roof of his apartment building, looked toward the heavens, and pledged not to rest until the loss of East Pakistan had been avenged and "the fatherland's defences were made impregnable." (*Dr. A.Q. Khan, From Kahuta to Hills of Chaghi*, p. 5.)
28. Weissman and Krosney, *The Islamic Bomb*, p. 176; Malik, *Dr. A.Q. Khan and the Islamic Bomb*, p. 53; William Langewiesche, "The Wrath of Khan," *The Atlantic Monthly*, November 2005; and Joop Boer, Henk van der Keur, Karl Koster, and Frank Slijper, "A.Q. Khan, Urenco and the Proliferation of Nuclear Weapons," May 2004 (available at http://www.stopwapenhandel.org/Ex_Summary_Khan_Urenco_report.html). Regarding Brabers, see chapter 4 in this book and "Musharraf Appoints Renowned Scientist as Rector

of GIKI," *The News* (Pakistan), April 5, 2005; "Biodata of Dr. A. Q. Khan," http://draqkhan.com.pk/dkbdata.pdf, available at http://draqkhan.com.pk/index1.htm; and Nuclear Threat Initiative, "Pakistan Profile: Missile Facilities," http://www.nti.org/e_research/profiles/Pakistan/Missile/3294_3325.html.

29. Malik, *Dr. A.Q. Khan and the Islamic Bomb*, pp. 53, 265.
30. Weissman and Krosney, *The Islamic Bomb*, p. 176.
31. Boer, et al., "A.Q. Khan, Urenco," p. 21.
32. Malik, *Dr. A.Q. Khan and the Islamic Bomb*, pp. 53–55; Weissman and Krosney, *The Islamic Bomb*, pp. 177–78.
33. Tweede Kamer der Staten-General, Zitting 1979–1980, 16 082, "Onderzoek zaak-Khan," p. 9.
34. Mark Hibbs, "Pakistan in 1976 Told Netherlands It Had Italian Centrifuge Design," *Nucleonics Week*, September 22, 2005; Douglas Frantz, "From Patriot to Proliferator," *Los Angeles Times*, September 23, 2005.
35. Weissman and Krosney, *The Islamic Bomb*, pp. 177–178.
36. Ibid., p. 177; Frantz, "From Patriot to Proliferator."
37. Weissman and Krosney, *The Islamic Bomb*, p. 178.
38. Ibid.
39. Ibid.; and Malik, *Dr. A.Q. Khan and the Islamic Bomb*, pp. 55–56.
40. Malik, *Dr. A.Q. Khan and the Islamic Bomb*, p. 58
41. Ibid., p. 6.
42. Ibid., p. 59.
43. Shahid-Ur-Rehman, *Long Road to Chagai*, p. 47; Malik, *Dr. A.Q. Khan and the Islamic Bomb*, p. 59. There is some discrepancy about the date of Khan's letter. Malik indicates Khan wrote it "in the middle of 1974." Shaid-Ur-Rehman reports that Khan wrote it on September 17, 1974.
44. Weissman and Krosney, *The Islamic Bomb*, p. 166.
45. Shahid-Ur-Rehman, *Long Road to Chagai*, pp. 2–3, 47–48.
46. Ibid., pp. 48, 50.
47. Kamran Khan and Molly Moore, "2 Nuclear Experts Briefed Bin Laden, Pakistanis Say," *Washington Post*, December 12, 2001; Pete Baker, "Pakistani Scientist Who Met Bin Laden Failed Polygraphs," *Washington Post*, March 3, 2002; Daniel Pearl and Steve LeVine, "Pakistan Has Ties to Groups It Vowed to Curb," *Wall Street Journal*, December 24, 2001; and Ron Suskind, *The One Percent Doctrine* (New York: Simon & Schuster, 2006), p. 122.
48. M.A. Chaudhri (alias), "Pakistan's Nuclear History: Separating Myth From Reality," *Defence Journal*, May 2006 (www.defencejournal.com).
49. Tweede Kamer der Staten-General, Zitting 1979–1980, 16 082, "Onderzoek zaak-Khan," p. 10; and Weissman and Krosney, *The Islamic Bomb*, p. 179. As the authors of *The Islamic Bomb* note, the technology in question, known as the vertical centrifuge method, had an unsettling history. "It went back to World War II, when Nazi scientists were rushing to forge the ultimate weapon for Hitler. The Russians later perfected the method with the help of some of the same Nazi scientists, and now the secret had returned to Germany, where scientists had developed it experimentally. Urenco hoped to incorporate the new method, and the report was part of the effort" (p. 179). Also see Jack Boureston, "Tracking the Technology," *Nuclear Engineering International*, August 31, 2004.
50. Mark Hibbs, "Pakistan in 1976 Told Netherlands It Had Italian Centrifuge Design," *Nucleonics Week*, September 22, 2005.
51. Tweede Kamer der Staten-General, Zitting 1979–1980, 16 082, "Onderzoek zaak-Khan," p. 19.
52. Weissman and Krosney, *The Islamic Bomb*, pp. 179–80. There is some discrepancy as to how long Khan's assignment at Urenco lasted. Weissman and Krosney report that he was posted there for three or four weeks. Malik writes that Khan was there four months. (*Dr. A.Q. Khan and the Islamic Bomb*, p. 56.) The Dutch government report is not clear on this point, saying only that Khan spent sixteen "work days" at the brainbox. (Tweede Kamer der Staten-General, Zitting 1979–1980, 16 082, "Onderzoek zaak-Khan," p. 10.)

53. Weissman and Krosney, *The Islamic Bomb*, p. 178.
54. Malik, *Dr. A.Q. Khan and the Islamic Bomb*, p. 60; Shahid-Ur-Rehman, *Long Road to Chagai*, p. 50. Also see Frantz, "From Patriot to Proliferator."
55. Shahid-Ur-Rehman, *Long Road to Chagai*, pp. 4, 50.
56. Malik, *Dr. A.Q. Khan and the Islamic Bomb*, p. 61.
57. Shahid-Ur-Rehman, *Long Road to Chagai*, p. 50.
58. "Ford Sees Bhutto and Hints US May Ease Pakistan Arms Curb," *New York Times*, February 6, 1975; and National Security Decision Memorandum 289, March 24, 1975.
59. Shahid-Ur-Rehman, *Long Road to Chagai*, pp. 50, 52; and Malik, *Dr. A.Q. Khan and the Islamic Bomb*, p. 69.
60. Shahid-Ur-Rehman, *Long Road to Chagai*, p. 52. Note that Shahid-Ur-Rehman quotes Khan referring to the uranium enrichment program as Project 726 (p. 54). But elsewhere it is widely referred to as Project 706. (See, for example, Weissman and Krosney, *The Islamic Bomb*, p.182.)
61. Shahid-Ur-Rehman, *Long Road to Chagai*, pp. 51, 1–2.
62. Gordon Corera, *Shopping for Bombs* (New York: Oxford University Press, 2006), p. 15.
63. William Langewiesche, "The Wrath of Khan," *Atlantic Monthly*, November 2005.
64. Shyam Bhatia, "Ex-Colleague Spills Beans on AQ Khan," www.rediff.com, January 29, 2004; and Frantz, "From Patriot to Proliferator." Also see David Rohde and Talat Hussain, "Delicate Dance for Musharraf in Nuclear Case," *New York Times*, February 8, 2004.
65. Weissman and Krosney, *The Islamic Bomb*, p. 180.
66. Langewiesche, "The Wrath of Khan."
67. Ibid.
68. Hibbs, "Pakistan in 1976 Told Netherlands It Had Italian Centrifuge Design."
69. William Broad and David Sanger, "As Nuclear Secrets Emerge in Khan Inquiry, More Are Suspected," *New York Times*, December 26, 2004; Frantz, "From Patriot to Proliferator."
70. BBC, "CIA 'let Atomic Expert Khan go,'" August 9, 2005, http://news.bbc.co.uk/1/hi/world/europe/4135998.stm. A former Congressional official with access to intelligence information about the case reportedly confirms Lubbers' account. See Broad and Sanger, "As Nuclear Secrets Emerge in Khan Inquiry."
71. "CIA Asked Us to Let Nuclear Spy Go, Ruud Lubbers Claims," *Expatica*, August 9, 2005.
72. "US Protected Pakistan Nuke Scientist — Dutch ex-PM," Reuters, August 9, 2005.
73. Boer, et al., "A.Q. Khan, Urenco," p.10; and Weissman and Krosney, *The Islamic Bomb*, p. 180.
74. Shahid-Ur-Rehman, *Long Road to Chagai*, p. 51.
75. Weissman and Krosney, *The Islamic Bomb*, p. 180; and Malik, *Dr. A.Q. Khan and the Islamic Bomb*, p. 56.
76. Mark Hibbs, "Using Catch-All Rule, the Hague Blocked 20 Exports Since 1996," *Nuclear Fuel*, March 15, 2004.

Chapter 4

1. Zahid Malik, *Dr. A.Q. Khan and the Islamic Bomb* (Islamabad: Hurmat Publications, 1992), p. 61.
2. Ibid. Malik reports that Bhutto asked Khan to "set up the uranium enrichment project," but this is clearly implausible, given that the program existed before Khan came on board.
3. Shahid-Ur-Rehman, *Long Road to Chagai* (Islamabad: Print Wise Publication, 1999), p. 53.
4. Ibid., p. 54.
5. Malik, *Dr. A.Q. Khan and the Islamic Bomb*, pp. 61–62; Shahid-Ur-Rehman, *Long Road to Chagai*, p. 51.
6. Ibid.
7. Ibid., pp. 51–52.
8. Malik, *Dr. A.Q. Khan and the Islamic Bomb*, p. 63.
9. Shahid-Ur-Rehman, *Long Road to Chagai*, p. 52.
10. M.A. Chaudhri (alias), "Pakistan's Nuclear History: Separating Myth from Reality," *Defence Journal*, May 2006 (available at www.defencejournal.com).

11. Shahid-Ur-Rehman, *Long Road to Chagai*, pp. 52–53.
12. Ibid., p. 53; and Malik, *Dr. A.Q. Khan and the Islamic Bomb*, p. 63.
13. Kamran Khan, "Business in Timbuktu," *The News International*, February 1, 2004.
14. Shahid-Ur-Rehman, *Long Road to Chagai*, p. 97; Malik, *Dr. A.Q. Khan and the Islamic Bomb*, p. 97; Steve Weissman and Herbert Krosney, *The Islamic Bomb* (New York: Times Books, 1981), p. 181.
15. Malik, *Dr. A.Q. Khan and the Islamic Bomb*, pp. 67–68.
16. Ibid., p. 70.
17. Don Oberdorfer, "Pakistan: The Quest for Atomic Bomb," *Washington Post*, August 27, 1979, p. A1; and Khalid Mahmud Arif, *Working with Zia: Pakistan's Power Politics 1977–1988* (New York: Oxford University Press, 1995), p. 366.
18. Weissman and Krosney, *The Islamic Bomb*, p. 162.
19. Dennis Kux, *The United States and Pakistan 1947–2000* (Baltimore: Johns Hopkins University Press, 2001), p. 221; and Weissman and Krosney, *The Islamic Bomb*, p. 149.
20. Weissman and Krosney, *The Islamic Bomb*, pp. 150, 145.
21. Ibid., p. 150.
22. Ibid.
23. State Department, Swartz (Eur/Rpe) to Hartman (Eur), "Demarche to Pakistan on Nuclear Fuel Reprocessing," January 30, 1976. Available at the National Security Archive, http://www.gwu.edu/~nsarchiv/NSAEBB/NSAEBB6/ipn21_1.htm; State Department, Memorandum of Conversation, "Proposed Cable to Tehran on Pakistani Nuclear Reprocessing," May 12, 1976. Available at the National Security Archive, http://www.gwu.edu/~nsarchiv/NSAEBB/NSAEBB114/index.htm, Document 3. Pakistani officials later confirmed that Qaddafi had offered to finance the French reprocessing plant "in return for access to the plutonium it produced." The officials claimed, however, that Pakistan rejected the offer. (Oberdorfer, "Pakistan: The Quest for Atomic Bomb.")
24. Jimmy Carter, *Why Not the Best?* (New York: Bantam Books, 1975), p. 178.
25. "On Nuclear Proliferation," speech by Jimmy Carter in San Diego, CA, September 25, 1976, in *The Presidential Campaign 1976, Vol. 1, Part II, Jimmy Carter* (Washington, DC: US Government Printing Office, 1978), p. 815.
26. Jimmy Carter, "Nuclear Non-Proliferation Message to the Congress," April 27, 1977, *Public Papers of the Presidents: Jimmy Carter, 1977* (Washington, DC: Government Printing Office), pp. 728–32.
27. Don Oberdorfer, "Arms Aid Bills Gain on Hill," *Washington Post*, May 12, 1976; Weissman and Krosney, *The Islamic Bomb*, p. 164.
28. Zulfikar Ali Bhutto, *If I Am Assassinated* (Sahibabad, India: Bell Books, 1979), p. 118; Weissman and Krosney, *The Islamic Bomb*, p. 163; Kux, *The United States and Pakistan*, pp. 222–23; Shahid-Ur-Rehman, *Long Road to Chagai*, prologue and p. 33. According to *Long Road to Chagai*, Kissinger later denied threatening Bhutto but acknowledged warning the prime minister that "down the road" the reprocessing plant could "create problems for Pakistan"
29. Weissman and Krosney, *The Islamic Bomb*, p. 163.
30. Ibid., p. 164.
31. "Pakistan Sticks to French Nuclear Deal," *Washington Post*, January 4, 1977; Milton Benjamin, "Pakistan Says France Killing Controversial Nuclear Deal," *Washington Post*, August 24, 1978. Also, Weissman and Krosney, *The Islamic Bomb*, p. 166.
32. Don Oberdorfer, "Pakistan: The Quest for Atomic Bomb."
33. A.Q. Khan, "Uranium Enrichment at Kahuta, a Decade of Pakistani Experience," *Dawn*, August 1, 1986, reprinted as "A Decade of Uranium Enrichment at Kahuta" in Zahid Malik, *Dr. A.Q. Khan and the Islamic Bomb*, pp. 90–99.
34. A.Q. Khan, "Uranium Enrichment at Kahuta."
35. Oberdorfer, "Pakistan: The Quest for Atomic Bomb."
36. "Pakistan Achieved Nuclear Capability in 1984: Dr. Abdul Qadeer," Associated Press of Pakistan, February 12, 1999.

37. Malik, *Dr. A.Q. Khan and the Islamic Bomb*, p. 83.
38. Ibid., p. 65.
39. Quoted in Weissman and Krosney, *The Islamic Bomb*, p. 182.
40. Ibid.; and Joop Boer, Henk van der Keur, Karl Koster, and Frank Slijper, "A.Q. Khan, Urenco and the Proliferation of Nuclear Weapons," May 2004, p.15 (available at http://www.stopwapenhandel.org/Ex_Summary_Khan_Urenco_report.html). Also see Frits Verrman and Jaques Ros, "Introduction," http://www.atoomspionage.com/introductie.htm.
41. Boer, et al., "A.Q. Khan, Urenco."; Federation of Atomic Scientists, "Government Entities Determined to be Involved in Nuclear or Missile Activities," http://www.fas.org/nuke/control/export/news/Ind-Pak2.htm; Federation of Atomic Scientists; "Ghulam Ishaq Khan Institute of Engineering Sciences and Technology," at http://www.fas.org/nuke/guide/pakistan/agency/giki.htm; "Musharraf Appoints Renowned Scientist as Rector of GIKI," *The News* (Pakistan), April 5, 2005; "Biodata of Dr. A. Q. Khan," http://draqkhan.com.pk/index1.htm; and Nuclear Threat Initiative, "Pakistan Profile: Missile Facilities," http://www.nti.org/e_research/profiles/Pakistan/Missile/3294_3325.html. GIKI also appears on a secret British list of companies of "proliferation concern" due to having "procured goods and/or technology for weapons of mass destruction programs." (UK Security Service [MI5] "Companies and Organizations of Proliferation Concern," April 2003.)
42. Shahid-Ur-Rehman, *Long Road to Chagai*, pp. 57–58.
43. Shyam Bhatia, "Yours Sincerely, Dr A Q Khan," February 11, 2004, http://www.rediff.com/news/2004/feb/11spec1.htm; and David Albright and Mark Hibbs, "Pakistan's Bomb: Out of the Closet," *Bulletin of the Atomic Scientists*, July-August 1992, available at http://www.thebulletin.org/article.php?art_ofn=ja92albright.
44. Shahid-Ur-Rehman, *Long Road to Chagai* (Islamabad: Print Wise Publication, 1999), p. 58.
45. Langewiesche, "The Wrath of Khan," *Atlantic Monthly*, November 2005; and Albright and Hibbs, "Pakistan's Bomb." Khan's defense attorney had his expenses paid for indirectly by BCCI (Dan Atkinson and John Wilcock, "BCCI Aided Nuclear Project," *The Guardian* (London), July 26, 1991).
46. "US Protected Pakistan Nuke Scientist — Dutch ex-PM," Reuters, August 9, 2005; "CIA asked the Netherlands 'Not to Arrest' Khan," Aljazeera.com, August 9, 2005.
47. "Former Netherlands Prime Minister Claims CIA Prevented Atom Spy Khan's Arrest," *NRC Handelsblad*, August 9, 2005.
48. Mark Hibbs, "Dutch Dossiers Show Ministries Dismissed Khan Role in Pakistan," *Nuclear Fuel*, August 1, 2005. Also see Weissman and Krosney, *The Islamic Bomb*, p. 184; and Boer, et al., "A.Q. Khan, Urenco." p. 11.
49. Tweede Kamer der Staten-General, Zitting 1979–1980, 16 082, "Onderzoek zaak-Khan"; and Albright and Hibbs, "Pakistan's Bomb."
50. Albright and Hibbs, "Pakistan's Bomb."
51. John J. Fialka, "Nuclear Club: Set to Explode?" *Wall Street Journal*, November 26,1984. Also see David Willis, "On the Trail of the A-Bomb Makers," *Christian Science Monitor*, December 1, 1981.
52. Shahid-Ur-Rehman, *Long Road to Chagai*, p. 56–57.
53. Ibid., p. 60; Weissman and Krosney, *The Islamic Bomb*, pp. 182–83; and Frederick Lamy, "Export Controls Violations and Illicit Trafficking by Swiss Companies and Individuals in the Case of the A.Q. Khan Network," paper for the Geneva Centre for Security Policy, August 19, 2004, available at www.pircenter.org.
54. Polis Diraja Malaysia, "Press Release by Inspector General of Police in Relation to Investigation on the Alleged Production of Components for Libya's Uranium Enrichment Programme," February 20, 2004; Lamy, "Export Controls Violations"; Craig Smith, "Roots of Pakistan Atomic Scandal Traced to Europe," *New York Times*, February 19, 2004; Ellen Nakashima and Alan Sipress, "Insider Tells of Nuclear Deals," *Washington Post*, February 21, 2004.

55. Weissman and Krosney, *The Islamic Bomb*, p. 183.
56. Lamy, "Export Controls Violations."
57. Weissman and Krosney, *The Islamic Bomb*, p. 183.
58. Ibid., p. 185. Regarding Leybold being a Urenco subcontractor, see Mark Hibbs, "European Suspects in KRL Case May Have Previous Involvement," *Nucleonics Week*, February 5, 2004.
59. Shahid-Ur-Rehman, *Long Road to Chagai*, pp. 61–62.
60. Lamy, "Export Controls Violations."
61. Juergen Dahlkamp et al., "Network of Death on Trial," *Der Spiegel*, March 13, 2006; Steve Coll, "The Atomic Emporium," *New Yorker*, August 7 and 14, 2006; James Markham, "Bonn Checks Report of Smuggling of Atomic Technology to Pakistan," *New York Times*, May 5, 1987; and Lamy, "Export Controls Violations."
62. Weissman and Krosney, *The Islamic Bomb*, p. 185.
63. Ibid., pp. 185–86; Gary Milhollin, "Asia's Nuclear Nightmare: The German Connection," *Washington Post*, June 10, 1990; and E.A. Wayne, "Secret Travels of Bombmaking Materials," *Christian Science Monitor*, July 8, 1988. Regarding Western companies (and the IAEA) assisting Pakistan's efforts to obtain uranium gas, see Weissman and Krosney, *The Islamic Bomb*, pp. 219–21; and Shahid-Ur-Rehman, *Long Road to Chagai*, pp. 67–71.
64. Boer, et al., "A.Q. Khan, Urenco," p. 10.
65. Weissman and Krosney, *The Islamic Bomb*, pp. 184–85. Regarding S.A. Butt placing the original order, see Boer, et al., "A.Q. Khan, Urenco," p. 10.
66. Boer, et al., "A.Q. Khan, Urenco," p. 11.
67. Leonard Spector, *The New Nuclear Nations* (New York: Vintage Books, 1985), p. 24; and Boer, et al., "A.Q. Khan, Urenco," p. 11. According to an independent Dutch report, the FDO sales manager told a Dutch journalist in 2004 that he had also visited Khan in 1976 "in cooperation with the Dutch security service BVD, to make clear that he 'should not bother us [FDO] anymore'" by asking former colleagues for assistance with sensitive information. (Boer, et al., "A.Q. Khan, Urenco," p. 11.)
68. A.Q. Khan, "Uranium Enrichment at Kahuta," in Malik, *Dr. A.Q. Khan and the Islamic Bomb*, p. 94.
69. William Broad and David Sanger, "As Nuclear Secrets Emerge in Khan Inquiry, More Are Suspected," *New York Times*, December 26, 2004.
70. Seymour Hersh, "On the Nuclear Edge," *New Yorker*, March 29, 2003.
71. Don Oberdorfer, "Pakistan: The Quest for Atomic Bomb."

Chapter 5

1. "How Pakistan Fooled the World and Got the Bomb," *8 Days* (London), June 23, 1979; Steve Weissman and Herbert Krosney, *The Islamic Bomb* (New York: Times Books, 1981), p. 185; Shahid-Ur-Rehman, *Long Road to Chagai* (Islamabad: Print Wise Publication, 1999), p. 63; and "Wanted. . . . Bomb Business: Nuclear Aid for Pakistan and India," transcript of a film by Egmont Koch, West German Broadcasting (WDR), 1986.
2. "How Pakistan Fooled the World and Got the Bomb"; Weissman and Krosney, *The Islamic Bomb*, pp. 182, 185; Zahid Malik, *Dr. A.Q. Khan and the Islamic Bomb* (Islamabad: Hurmat Publications, 1992), p. 62.
3. "How Pakistan Fooled the World and Got the Bomb"; Weissman and Krosney, *The Islamic Bomb*, p. 185; "Wanted. . . . Bomb Business,"; and Don Oberdorfer, "Pakistan: The Quest for Atomic Bomb," *Washington Post*, August 27, 1979, p. A1.
4. "How Pakistan Fooled the World and Got the Bomb." Also see Oberdorfer, "Pakistan: The Quest for Atomic Bomb." Pakistan's first attempt to obtain inverters came while A.Q. Khan was still at FDO. According to Weissman and Krosney, in August 1975, the Pakistani embassy in Brussels, where S.A. Butt was posted, made inquiries with a Dutch firm about the devices. (Weissman and Krosney, *The Islamic Bomb*, pp. 181, 186.) Documents released by the Dutch government in 2005 show that the Netherlands learned in the late 1970s that

a Dutch firm had provided Pakistan's nuclear program with an inverter in 1976. (Mark Hibbs, "Dutch Dossiers Show Ministries Dismissed Khan Role in Pakistan," *Nuclear Fuel*, August 1, 2005.)

5. "How Pakistan Fooled the World and Got the Bomb"; and Press Trust of India, "German Jailed for N-arms Equipment Sale to Pak," *The Tribune* (India), July 22, 1998. Also see Weissman and Krosney, *The Islamic Bomb*, pp. 186–87. In 1998, a German court convicted Piffl of illegally exporting centrifuge parts to Pakistan between 1988 and 1993. He was sentenced to three years and nine months imprisonment and ordered to pay a $240,000 fine. Despite Piffl's public revelation at the end of the trial that he was cooperating with German intelligence, the Khan network attempted to purchase parts from Team Industries in both 2000 and 2001. (Mark Hibbs, "The Unmaking of a Nuclear Smuggler," *Bulletin of the Atomic Scientists*, November/December 2006; and Press Trust of India, "German Jailed for N-arms Equipment Sale to Pak." Also see Mark Hibbs, "Report Suggests Pakistan Bought Components for Two Steel Centrifuges," *Nuclear Fuel*, July 4, 2005.)
6. Weissman and Krosney, *The Islamic Bomb*, p. 187.
7. "How Pakistan Fooled the World and Got the Bomb."
8. Weargate corporate records, Companies House, London; Weissman and Krosney, *The Islamic Bomb*, p. 186; author's telephone interview with Scimitar director Robert Phillips, May 11, 2007; and Simon Henderson, "Islamabad Continues Its Quest for Nuclear Might," *Financial Times*, December 22, 1980.
9. Weissman and Krosney, *The Islamic Bomb*, p. 186. Regarding Salam identifiying himself as an engineer, see Weargate's corporate records, Companies House, London.
10. "How Pakistan Fooled the World and Got the Bomb."
11. Ibid.
12. Oberdorfer, "Pakistan: The Quest for Atomic Bomb."
13. "How Pakistan Fooled the World and Got the Bomb." It is not clear how Allaun learned of the Pakistani inverter deal and he refused to say. A.Q. Khan later suggested that Team Industries, angry at not getting the second order, had informed Allaun. See Leonard Spector, *The New Nuclear Nations* (New York: Vintage Books, 1985), p. 257; and Shahid-Ur-Rehman, *Long Road to Chagai*, pp. 62–63.
14. Weissman and Krosney, *The Islamic Bomb*, p. 187.
15. Ibid.; and Oberdorfer, "Pakistan: The Quest for Atomic Bomb."
16. "How Pakistan Fooled the World and Got the Bomb."
17. Weissman and Krosney, *The Islamic Bomb*; and Oberdorfer, "Pakistan: The Quest for Atomic Bomb." A.Q. Khan mentions the requested modifications in the article "Uranium Enrichment at Kahuta, a Decade of Pakistani Experience," *Dawn*, August 1, 1986, reprinted as "A Decade of Uranium Enrichment at Kahuta" in Malik, *Dr. A.Q. Khan and the Islamic Bomb*, pp. 95–96; and Kuldip Nayar, "Has Pakistan Really Joined the Nuclear Club?" *Globe and Mail*, March 2, 1987, reprinted from the *Observer* (London) March 1, 1987.
18. Weissman and Krosney, *The Islamic Bomb*, p. 187–88; and Oberdorfer, "Pakistan: The Quest for Atomic Bomb." Regarding Pakistan reverse engineering the inverters, also see Shahid-Ur-Rehman, *Long Road to Chagai*, p. 63; and John J. Fialka, "Nuclear Club: Set to Explode?" *Wall Street Journal*, November 26, 1984.
19. Milton Benjamin, "Pakistan Says France Killing Controversial Nuclear Deal," *Washington Post*, August 24, 1978; Milton Benjamin, "US Officials View Pakistan as the Leading Threat to Join the Nuclear Club," *Washington Post*, December 8, 1978; Oberdorfer, "Pakistan: The Quest for Atomic Bomb."; and Weissman and Krosney, *The Islamic Bomb*, p. 167.
20. Don Oberdorfer, "US to Renew Aid to Pakistan," *Washington Post*, August 25, 1978, p. A23.
21. Oberdorfer, "Pakistan: The Quest for Atomic Bomb."
22. "US to Renew Aid," *Facts on File World News Digest*, November 3, 1978.
23. Don Oberdorfer, "Arms Sales to Pakistan Urged to Stave Off A-Bomb There," *Washington Post*, August 6, 1979, p. A7.
24. Oberdorfer, "US to Renew Aid."

25. Oberdorfer, "Pakistan: The Quest for Atomic Bomb."
26. Hibbs, "Dutch Dossiers Show Ministries Dismissed Khan Role in Pakistan."
27. Don Oberdorfer, "US Cutting Aid to Pakistan in A-Facility Dispute," *Washington Post*, April 7, 1979, pA1.
28. Weissman and Krosney, *The Islamic Bomb*, pp. 168, 192. Also see Salmaan Taseer, *Bhutto: A Political Biography* (London: Ithaca Press, 1979), p. 154; and Jamshed Nazar, "A History of US–Pakistan Relations," http://www.chowk.com, December 12, 2003.
29. Zulfikar Ali Bhutto, *If I Am Assassinated* (Sahibabad, India: Bell Books, 1979), p. 118
30. Weissman and Krosney, *The Islamic Bomb*, p. 161.
31. Dennis Kux, *The United States and Pakistan 1947–2000* (Baltimore: Johns Hopkins University Press, 2001), p. 239; Oberdorfer, "US Cutting Aid to Pakistan in A-Facility Dispute."
32. Oberdorfer, "Pakistan: The Quest for Atomic Bomb."
33. "US Aid to Pakistan Is Ended over A-plant," *Globe and Mail* (Canada), Reuters, April 7, 1979; Oberdorfer, "US Cutting Aid to Pakistan in A-Facility Dispute"; and Richard Burt, "US Aid to Pakistan Cut After Evidence of Atom Arms Plan," *New York Times*, April 7, 1979, p. 1.
34. Oberdorfer, "Pakistan: The Quest for Atomic Bomb."
35. Weissman and Krosney, *The Islamic Bomb*, p. 190; and Burt, "US Aid to Pakistan Cut After Evidence of Atom Arms Plan."
36. Burt, "US Aid to Pakistan Cut After Evidence of Atom Arms Plan."
37. Stuart Auerbach, "Panel Told Pakistan Gained A-Weapons Ability by 'End Runs,'" *Washington Post*, May 2, 1979. A.Q. Khan was told of Pickering's testimony in a letter from an associate in Canada. See Spector, *The New Nuclear Nations*, p. 257.
38. Shahid-Ur-Rehman, *Long Road to Chagai*, pp. 58–59.
39. Kux, *The United States and Pakistan 1947–2000*, p. 240; Weissman and Krosney, *The Islamic Bomb*, p. 193; and Simon Henderson, "We Can Do It Ourselves," *Bulletin of the Atomic Scientists*, September 1993. Henderson also reports that two weeks before the incident with the French ambassador, an American State Department official, Robert Gallucci, had defiantly driven by the prohibited Kahuta facility. When Zia heard that the French ambassador had been beaten, he reportedly remarked, "I wish it had been the American bastard."
40. Weissman and Krosney, *The Islamic Bomb*, p. 193; and Henderson, "We Can Do It Ourselves." Henderson notes that at the time of the assault, Khan was reportedly in Britain "visiting his network of suppliers."
41. Richard Burt, "US Will Press Pakistan to Halt A-Arms Plant," *New York Times*, August 12, 1979, p. 1.
42. Weissman and Krosney, *The Islamic Bomb*, p. 193; and Kux, *The United States and Pakistan 1947–2000*, p. 240.
43. Simon Henderson, "Delhi Attack on Atomic Facility Feared," *Financial Times* (London), August 14, 1979; and Nuclear Threat Initiative, "Pakistan Profiles: Nuclear Chronology," entry for 14 August 1979, http://www.nti.org/e_research/profiles/Pakistan/Nuclear/5593_5678.html.
44. Kux, *The United States and Pakistan 1947–2000*, p. 241.
45. Steve Coll, *Ghost Wars* (New York: Penguin Press, 2004), p. 44.
46. Ibid., pp. 42–44.
47. Robert Gates, *From the Shadows* (New York, Simon and Schuster, 1996), p. 144. Gates also notes that, "a senior [Saudi] official also had raised the prospect of a Soviet setback in Afghanistan and said that his government was considering officially proposing that the United States aid the rebels."
48. Coll, *Ghost Wars*, pp. 45–46.
49. Jimmy Carter, *Keeping Faith* (New York: Bantam Books, 1982), p. 143.
50. John Cooley, *UnHoly Wars* (Sterling, VA: Pluto Press, 2000), pp. 23–24.
51. Gates, *From the Shadows*, p. 146. Gates indicates the initial allotment of funds had all been drawn within the first six weeks of the operation (p. 146).

52. Coll, *Ghost Wars*, p. 46.
53. John Cooley, *UnHoly Wars*, pp. 19–20. The Carter Library was unable to locate the memo Brzezinski described (e-mail, Keith Shuler to author, July 6, 2007).
54. Ibid., p. 16.
55. Gates, *From the Shadows*, p. 146.
56. "Pakistan President Reaffirms Peaceful Nuclear Programme," Xinhua General Overseas News Service, July 29, 1979.
57. State Department Bureau of Intelligence and Research (BIS) report (Secret), "Pakistan's Short-Term Prospects," August 24, 1979, in *Documents from the US Espionage Den* vol. 46-2, p. 54. Also available in the collection: *US Nuclear Non-Proliferation Policy, 1945–1991*, Virginia I. Foran, Project Director (Alexandria, VA: The National Security Archive and Chadwyck-Healey Inc., 1991).
58. Kux, *The United States and Pakistan 1947–2000*, p. 240.
59. "Pakistan Said to Stop Work at Uranium Enriching Plant," World News Briefs, *New York Times*, Reuters, October 12, 1979, p. A4.
60. Kux, *The United States and Pakistan 1947–2000*, pp. 240–41. After meeting with Vance, Shahi reportedly told members of Congress that Zia would "be willing to give the United States a 'no-explosion' pledge during the life of this current government." (Don Oberdorfer, "Effort to Block Pakistan From A-Bomb Faltering," *Washington Post*, October 20, 1979, p. A3.) Regarding Zia establishing a "genuine Islamic order," see Coll, *Ghost Wars*, p. 27.
61. Stuart Auerbach, "Pakistan Holds A-Option Open," *Washington Post*, October 27, 1979, p. A17.
62. Ibid.
63. Lee Daniels, "Medical Tests in Manhattan," *New York Times*, October 24, 1979, p. A1.
64. Peter Bergen, *Holy War Inc.* (New York: Touchstone, 2002), p. 51.
65. Coll, *Ghost Wars*, pp. 24–29; and Kux, *The United States and Pakistan 1947–2000*, pp. 242–43.
66. Kux, *The United States and Pakistan 1947–2000*, pp. 242–43.
67. Ibid., pp. 243–45.
68. Coll, *Ghost Wars*, p. 36.
69. Kux, *The United States and Pakistan 1947–2000*, pp. 244-245. Regarding Zia's decsion to let the riot run its course, see Coll, *Ghost Wars*, pp. 31–32.
70. Kux, *The United States and Pakistan 1947–2000*, p. 245.
71. Lohbeck, *Holy War, Unholy Victory* (Washington, DC: Regnery Gateway, 1993), p. 43.
72. Memoranda from Zbigniew Brzezinski to the President, December 26, 1979, "Reflections on Soviet Intervention of Afghanistan," published by the Cold War International History Project (CWIHP), in Vol. 1 of the Document Reader prepared for the international conference "Toward an International History of the War in Afghanistan, 1979–1989" (April 2002), edited by Christian Ostermann and Mircea Munteanu.
73. Ibid.
74. Kux, *The United States and Pakistan 1947–2000*, p. 247; and Terrance Smith, "Warns on Olympics," *New York Times*, January 5, 1980, p. 1.
75. George Crile, *Charlie Wilson's War* (New York: Atlantic Monthly Press, 2003), p. 14.
76. Kux, *The United States and Pakistan 1947–2000*, p. 247; and Smith, "Warns on Olympics."
77. Cooley, *UnHoly Wars*, p. 30.
78. Kux, *The United States and Pakistan 1947–2000*, p. 252.
79. Ibid., p. 247.
80. Smith, "Warns on Olympics."
81. Kux, *The United States and Pakistan 1947–2000*, pp. 248–49.
82. Ibid., p. 249.
83. Don Oberdorfer, "Pakistan Offered $400 Million Aid," *Washington Post*, January 15, 1980, p. A1.
84. Kux, *The United States and Pakistan 1947–2000*, p. 249.

85. Ibid., p. 250.
86. Ibid.
87. Richard Burt, "Pakistani Nuclear Effort May Imperil US Aid Plan," *New York Times*, February 28, 1980.
88. James Dorsey, "Afghanistan Crisis Yields Critical Policy," *Christian Science Monitor*, February 4, 1980.
89. Kux, *The United States and Pakistan 1947–2000*, p. 251.
90. Robert Lindsey, "Reagan Says America Should Not Bar Others from A-Bomb Output," *New York Times*, February 1, 1980.
91. Burt, "Pakistani Nuclear Effort May Imperil US Aid Plan."
92. John Cooley, "US-India Nuclear Transaction Watched," *Christian Science Monitor*, June 11, 1980. In fact, the administration had been trying to win international acceptance for what a report on the White House's nonproliferation policy described as "the position that access to weapons-grade materials is appropriate for industrialized countries but not for developing ones" (Richard Burt, "US Said to Ease Ban on Export of Atom Materials," *New York Times*, October 25, 1979).
93. Kux, *The United States and Pakistan 1947–2000*, p. 251.
94. Ibid.
95. "Pakistan President on Nuclear Development," Xinhau General Overseas News Service, March 11, 1980.
96. Khalid Mahmud Arif, *Working with Zia* (New York: Oxford University Press, 1995), p. 368.
97. Gates, *From the Shadows*, p. 148.
98. Kux, *The United States and Pakistan 1947–2000*, p. 252.
99. Gates, *From the Shadows*, p. 148.
100. Coll, *Ghost Wars*, p. 65; and Bergen, *Holy War Inc.*, p. 71.
101. See, for example Walter Pincus, "Panels to Probe Afghan Arms Fund," *Washington Post*, January 13, 1987; Coll, *Ghost Wars*, pp. 66–67; and Robert Manning, et al., "Casey's CIA: New Clout, New Danger," *US News & World Report*, June 16, 1986.

Chapter 6

1. John J. Fialka, "Nuclear Club: Set to Explode?" *Wall Street Journal*, November 26, 1984; and David Albright and Mark Hibbs, "Pakistan's Bomb: Out of the Closet," *The Bulletin of the Atomic Scientists*, July-August 1992.
2. "Pakistan Said to Stop Work at Uranium Enriching Plant," World News Briefs, *New York Times*, Reuters, October 12, 1979, p. A4. Interestingly, these reports appeared in early October 1979, on the eve of US talks with Pakistan aimed at getting Islamabad to back off its nuclear program while still keeping it on board the fledging Afghan initiative. (See chapter five).
3. Richard Burt, "US Will Press Pakistan to Halt A-Arms Plant," *New York Times*, August 12, 1979, p. 1. (Also see chapter five).
4. Gordon Corera, *Shopping for Bombs* (New York: Oxford University Press, 2006), p. 28.
5. Simon Henderson, "Delhi Attack on Atomic Facility Feared," *Financial Times* (London), August 14, 1979; and Nuclear Threat Initiative, "Pakistan Profiles: Nuclear Chronology," entry for 14 August 1979, http://www.nti.org/e_research/profiles/Pakistan/Nuclear/5593_5678.html.
6. Memoranda from Zbigniew Brzezinski to the President, December 26, 1979, "Reflections on Soviet Intervention of Afghanistan," published by the Cold War International History Project (CWIHP) in Volume One of the Document Reader prepared for the international conference "Toward an International History of the War in Afghanistan, 1979–1989" (April 2002), edited by Christian Ostermann and Mircea Munteanu.
7. "Pak Nuclear Capability Indebted to Afghan war," *Dawn*, May 27, 2001.
8. Simon Henderson, "UK Deals May Aid Pakistan Atom Bid," *Financial Times* (London), August 22, 1979; and Simon Henderson, "Pakistan's Nuclear Shopping List: The British Link," *Financial Times* (London), August 22, 1979.

9. Simon Henderson, "Musharraf's Mess," *National Review*, February 12, 2004.
10. Regarding mandrels and bellows, see, for example, "E-book Glossary," Institute for Science and International Security, www.exportcontrols.org/glossary.html.
11. E-mail, Student Records Officer, Oxford Brookes University, to author, January 29, 2007; Owen Bowcott, et al., "Businessman Under Scrutiny 25 Years Ago after Ordering Unusual Supplies," *The Guardian* (London), March 5, 2004; and Weargate corporate records, Companies House, London.
12. Steve Coll, "The Atomic Emporium," *New Yorker*, August 7 and 14, 2006; Corera, *Shopping for Bombs*, p. 26.
13. Corporate records of Weargate and other Griffin-Salam companies, Companies House, London.
14. Shahid-Ur-Rehman, *Long Road to Chagai*, p. 63. In response to an inquiry from one of the authors of this book, Shahid-Ur-Rehman explained that the information about Salam "was revealed in background interviews by Dr. A.Q. Khan himself" and one of Khan's associates. (E-mail, Shahid-Ur-Rehman to author, August 3, 2006.)
15. Corera, *Shopping for Bombs*, p. 26; Coll, "The Atomic Emporium." Regarding Griffin's father, see Griffin's marriage certificate, Register Office, Swansea, County of West Glamorgan, UK.
16. Simon Henderson, "Pakistan's nuclear shopping list"; and Simon Henderson, "Islamabad Continues Its Quest for Nuclear Might," *Financial Times* (London), December 22, 1980.
17. Source Reliance International's corporate records, Companies House, London.
18. William J. Broad, et al., "A Tale of Nuclear Proliferation: How Pakistani Built His Network," *New York Times*, February 12, 2004, p. 1.
19. Bowcott, et al., "Businessman Under Scrutiny."
20. Farah Stockman, "US Prods UN for a Nuclear Export Rule," *Boston Globe*, April 4, 2004.
21. See, for example, Coll, "The Atomic Emporium."
22. Henderson, "Islamabad Continues Its Quest."
23. Henderson, "Islamabad Continues Its Quest"; Weissman and Krosney, *The Islamic Bomb*, p. 216; Fialka, "Nuclear Club: Set to Explode?"; Shahid-Ur-Rehman, *Long Road to Chagai*, p. 62; Leonard Spector, *The New Nuclear Nations* (New York: Vintage Books, 1985), p. 257; and Court Records, Her Majesty the Queen, Petitioner, v. Serabit Electronics Ltd., Salam El Menyawi, Mohammad Ahmed, Abdul Aziz Khan, Respondents, Superior Court (Criminal Jurisdiction), Province of Quebec, District of Montreal.
24. Henderson, "Islamabad Continues Its Quest"; and Fialka, "Nuclear Club: Set to Explode?"
25. Shahid-Ur-Rehman, *Long Road to Chagai*, pp. 57–58.
26. Henderson, "Islamabad Continues Its Quest"; Weissman and Krosney, *The Islamic Bomb*, p. 216; Fialka, "Nuclear Club: Set to Explode?"; and David Willis, "On the Trail of the A-Bomb Makers," *Christian Science Monitor*, December 1, 1981.
27. Fialka, "Nuclear Club: Set to Explode?"; David Albright and Mark Hibbs, "Pakistan's Bomb: Out of the Closet," *The Bulletin of the Atomic Scientists*, July-August 1992; Willis, "On the Trail of the A-Bomb Makers"; and Spector, *The New Nuclear Nations*, p. 257.
28. Fialka, "Nuclear Club: Set to Explode?".
29. Shahid-Ur-Rehman, *Long Road to Chagai*, pp. 58–59.
30. Spector, *The New Nuclear Nations*, p. 257.
31. Fialka, "Nuclear Club: Set to Explode?"
32. Ibid.; Willis, "On the Trail of the A-Bomb Makers"; and Henderson, "Islamabad Continues Its Quest." A transcript of a 1986 German documentary about Pakistan's nuclear program states that after leaving his post in Bonn, I.H. Khan became a director of "the cover firm Technical Equipment." It gives no other information. ("Wanted. . . . Bomb Business: Nuclear Aid for Pakistan and India," transcript of a film by Egmont Koch, West German Broadcasting (WDR), 1986.)
33. Fialka, "Nuclear Club: Set to Explode?"; and Weissman and Krosney, *The Islamic Bomb*, pp. 216–17. In a November 1979 letter to A.Q. Khan, A.A. Khan indicated that he would obtain a copy of a paper A.Q. Khan wanted (Spector, *The New Nuclear Nations*, p. 257).

34. Willis, "On the Trail of the A-Bomb Makers"; and Weissman and Krosney, *The Islamic Bomb*, pp. 216–17.
35. Fialka, "Nuclear Club: Set to Explode?"
36. Campagne tegen Wapenhandel, "The Dutch Connnection: A.Q. Khan, Nuclear Proliferation and the Netherlands (1963–2004), Part 2: Khan's Support Network," http://www.stopwapenhandel.org/Khan_part2.html.
37. Fialka, "Nuclear Club: Set to Explode?"
38. Court Records, Her Majesty the Queen, Petitioner, v. Serabit Electronics Ltd., Superior Court (Criminal Jurisdiction), Province of Quebec, District of Montreal.
39. Author's telephone interview with Guy Gilbert, April 10, 2006.
40. "Two Engineers Guilty on One Export Charge," *Globe and Mail* (Canada), July 14, 1984; "Engineers, Firm Fined for Deal with Pakistan," *Globe and Mail* (Canada), August 9, 1984; Fialka, "Nuclear Club: Set to Explode?"
41. Interview with Guy Gilbert; and Irwin Block, "'Great' Features in Recollections of Ryan," *The Gazette* (Montreal), February 12, 2004.
42. Interview with Guy Gilbert.
43. Quoted in Weissman and Krosney, *The Islamic Bomb*, p. 217.
44. "Chehlum of Imtiza Bhatti Sitara-i-Imtiaz to be observed on Feb 3," Pakistan Newswire, February 1, 2002; and e-mail, Shahid-ur-Rehman to author, January 17, 2007.
45. Leonard Spector and Haider Nizamani, "New Head of Pakistan Atomic Energy Commission Apparently Tied to 1980s Nuclear Smuggling," *WMD Insights*, Center for Nonproliferation Studies, May 4, 2006.
46. Quoted in Henderson, "Islamabad Continues Its Quest."
47. Henderson, "Musharraf's Mess."
48. Florida Department of State corporate records, Tallahassee; "Extraordinary Resolutions of Source Reliance International Limited," July 7, 1982, Source Reliance International Ltd. corporate records, Companies House, London.
49. Author's telephone interview with David Reed, June 29, 2006.
50. Author's telephone interview with Julio Bagiardi, June 29, 2006.
51. Ibid.
52. Ibid.
53. Interview with David Reed.
54. William Carley, "Paying for Peace: To Finance Terrorism, Abu Nidal Is Believed to Shake Down Arabs," *Wall Street Journal*, August 9, 1991; Simon Pincombe, "Price Waterhouse Faced Sack as BCCI Auditors," *The Independent* (London), July 30, 1991.
55. Peter Truell and Larry Gurwin, *False Profits* (Boston: Houghton Mifflin, 1992), pp. 247–48; and Jeff Testerman, "Entangled in the Web," *St. Petersburg Times*, October 20, 1991.
56. Centaur Impex corporate records, Florida Department of State, Tallahassee.
57. Interview with David Reed.
58. Centaur Impex and Natasha International corporate records, Florida Department of State, Tallahassee. At some point during the 1990s, Natasha filed paperwork to do business as Centaur Computer Systems. Centaur Computer Systems also kept a low public profile.
59. See, for example Natasha's company profile at Electric Commerce Europe (http://www.eceurope.com/natasha) and at Tradeport (http://mbita.worldbid.com/natasha).
60. http://www.ellafinance.com/35235.php.
61. Ibid.
62. EllaFinance Conference Room, Confidential News Flash, "Leasing of Investment Power," http://www.ellafinance.com/conf/start01.htm. (Accessed June 22, 2006; login was required to view these pages.)
63. EllaFinance "Provider Section," http://ellafinance.com/. (Accessed June 15, 2006.)
64. Author's telephone interview with Jack Blum, April 13, 2007.
65. EllaFinance of Barbados Web site, http://www.ellafinance.com/35235.php.
66. The Central Bank of Barbados, "Warning Notice," http://www.centralbank.org.bb/Financial/warning_notice.shtml/.

67. Letter from Marlene E.A. Bayne, director, Bank Supervision Dept., Central Bank of Barbados, to author, July 31, 2006.
68. E-mail, "Abdul" Salam to author, July 14, 2006.
69. EllaFinance of Barbados Web site, http://www.ellafinance.com/35235.php.

Chapter 7

1. Memo, Shultz to Reagan, "Visit of Zia-ul-Haq," November 29, 1982, in Cold War International History Project Briefing Book *Towards an International History of the War in Afghanistan, 1979–1989,* vol. 1, edited by Christian Ostermann and Mircea Munteanu (Washington, DC: Cold War International History Project, 2002).
2. Dennis Kux, *The United States and Pakistan 1947–2000* (Baltimore: Johns Hopkins University Press, 2001), pp. 256, 259–260.
3. Kux, *The United States and Pakistan 1947–2000,* p. 257.
4. Quoted in Kux, *The United States and Pakistan 1947–2000,* p. 257.
5. Steve Coll, *Ghost Wars* (New York: Penguin Press, 2004), p. 65.
6. Edith Lenart, "Saudis Offer to Help Zia Build H-bomb," *Sunday Times* (London), January 18, 1981. According to the article, the offer was ostensibly contingent on Pakistan not sharing its nuclear technology with Iraq or Libya. A.Q. Khan was said to have participated in the negotiations.
7. See, for example, Arnaud de Borchrave, "Pakistan and Saudi Arabia Have Concluded a Secret Agreement on Nuclear Cooperation," United Press International, October 20, 2003; and Thomas Wilson, "The Sino-Soviet Connection," The Jamestown Foundation, October 24, 2002.
8. Kux, *The United States and Pakistan 1947–2000,* p. 260; and Stuart Auberbach, "US Aid as Deterrent," *Washington Post,* December 8, 1981.
9. Auberbach, "US Aid as Deterrent."
10. State Department cable, American Embassy Islamabad to Secretary of State, Confidential Islamabad 21106, E.O. 12356: DECL: OADR, "Draft Letter to Congressman Solarz," October 1987.
11. Kux, *The United States and Pakistan 1947–2000,* p. 261.
12. Coll, *Ghost Wars,* pp. 57, 69. Also see George Crile, *Charlie Wilson's War* (New York: Atlantic Monthly Press, 2003), pp. 463–464.
13. Auberbach, "US Aid as Deterrent." Also see Judith Miller, "US Says Pakistan's Nuclear Potential is Growing," *New York Times,* January 24, 1982.
14. State Department Briefing Paper, "The Pakistani Nuclear Program," June 23, 1983, available through the National Security Archives, http://www.gwu.edu/~nsarchiv/NSAEBB/NSAEBB6/index.html, Document 22. Regarding being based on CIA intelligence, see Tim Weiner, "Nuclear Anxiety; The Know-How; US and China Helped Pakistan Build Its Bomb," *New York Times,* June 1, 1998.
15. State Department Memo No. 8238186, "Pakistan President Zia in US Visit," December 13, 1982, and State Department Memo No. 348949, "State Visit of President Mohammad Zia-ul-Haq," December 1982, both in *Towards an International History of the War in Afghanistan, 1979-1989.*
16. Crile, *Charlie Wilson's War,* p. 464.
17. Memo, Shultz to Reagan, "Visit of Zia-ul-Haq."
18. David Ignatius, "US Pressuring Pakistan to Abandon Controversial Nuclear-Arms Program," *Wall Street Journal,* October 25, 1984.
19. Zahid Malik, *Dr. A.Q. Khan and the Islamic Bomb* (Islamabad: Hurmat Publications, 1992), pp. 87, 255.
20. Seymour Hersh, "On the Nuclear Edge," *New Yorker,* March 29, 1993.
21. Shahid-Ur-Rehman, *Long Road to Chagai* (Islamabad: Print Wise Publication, 1999), pp. 78–79.
22. "Pakistan's Nuclear Chief Says It Could Build a Bomb," *Washington Post,* February 10, 1984.

23. Shahid-Ur-Rehman, *Long Road to Chagai*, p. 33; and State Department Briefing Paper, "The Pakistani Nuclear Program," June 23, 1983, available through the National Security Archives, http://www.gwu.edu/~nsarchiv/NSAEBB/NSAEBB6/index.html, Document 22.
24. George Perkovich, *India's Nuclear Bomb* (Berkeley: University of California Press, 1999), pp. 196–97.
25. State Department, "The Pakistani Nuclear Program."
26. William Broad and David Sanger, "As Nuclear Secrets Emerge in Khan Inquiry, More Are Suspected," *New York Times*, December 26, 2004; Tim Weiner, "Nuclear Anxiety; The Know-How; US and China Helped Pakistan Build Its Bomb," *New York Times*, June 1, 1998. Also see Hedrick Smith, "A Bomb Ticks in Pakistan," *New York Times Magazine*, March 6, 1988.
27. State Department, "The Pakistani Nuclear Program."
28. Ibid.
29. Seymour Hersh, "Pakistani in US Sought to Ship A-Bomb Trigger," *New York Times*, February 25, 1985; Rick Atkinson, "Use in Arms Feared: Nuclear Parts Sought by Pakistan," *Washington Post*, July 21, 1984.
30. Hersh, "Pakistani in US Sought to Ship A-Bomb Trigger."
31. Ibid.; Atkinson, "Use in Arms Feared: Nuclear Parts Sought by Pakistan"; and Affidavit of Special Agent Justice A. McCalley, US v. Nazir Ahmed Vaid, US District Court, Southern District of Texas, Houston Division, Houston.
32. Atkinson, "Use in Arms Feared: Nuclear Parts Sought by Pakistan."
33. Affidavit of Special Agent McCalley; Atkinson, "Use in Arms Feared: Nuclear Parts Sought by Pakistan"; and Hersh, "Pakistani in US Sought to Ship A-Bomb Trigger."
34. Atkinson, "Use in Arms Feared: Nuclear Parts Sought by Pakistan"; and Hersh, "Pakistani in US Sought to Ship A-Bomb Trigger."
35. Affidavit of Special Agent McCalley.
36. Ibid.; Hersh, "Pakistani in US Sought to Ship A-Bomb Trigger"; Atkinson, "Use in Arms Feared: Nuclear Parts Sought by Pakistan."
37. Affidavit of Special Agent McCalley; Atkinson, "Use in Arms Feared: Nuclear Parts Sought by Pakistan."
38. Affidavit of Special Agent McCalley; Hersh, "Pakistani in US Sought to Ship A-Bomb Trigger."
39. Affidavit of Special Agent McCalley; Hersh, "Pakistani in US Sought to Ship A-Bomb Trigger."
40. Affidavit of Special Agent McCalley; Hersh, "Pakistani in US Sought to Ship A-Bomb Trigger"; Atkinson, "Use in Arms Feared: Nuclear Parts Sought by Pakistan." Regarding the krytrons being "materially altered," see "Defendant's Motion to Suppress Fruits of Search and Seizure," August 24, 1984, p. 4, US v. Nazir Ahmed Vaid, US District Court, Southern District of Texas, Houston Division, Houston.
41. Affidavit of Special Agent McCalley; Hersh, "Pakistani in US Sought to Ship A-Bomb Trigger"; and "Buying the Bomb," PBS *Frontline,* March 5, 1985.
42. Affidavit of Special Agent McCalley; Hersh, "Pakistani in US Sought to Ship A-Bomb Trigger"; Atkinson, "Use in Arms Feared: Nuclear Parts Sought by Pakistan."
43. Hersh, "Pakistani in US Sought to Ship A-Bomb Trigger"; "Buying the Bomb"; and Nancy Stancill, "Lawyers: US Didn't Probe Clue That Man Possible Spy," *Houston Chronicle*, February 26, 1985. Regarding Butt, also see Steve Weissman and Herbert Krosney, *The Islamic Bomb* (New York: Times Books, 1981), p. 182 (and elsewhere) and Shahid-Ur-Rehman, *Long Road to Chagai.*
44. State Department Cable, American Embassy Islamabad (Hummel) to Secretary of State, et al., Secret Islamabad 03745, "Pakistani Efforts to Obtain Zircalloy Tubes/Production Equipment," E.O. 12065: RDS-2, March 1981.
45. Hersh, "Pakistani in US Sought to Ship A-Bomb Trigger"; "Buying the Bomb"; and Stancill, "Lawyers: US Didn't Probe Clue that Man Possible Spy."
46. Hersh, "Pakistani in US Sought to Ship A-Bomb Trigger"; Atkinson, "Use in Arms Feared: Nuclear Parts Sought by Pakistan"; Burke Watson, "3 Indicted in Plot to Ship Parts for

N-arms to Pakistan," *Houston Chronicle*, July 18, 1984; Jim Barlow, "US Magistrate Refuses to Lower $200,000 Bond," *Houston Chronicle*, June 24, 1984; and author's telephone interview with Sam Longoria, September 27, 2004.

47. Watson, "3 Indicted in Plot to Ship Parts for N-arms to Pakistan." The court apparently turned down a prosecution motion to increase Vaid's bail. (See State Department Memorandum, Secret, James Larocco to Howard Schaffer, "Pakistani Nationals on Trial in Houston for Attempting ot Smuggle Nuclear Explosive Components," July 12, 1984.)
48. State Department Memorandum Arnold Raphel to Michael Armacost, "Pakistanis Arrested for Smuggling Nuclear Weapons Components," July 20, 1984; and Archives of the Ronald Reagan Presidential Library: Speeches, "Nomination of Arnold Lewis Raphel To Be United States Ambassador to Pakistan," http://www.reagan.utexas.edu/archives/speeches/1986/101586b.htm.
49. State Department Memorandum, Secret, James Larocco to Howard Schaffer, "Pakistani Nationals on Trial in Houston for Attempting ot Smuggle Nuclear Explosive Components," July 12, 1984. Also see Larocco through Schaffer to Richard Murphy, "Possible Press Attention to Case of Pakistani Arrested for Attempting to Smuggle Nuclear Weapons Components," June 28, 1984.
50. State Department Memorandum, James Larocco to Howard Schaffer, "Paks Nuke Smugglers," July 31, 1984. In Pakistan, some saw dark hands behind the incident. During a meeting in Islamabad shortly before Vaid's trial was to begin, Pakistan's Foreign Secretary Niaz Naik told US ambassador Dean Hinton that Gen. Zia had "personally directed an on-going investigation" into how and by whom the "unauthorized krytron caper had been initiated." While the investigation had so far been "inconclusive," Naik told Hinton that it left open the "possibility that [the] individuals arrested in the US were in the pay of someone wanting to discredit Pakistan." (Hinton to Shultz, Secret, "Nuclear Weapons Problem," "Islama 18574 130718Z," September 1984.)
51. Hersh, "Pakistani in US Sought to Ship A-Bomb Trigger."
52. Watson, "3 Indicted in Plot to Ship Parts for N-arms to Pakistan"; and Indictment, July 16, 1984, US v. Nazir Ahmed Vaid, et al., US District Court, Southern District of Texas, Houston Division, Houston.
53. Author's telephone interview with Sam Longoria.
54. Hersh, "Pakistani in US Sought to Ship A-Bomb Trigger"; and Indictment, US v. Nazir Ahmed Vaid, et al.
55. "Buying the Bomb," PBS/Frontline, March 5, 1985.
56. "Motion to Restrict Juror's Access to Publicity," filed August 24, 1984, US v. Nazir Ahmed Vaid, et al., US District Court, Southern District of Texas, Houston Division, Houston.
57. Hersh, "Pakistani in US Sought to Ship A-Bomb Trigger"; "Motion to Strike Surplusage from the Indictment," filed August 24, 2004, "Government's Response to Defendant's Motion to Strike Surplusage from Indictment," 37, September 5, 1984, and Courtroom Minutes — Motion Hearing, Pre-Trial Hearing, September 12, 1984, all in US v Vaid, et al., US District Court, Southern District of Texas, Houston Division, Houston.
58. Hersh, "Pakistani in US Sought to Ship A-Bomb Trigger."
59. Author's interviews with Peter Williamson, July 21, 2004, and September 23, 2004; and e-mail, Williamson to Armstrong, July 21, 2004.
60. Author's interviews with Peter Williamson.
61. Author's interview with James DeAnda, 9/23/04; Dennis Heveshi, "James deAnda, 81, Lawyer in Case of Hispanic Jurors," *New York Times*, September 9, 2006.
62. "Buying the Bomb."
63. Author's interview with Sam Longoria.
64. Crile, *Charlie Wilson's War*, p. 463.
65. Author's telephone interview with Charlie Wilson, November 2, 2005.
66. Docket, US v Vaid, et al., US District Court, Southern District of Texas, Houston Division, Houston.
67. Hersh, "Pakistani in US Sought to Ship A-Bomb Trigger."

68. Copies of telexes accompanying Hersh, "Pakistani in US Sought to Ship A-Bomb Trigger."
69. Hersh, "Pakistani in US Sought to Ship A-Bomb Trigger."
70. State Department Cable, American Embassy, Paris (Friedman) to Secretary of State, et al., Confidential Paris 02246, E.O. 12065: GDS-1, "Pakistan Nuclear Affairs," January 1982. Also see State Department Cable, American Embassy, Islamabad (Hummel) to Secretary of State, et al., Secret Islamabad 03745, "Pakistani Efforts to Obtain Zircalloy Tubes/ Production Equipment," E.O. 12065: RDS-2, March 1981; and State Department Cable, American Embassy, Tokyo to Secretary of State, Secret Tokyo 15669, E.O. 12356: DECL: OADR, "Japanese Assistance to the Indian and Pakistani Nuclear Programs," August 1985.
71. Hersh, "Pakistani in US Sought to Ship A-Bomb Trigger." Also, interview with Sam Longoria.
72. Hersh, "Pakistani in US Sought to Ship A-Bomb Trigger"; "Government's Motion to Dismiss," September 5, 2004; Order [to Dismiss], #56, September 18, 2004, and Docket, all in US v Vaid, et al., US District Court, Southern District of Texas, Houston Division, Houston.
73. Author's interview with James DeAnda, September 23, 2004.
74. "Transcript of Proceedings," 70, October 22, 1984, US v. Vaid, et al., US District Court, Southern District of Texas, Houston Division, Houston.
75. Author's interview with James DeAnda.
76. "Transcript of Proceedings," 70, US v. Vaid, et al.
77. Ibid.
78. Author's interview with Peter Williamson, September 23, 2004.
79. Hersh, "Pakistani in US Sought to Ship A-Bomb Trigger"; and "Buying the Bomb."
80. Ibid.
81. Author's interview with Peter Williamson.
82. "Buying the Bomb."
83. Hersh, "Pakistani in US Sought to Ship A-Bomb Trigger."
84. Author's telephone interview with Sam Longoria.
85. Author's inverview with James DeAnda.
86. Hersh, "Pakistani in US Sought to Ship A-Bomb Trigger"; "Buying the Bomb."
87. Texas Secretary of State corporate records, Austin.
88. "Criminal Complaint," June 22, 1984, "Commitment," June 25, 1984, and "Magistrate Criminal Minutes," June 25, 1984, all in US v. Nazir Ahmed Vaid aka Nazir Ahmed.
89. Texas Secretary of State corporate records, Austin.
90. Author's telephone interview with Nazir Vaid, June 24, 2004. As he had done with Seymour Hersh years earlier, Vaid invited the News Service reporter to speak with his lawyer about his visa applications. Vaid's "lawyer" turned out to be a paralegal working for an unusual Houston law firm that specializes in immigration work. The paralegal, who had helped Vaid set up his new Texas companies and use them to apply for visas, was a naturalized American citizen, originally from Pakistan, who had himself been sentenced to three years in prison during the 1990s for conspiring to illegally export individual chemical agent detector units to Pakistan. Other employees of the law firm for which the paralegal works have colorful histories as well. Another paralegal at the firm, who was also involved in setting up Vaid's companies, was convicted in 1985 on a variety of immigration fraud charges. He had been sentenced to eighteen months in prison and was forced to surrender his license to practice law. And one of the lawyers at the firm is a former Pakistani civil servant who became known as a member of the gang that surrounded deposed Pakistani prime minister Nawaz Sharif.
91. Webarchive.org has remnants of the Finatra Group of Companies' Web sites; http://web.archive.org/web/19981111185327/http://finatra.com/.
92. "Pakistan Telecom to Start Pre-Paid Calling Card Service," *Asia Pulse*, September 3, 1998.
93. Finatra Communications and Finatra Group of Companies corporate records, Texas Secretary of State, Austin. *Note:* The technical contact for Finatra's defunct Web site, www.finatra.com, is Catalyst Communication, another of Vaid's companies.

94. Author's telephone interview with Nazir Vaid, July 6, 2004.
95. Public Notice, "International Authorizations Granted," DA No. 01-2951, December 20, 2001, Federal Communications Commission, Washington, DC; available at www.cc.gov/bureaurs/international/public_notices/2001/da012951.pdf.
96. Catalyst Communication USA Inc. corporate records, Texas Secretary of State, Austin.
97. www.pseb.org.pk/profile.php?id=680 (accessed July 20, 2005). The Pakistan Software Export Board listing for Catalyst Communication (PVT) Ltd. shows the CEO as "Dr. Nazir Ahmed Vaid." Regarding the Pakistani and American companies being related, see www.web.archive.org/web20041126031 92/http://catcomntc.com/; click on "About CATCOM" and scroll to the bottom of the page. Under the heading "b) USA," it states: "CATCOM [the Pakistani entity] is registered in USA as Catalyst Communication USA (Inc.)."
98. E-mail, Nazir Vaid to Armstrong, June 12, 2006

Chapter 8

1. Associated Press, "Pakistan Calls Atom Program Peaceful," *New York Times*, February 26, 1985.
2. Mary Anne Weaver, "Pakistan 'Over the Hump' – Zia interview," *Christian Science Monitor*, March 1, 1985.
3. George Crile, *Charlie Wilson's War* (New York: Atlantic Monthly Press, 2003), p. 463.
4. Ibid., p. 464.
5. Ibid.; and Patricia Sullivan, "CIA Agent Gust L. Avrakotos Dies at Age 67," *Washington Post*, December 25, 2005.
6. Joanne Omang, "Lawmakers Exercise Foreign Aid Initiative," *Washington Post*, March 21, 1985; Robert Pear, "Legislators Move on Atom Exports," *New York Times*, March 26, 1985; Joanne Omang, "House Panel Votes Aid for Cambodian Forces," *Washington Post*, April 4, 1985; Leonard Spector, *The New Nuclear Nations* (New York: Vintage Books, 1985), pp. 119–20; Dennis Kux, *The United States and Pakistan 1947–2000* (Baltimore: Johns Hopkins University Press, 2001), pp. 277–78.
7. Kux, *The United States and Pakistan 1947-2000*, p. 277; Omang, "House Panel Votes Aid for Cambodian Forces"; and Spector, *The New Nuclear Nations*, pp. 119–20.
8. *World News Tonight*, ABC News, July 11, 1985.
9. David Ottaway, "US Expediting Missiles for Pakistan's Defenses," *Washington Post*, July 11, 1985.
10. Hedrick Smith, "A Bomb Ticks in Pakistan," *New York Times Magazine*, March 6, 1988; and Kux, *The United States and Pakistan 1947–2000*, p. 276. Also see David Ignatius, "US Pressuring Pakistan to Abandon Controversial Nuclear-Arms Program," *Wall Street Journal*, October 25, 1984; Simon Henderson, "US Warns Pakistan on Enriching Uranium," *Financial Times* (London), December 7, 1984; "Pakistanis Admit Gains in Uranium Technology," *New York Times*, February 25, 1985; and Mary Anne Weaver, "Pakistan 'Over the Hump' – Zia Interview," *Christian Science Monitor*, March 1, 1985
11. Quoted in George Perkovich, *India's Nuclear Bomb* (Berkeley: University of California Press, 1999), p. 264.
12. Smith, "A Bomb Ticks in Pakistan"; and Don Oberdorfer, "Two Reportedly Agree on Talks," *Washington Post*, October 24, 1985.
13. Portions of the Congressional Record posted by Federation of American Scientists, "The Pressler Amendment and Pakistan's Nuclear Weapons Program," http://www.fas.org/news/pakistan/1992/920731.htm; see "Letters to Congress from Presidents Reagan & Bush, 1985–1989, required under sec. 620E(e) of Foreign Assistance Act (Pressler Amendment)."
14. Smith, "A Bomb Ticks in Pakistan,"; Bob Woodward, "Pakistan Reported Near Atom Arms Production," *Washington Post*, November 4, 1986; and Kux, *The United States and Pakistan 1947–2000*, p. 283.
15. Woodward, "Pakistan Reported Near Atom Arms Production."
16. Author's interviews with Richard Barlow, August 27, 2006, and February 2, 2007. Although President Bush continued withholding certification in 1991 and 1992, he nevertheless

allowed Pakistan to purchase American-made weapons, thereby undercutting the effect of the Pressler amendment sanctions (Seymour Hersh, "On the Nuclear Edge," *New Yorker*, March 29, 1993.)

17. Kux, *The United States and Pakistan 1947–2000*, p. 279; and Hersh, "On the Nuclear Edge." Kux reports that the blueprints shown to Zia had been stolen from A.Q. Khan's hotel room by a "friendly foreign intelligence service" (Kux, p. 417, footnote 98). The *Washington Post* reported that US intelligence swiped the plans (R. Jeffrey Smith, "US Aides See Troubling Trend in China-Pakistan Nuclear Ties," *Washington Post*, April 1, 1996). According to the *New York Times*, two former Pakistani military intelligence officers said that when the US confronted Zia with evidence of his country's bomb he "indicated his belief that he had the blessings of President Reagan and the Director of Central Intelligence, William J. Casey, to go ahead and build the bomb" (Tim Weiner, "Nuclear Anxiety; The Know-How; US and China Helped Pakistan Build Its Bomb," *New York Times*, June 1, 1998). The mock-up of the bomb shown to Benazir Bhutto (and to Pakistan's foreign minister) was built by weaponeers at the Lawrence Livermore National Laboratory, based on the Chinese designs purloined from Khan (Smith, "US Aides See Troubling Trend in China-Pakistan Nuclear Ties").
18. Weiner, "Nuclear Anxiety; The Know-How; US and China Helped Pakistan Build Its Bomb."
19. Michael Gordon, "Pakistani Seized by US in Plot on A-Arms Alloy," *New York Times*, July 15, 1987; Smith, "A Bomb Ticks in Pakistan"; Hersh, "On the Nuclear Edge"; and House Subcommittees on Asian and Pacific Affairs on International Economic Policy and Trade of the Committee on Foreign Affairs, hearings on Pakistan's Illegal Nuclear Procurement in the United States, 100th Cong., 1st sess., July 22, 1987.
20. Anna Tomforde and Dan Atkinson, "US sought the arrest of BCCI-backed Bomb Buyer," *Guardian* (London) August 7, 1991; and Gordon, "Pakistani Seized by US in Plot on A-Arms Alloy." According to a Pakistani press report, ul-Haq's company, Multinational Incorporated, also had branches in "Houston, Texas, Northampton and Dubai." (*The Nation* [Lahore], July 27, 1987, quoted in State Department Cable, American Consulate Lahore (Thibault) to Secretary of State (Shultz), "Lahore paper reports that Lahore nuclear importer Brig. Inam ul-Haq has vanished," document number 87Lahore02142, July 1987.)
21. Smith, "A Bomb Ticks in Pakistan." Also, Andre Picard, "RCMP Charges Metro Man on Trial in US Accused of Seeking A-bomb Material," *Globe and Mail* (Toronto), December 12, 1987.
22. "Nuclear Weapons Proliferation Concerns and United States Assistance to Pakistan," S 9885, 100th Cong. 1st sess., *Cong. Rec.* 133 no. 116, (July 14, 1987); Smith, "A Bomb Ticks in Pakistan"; and Gordon, "Pakistani Seized by US in Plot on A-Arms Alloy."
23. Mark Hibbs, "US Will Charge Pakistani in BCCI-financed Nuclear Case," *Nucleonics Week*, August 8, 1991; Marcus Brauchli and John Fialka, "Arrest of Retired General May Help US Link BCCI to Pakistani Nuclear Weapons," *Wall Street Journal*, August 5, 1991.
24. Don Oberdorfer, "Pakistani Native Charged with Nuclear Export Plot," *Washington Post*, July 15, 1987.
25. Ibid.; Hersh, "On the Nuclear Edge"; "Nuclear Weapons Proliferation Concerns and United States Assistance to Pakistan," S 9885 *Cong. Rec.*, 133; and Michael Gordon, "US Pressing Pakistan on Export Plot," *New York Times*, July 16, 1987.
26. Author's interview with Richard Barlow, February 2, 2007.
27. Crile, *Charlie Wilson's War*, p. 479. Solarz's Asian Affairs subcommittee held the hearing in conjunction with the House Subcommittee on International Economic Policy and Trade.
28. Author's interview with Barlow; and Hersh, "On the Nuclear Edge."
29. Author's interview with Barlow.
30. Ibid.
31. Hersh, "On the Nuclear Edge."
32. Author's interview with Barlow.
33. Hersh, "On the Nuclear Edge."

34. Author's interview with Barlow.
35. Hersh, "On the Nuclear Edge."
36. House Subcommittees on Asian and Pacific Affairs on International Economic Policy and Trade of the Committee on Foreign Affairs, Hearing on Pakistan's Illegal Nuclear Procurement in the United States, 100th Cong., 1st sess., July 22, 1987; and Michael Gordon, "US Indicts 3 in the Export of Equipment to Pakistan," *New York Times*, July 18, 1987.
37. Author's interview with Barlow.
38. Crile, *Charlie Wilson's War*, p. 479.
39. Cable, American Embassy Islamabad to Secretary of State, Confidential Islamabad 21106, E.O. 12356: DECL: OADR, "Draft Letter to Congressman Solarz," October 1987.
40. Smith, "A Bomb Ticks in Pakistan."
41. Crile, *Charlie Wilson's War*, pp. 479–80.
42. Ibid., p. 480.
43. Ibid., p. 481.
44. Ibid.
45. Ibid.
46. Kux, *The United States and Pakistan 1947–2000*, p. 286.
47. Michael Gordon, "Businessman Convicted in Pakistan Nuclear Plot," *New York Times*, December 18, 1987.
48. Smith, "A Bomb Ticks in Pakistan."
49. Gordon, "Businessman Convicted in Pakistan Nuclear Plot."
50. "The Pressler Amendment and Pakistan's Nuclear Weapons Program," http://www.fas.org/news/pakistan/1992/920731.htm; see "Letters to Congress from Presidents Reagan & Bush, 1985–1989." Dennis Kux notes that Reagan's formal declaration stated that the certification rested on the "statutory standard as legislated by Congress" of whether Pakistan "possesses a nuclear explosive device, not whether Pakistan is attempting to develop or has developed various relevant capacities." Yet earlier in the year, Kux writes, "State Department legal advisor Abraham Sofaer found that a country could 'possess' a nuclear device even if it had not assembled all the elements." (Kux, *The United States and Pakistan 1947–2000*, p. 286 and footnote, p. 418.)
51. Hersh, "On the Nuclear Edge"; "The Pressler Amendment," http://www.fas.org/news/pakistan/1992/920731.htm; see "From Myth to Reality: Evidence of Pakistan's 'Nuclear Restraint'"; Smith, "A Bomb Ticks in Pakistan."
52. Gordon, "Pakistani Seized by US in Plot on A-Arms Alloy.".
53. Ibid.; and Robert MacLeod, "US Frees Man Held Over Bid to Export Nuclear Material," *Globe and Mail* (Toronto), April 7, 1990.
54. "Retired Pakistani General Convicted," United Press International, July 8, 1992; and "Pakistani General Sentenced to Time Served in Arms Case," Associated Press, September 28, 1992. At the time of Inam ul-Haq's arrest, The *Guardian* newspaper in London reported that the retired brigadier was "a key figure in a three-nation cartel that, since the late 1970s, has surreptitiously been attempting to buy the components for an atomic bomb. The cartel comprised Argentina, Libya and Pakistan." (Tomforde and Atkinson, "US Sought the Arrest of BCCI-backed Bomb Buyer.")
55. Government of Pakistan Board of Investment; and e-mail, Brig. Inam ul-Haq to author, April 30, 2007.
56. Hersh, "On the Nuclear Edge"; and author's interview with Richard Barlow, August 27, 2006.

Chapter 9

1. Kuldip Nayar, "Has Pakistan Really Joined the Nuclear Club?" *Globe and Mail* (Toronto), March 2, 1987, reprinted from the *Observer* (London), March 1, 1987.
2. Nayar, "Has Pakistan Really Joined the Nuclear Club?"; and Steven R. Weisman, "Report of Pakistani A-Bomb," *New York Times*, March 2, 1987.

3. Weisman, "Report of Pakistani A-Bomb." Also see Nayar, "Has Pakistan Really Joined the Nuclear Club?"; and Dennis Kux, *The United States and Pakistan 1947–2000* (Baltimore: Johns Hopkins University Press, 2001), p. 284.
4. William Doerner, "Pakistan Knocking at the Nuclear Door," *Time*, March 30, 1987.
5. Kux, *The United States and Pakistan 1947-2000*, p. 284; and Doerner, "Pakistan Knocking at the Nuclear Door.".
6. Kux, *The United States and Pakistan 1947–2000*, pp. 281–82.
7. Ibid., p. 282.
8. Leonard Weiss, "Pakistan, It's Déjà Vu All Over Again," *Bulletin of the Atomic Scientists*, May/June 2004.
9. Egmont Koch, "Der Physiker der Mullahs," *FrankfurterRundschau On-line*, http://www.fr-online.de/in_und_ausland/dokumentation/?em_cnt=1053209.
10. Pierre Goldschmidt, "DDG-Sg's Statement to the Board of Governors Meeting Commencing 28 February 2005," International Atomic Energy Agency, March 1, 2005; Report by the Director General, "Implementation of the NPT Safeguards Agreement in the Islamic Republic of Iran," International Atomic Energy Agency, GOV/2005/87, November 18, 2005 (http://www.iaea.org/Publications/Documents/Board/2005/gov2005-87.pdf); Report by the Director General, "Implementation of the NPT Safeguards Agreement in the Islamic Republic of Iran," International Atomic Energy Agency, GOV/2005/67, September 2, 2005, (http://www.iaea.org/Publications/Documents/Board/2005/gov2005-67.pdf); and Dafna Linzer, "Iran was Offered Nuclear Parts," *Washington Post*, February 27, 2005.
11. Elaine Sciolino and David Sanger, "Pressed, Iran Admits It Discussed Acquiring Tools for Nuclear Arms," *New York Times*, February 28, 2005.
12. Steve Weissman and Herbert Krosney, *The Islamic Bomb* (New York: Times Books, 1981), pp. 64–65, 162; and Jim Hoagland, "Briefing to Yeltsin on Iran," *Washington Post*, May 17, 1995.
13. David Segal, "Atomic Ayatollahs," *Washington Post*, April 12, 1987.
14. Leonard Spector, *Nuclear Ambitions* (Boulder, CO: Westview Press, 1990), p. 206; and Leonard Spector, *Going Nuclear: The Spread of Nuclear Weapons 1986–1987* (Cambridge, MA: Ballinger Publishing Company, 1987), pp. 50–51.
15. Jack Boureston and Charles Ferguson, "Schooling Iran's Atom Squad," *Bulletin of the Atomic Scientists*, May/June 2004.
16. David Segal, "Atomic Ayatollahs."
17. Farzad Bazoft, "Iran Signs Secret Atom Deal," *Observer* (London), June 12, 1988.
18. Report by the Director General, "Implementation of the NPT Safeguards Agreement in the Islamic Republic of Iran," International Atomic Energy Agency, GOV/2004/83, November 15, 2004 (http://www.iaea.org/Publications/Documents/Board/2004/gov2004-83.pdf); and David Albright and Corey Hinderstein, "The Centrifuge Connection," *Bulletin of the Atomic Scientists*, March/April 2004.
19. Bazoft, "Iran Signs Secret Atom Deal"; David Segal, "Atomic Ayatollahs."
20. Bazoft, "Iran Signs Secret Atom Deal"; Kenneth Timmerman, *Countdown to Crisis* (New York: Crown Forum, 2005), p. 39.
21. Timmerman, *Countdown to Crisis*, p. 39. An Iranian opposition group claims that one of the officials with whom Khan met in 1986 and 1987 was Briga. Gen. Mohammad Eslami, a top commander in the Iranian Revolutionary Guard. (Statement by Alireza Jafarzadeh, "Iran Building Nuclear Capable Missiles in Underground Secret Tunnels," National Press Club, Washington, DC, November 21, 2005; "Iranian Dissident Fires Ukraine," Agence France-Presse, August 26, 2005; and "Opposition Claims Iran 'On Verge' of Nuclear Bomb," Agence France-Presse, November 22, 2005.) Two former Iranian diplomats told the *Los Angeles Times* in 2002 that Khan made several trips to Iran in 1987 to assist Iran's nuclear program. One of the diplomats said Khan was given a villa on the Caspian Sea in exchange for his help (Douglas Frantz, "Iran-Pakistan Atomic Link Seen," *Los Angeles Times*, November 28, 2003.)

22. Koch, "Der Physiker der Mullahs"; and author's telephone interview with Masud Naraghi, January 29, 2007. Naraghi himself reportedly had extensive dealings with some of Khan's alleged suppliers. According to a 2007 German documentary, over the past several years, Naraghi had paid repeated visits to the German firm Leybold Heraeus, a Urenco subcontractor and one of the companies from which Pakistan bought vacuum pumps and uranium gas conversion and purification technology, starting in the mid-1970s. (Koch, "Der Physiker der Mullahs"; Weissman and Krosney, *The Islamic Bomb*, p. 185; Shahid-Ur-Rehman, *Long Road to Chagai* (Islamabad: Print Wise Publication, 1999), pp. 61–62.) During his visits to Leybold to purchase vacuum equipment for what he says were conventional military applications, Naraghi reportedly became acquainted with two of the firm's employees, Gotthard Lerch, a division manager and salesman, and Otto Heilingbrunner, Leybold's export manager. (Koch, "Der Physiker der Mullahs.") Lerch had been involved in Leybold's dealings with Pakistan and through them had reportedly come to know A.Q. Khan. (Juergen Dahlkamp et al., "Network of Death on Trial," *Der Spiegel*, March 13, 2006.) Heilingbrunner reportedly had extensive dealings with Naraghi and visited the physicist in Tehran several times. (Koch, "Der Physiker der Mullahs.") In 1985, Lerch left Leybold to start his own firm in Switzerland. (Steve Coll, "The Atomic Emporium," *New Yorker*, August 7 and 14, 2006.) In 1987, German authorities named Lerch and Heilingbrunner as suspects in the theft of Urenco blueprints for uranium-enrichment equipment from Leybold. Neither Lerch nor Heilingbrunner were ever formally charged. (Mark Hibbs, "European Suspects in KRL Case May Have Previous Involvement," *Nucleonics Week*, February 5, 2004.) German authorities would later accuse Lerch of involvement in the Khan network's sale of nuclear technology to Libya (see epilogue). His first trial ended in a mistrial in 2006. He is scheduled to be retried. ("Retrial Ordered for German in Libya Arms Deal," www.expatica.com, July 26, 2006.)
23. Report by the Director General, "Implementation of the NPT Safeguards Agreement in the Islamic Republic of Iran," GOV/2005/87.
24. Coll, "The Atomic Emporium."
25. Albright and Hinderstein, "The Centrifuge Connection."
26. Dafna Linzer, "Iran Was Offered Nuclear Parts," *Washington Post*, February 27, 2005; Westmont Court (Ealing) Ltd. corporate records, Companies House, London; Foremost Trading Ltd. corporate records, Companies House, London; Summary of Interviews, R v. Abu Bakr Siddiqui, Southwark Crown Court, London; and Raymond Bonner, "Salesman on Nuclear Circuit Casts Blurry Corporate Shadow," *New York Times*, February 18, 2004.
27. Stephen Fidler, "Fresh Clues on Smuggling Network . . . ," *Financial Times* (London) March 12, 2005.
28. Koch, "Der Physiker der Mullahs"; and author's telephone interview with Naraghi. Naraghi claimed in the interview that even at the time of the meeting he did not know the nationalities of the other participants.
29. Louis Charbonneau, "Iran Stalls in Probe of Nuke Smuggling," Reuters, April 18, 2005.
30. Linzer, "Iran Was Offered Nuclear Parts"; and "Wanted. . . . Bomb Business: Nuclear Aid for Pakistan and India," transcript of a film by Egmont Koch, West German Broadcasting (WDR), 1986. Mebus had been an employee of the German firm CES Kalthof in the late 1970s, when the company supplied Pakistan with a complete uranium gas–enrichment facility. The company's owner, Albrecht Migule, was sentenced to eight months in prison for violating Germany's foreign trade law. In 1981, a letter bomb, possibly sent by Israeli intelligence, blew up Mebus's home. Mebus, who at the time was suspected of supplying Iraq's nuclear program, escaped and allegedly continued to have dealings with Khan. He died in 1992. ("More Scandals," *Power Europe*, January 19, 1989; and "Wanted. . . . Bomb Business: Nuclear Aid for Pakistan and India.") Tahir later told Malaysian police that Mebus had been involved in discussions between Khan and Iran to supply centrifuge designs "about 1984, '85." (Polis Diraja Malaysia, "Press Release by Inspector General of Police in Relation to Investigation on the Alleged Production of Components for Libya's Uranium Enrichment Programme," February 20, 2004).

31. Koch, "Der Physiker der Mullahs."
32. Ibid.
33. Report by the Director General, "Implementation of the NPT Safeguards Agreement in the Islamic Republic of Iran," GOV/2005/67.
34. Linzer, "Iran Was Offered Nuclear Parts"; and Report of the Director General, "Implementation of the NPT Safeguards Agreement in the Islamic Republic of Iran," International Atomic Energy Agency, GOV/2006/15 February 27, 2006.
35. Albright and Hinderstein, "The Centrifuge Connection."
36. Koch, "Der Physiker der Mullahs"; and Coll, "The Atomic Emporium."
37. Report by the Director General, "Implementation of the NPT Safeguards Agreement in the Islamic Republic of Iran," GOV/2005/67; and Dafna Linzer, "Strong Leads and Dead Ends in Nuclear Case Against Iran," *Washington Post*, February 10, 2006.
38. Bazoft, "Iran Signs Secret Atom Deal."
39. John Lancaster and Kamran Khan, "Pakistanis Say Nuclear Scientists Aided Iran," *Washington Post*, January 24, 2004.
40. Ohad Gozani, "Pakistan 'Aiding Iran' in Nuclear Weapons Venture," *Daily Telegraph* (London), November 26, 1988.
41. Seymour Hersh, "On the Nuclear Edge," *New Yorker*, March 29, 1993.
42. Kux, *The United States and Pakistan 1947–2000*, p. 291; and Edward Jay Epstein, "Who Killed Zia?" *Vanity Fair*, September 1989.
43. Kux, *The United States and Pakistan 1947–2000*, p. 291; and Epstein, "Who Killed Zia?"
44. Kux, *The United States and Pakistan 1947–2000*, pp. 291–92; Hersh, "On the Nuclear Edge"; and Elizabeth Rubin, "In the Land of the Taliban," *New York Times Magazine*, October 22, 2006.
45. Kux, *The United States and Pakistan 1947–2000*, pp. 291–93; Hersh, "On the Nuclear Edge."
46. Hersh, "On the Nuclear Edge"; author's interview with Richard Barlow, August 27, 2006.
47. "I Refused to Sell Pakistan's Nuclear Technology: Benazir," *Daily Times* (Pakistan), February 25, 2004; and Ian Black and Richard Norton-Taylor, "Saudis 'Trying to Buy Nuclear Weapons,'" *The Guardian* (London), August 4, 1999.
48. Lancaster and Khan, "Pakistanis Say Nuclear Scientists Aided Iran."
49. Ibid.; John Lancaster and Kamran Khan, "Musharraf Named in Nuclear Probe," *Washington Post*, February 3, 2004; Douglas Frantz, "Pakistan's Role in Scientists Nuclear Trafficking Debated," *Los Angeles Times*, May 16, 2005.
50. Federation of American Scientists, "The Pressler Amendment and Pakistan's Nuclear Weapons Program," http://www.fas.org/news/pakistan/1992/920731.htm. See "Letters to Congress from Presidents Reagan & Bush, 1985–1989."
51. Foremost Trading Ltd., corporate records, Companies House, London. Summary of Interviews, R v. Abu Bakr Siddiqui, Southwark Crown Court, London; Testimony of Abu Bakr Siddiqui, Southwark Crown Court, London, August 16, 2001; Polis Diraja Malaysia, "Press Release by Inspector General of Police in Relation to Investigation on the Alleged Production of Components for Libya's Uranium Enrichment Programme," February 20, 2004, available at www.webarchive.org by entering www.rmp.gov.my/rmp03/040220scomi_eng.htm.; and www.smbcomputers.com/english (accessed January 28, 2005).
52. Summary of Interviews, R v. Abu Bakr Siddiqui, Southwark Crown Court, London; Testimony of Abu Bakr Siddiqui; Reddy Siddiqui, www.reddysiddiqui.com; and Abdul Mabood Siddiqui, *Timbuktu, City in the Middle of Nowhere*, 2nd ed. (Islamabad: Agha Jee Printers, 2003).
53. Testimony of Abu Bakr Siddiqui; Foremost Trading Ltd., corporate records, Companies House, London; and Summary of Interviews, R v. Abu Bakr Siddiqui. Farook and Tahir would maintain an apartment near the Siddiquis' home in Ealing, West London, that they used during their regular visits to the UK. Corporate records for the apartment list Farook's nationality as "British." (Testimony of Abu Bakr Siddiqui; and Westmont Court (Ealing) Ltd. corporate records, Companies House, London.)

54. Foremost Trading Ltd., corporate records, Companies House, London; Testimony of Abu Bakr Siddiqui; Summary of Interviews, R v. Abu Bakr Siddiqui; Lucy Rock, "Nuke Test Smuggler Goes Free," *The Mirror* (London), October 9, 2001. Siddiqui later told investigators that he asked Tahir and Farook whether "the UK operations" would close because of a change in the Pakistani government. Siddiqui was assured that "the organization would continue" (Summary of Interviews, R v Abu Bakr Siddiqui).
55. Simon Henderson, "We Can Do It Ourselves," *Bulletin of the Atomic Scientists*, September 1993.
56. Author's telephone interview with Rifat Chowdhury, February 2, 2005.
57. Foremost Trading Ltd., corporate records, Companies House, London; Testimony of Abu Bakr Siddiqui; Summary of Interviews, R v. Abu Bakr Siddiqui.
58. SMB Europe Ltd. corporate records, Companies House, London; Summary of Interviews, R v. Abu Bakr Siddiqui; Testimony of Abu Bakr Siddiqui, Southwark Crown Court, London, August 16, 2001; Testimony of Abu Bakr Siddiqui, Southwark Crown Court, London, August 17, 2001; Ellen Nakashima and Alan Sipress, "Insider Tells of Nuclear Deals, Cash," *Washington Post*, February 21, 2004; www.smbcomputers.com/english. SMB Middle East shared a post office box with SMB Group. See "EC Yellow Pages," http://www.ecemirates.com/yp/list.asp?CompanyType=Readymade%20Garments %20Trading&page=289. Both PSM and UETC appear on a 2003 list complied by MI5 of companies of "proliferation concern," because they "procured goods and/or technology for weapons of mass destruction programmes." (UK Security Service [MI5], "Companies and Organizations of Proliferation Concern," April 2003.) A twenty-fifth anniversary booklet published by KRL credits A.Q. Khan with having "rehabilitated" PSM during the 1990s, transforming it into a modern producer of high-quality steel alloys for "the local industry." (KRL, "Dr. A.Q. Khan Research Laboratories, 1976–2001: 25 Years of Excellence," July 31, 2001.)
59. Report by the Director General, "Implementation of the NPT Safeguards Agreement in the Islamic Republic of Iran," GOV/2005/67; Polis Diraja Malaysia, "Press Release by Inspector General of Police in Relation to Investigation on the Alleged Production of Components for Libya's Uranium Enrichment Programme"; Gordon Corera, *Shopping for Bombs* (New York: Oxford University Press, 2006), pp. 68–69; and Report by the Director General, "Implementation of the NPT Safeguards Agreement in the Islamic Republic of Iran," GOV/2004/83, November 15, 2004.
60. Report by the Director General, "Implementation of the NPT Safeguards Agreement in the Islamic Republic of Iran," GOV/2005/67; Polis Diraja Malaysia, "Press Release by Inspector General of Police in Relation to Investigation on the Alleged Production of Components for Libya's Uranium Enrichment Programme"; and Corera, *Shopping for Bombs*, pp. 68–69.
61. Report by the Director General, "Implementation of the NPT Safeguards Agreement in the Islamic Republic of Iran," GOV/2005/87; Report by the Director General, "Implementation of the NPT Safeguards Agreement in the Islamic Republic of Iran," GOV/2005/67; Albright and Hinderstein, "The Centrifuge Connection"; and Linzer, "Strong Leads and Dead Ends in Nuclear Case Against Iran."
62. Albright and Hinderstein, "The Centrifuge Connection"; and Linzer, "Strong Leads and Dead Ends in Nuclear Case Against Iran."
63. Albright and Hinderstein, "The Centrifuge Connection"; and Hoagland, "Briefing to Yeltsin on Iran." According to *The Guardian*, in 2004 the magazine *Der Spiegel* reported that a German intelligence report from the mid-1990s stated that "there is said to be cooperation between Iran's atomic energy organization and Pakistan's Khan laboratories." (Quoted in Ian Traynor, "Pakistan's Nuclear Hero Throws Open Pandora's Box, *The Guardian* [London], January 31, 2004.)
64. Carla Ann Robbins, "As Evidence Grows Of Iran's Program, US Hits Quandary," *Wall Street Journal*, March 18, 2005.
65. Hoagland, "Briefing to Yeltsin on Iran."

66. David Albright and Corey Hinderstein, "Documents Indicate A.Q. Khan Offered Nuclear Designs to Iraq in 1990," Institute for Science and International Security, February 4, 2004, http://www.isis-online.org/publications/southasia/khan_memo.html; Allan Urry, "Britain 'Knew About Nuclear Network,'" *BBC News*, August 17, 2004; Kamran Khan, "Pakistanis Exploited Nuclear Network," *Washington Post*, January 28, 2004.
67. David E. Sanger, "The Khan Network," presented at the Conference on South Asia and the Nuclear Future, Stanford Institute for International Studies, Stanford University, June 4–5, 2004, available at http://iis-db.stanford.edu/evnts/3889/Khan_network-paper.pdf; "Research and Publications contributed by Dr. A. Q. Khan, Ni & Bar, Hi," http://draqkhan.com.pk/respub.htm; Andrew Koch, "The Nuclear Network," *Jane's Defense Weekly*, February 26, 2004; "Dr. A.Q. Khan Research Laboratories, 1976–2001"; David Albright and Holly Higgins, "A Bomb for the Ummah," *Bulletin of the Atomic Scientists*, March/April 2003; Zahid Malik, *Dr. A.Q. Khan and the Islamic Bomb* (Islamabad: Hurmat Publications, 1992); and Victoria Burnett and Stephen Fidler, "Connections Across the World Were Involved . . . ," *Financial Times* (London), April 7, 2004.
68. Bruce Odessey, "Sites in 5 Countries Identified with Weapons Proliferation," US Information Agency, June 30, 1997.
69. Polis Diraja Malaysia, "Press Release by Inspector General of Police in Relation to Investigation on the Alleged Production of Components for Libya's Uranium Enrichment Programme"; Bill Powell and Tim McGirk, "The Man Who Sold the Bomb," *Time*, February 6, 2005; Raymond Bonner and Craig Smith, "Pakistani Said to Have Given Libya Uranium," *New York Times*, February 21, 2004; Corera, *Shopping for Bombs*, pp. 107–108.
70. Polis Diraja Malaysia, "Press Release by Inspector General of Police in Relation to Investigation on the Alleged Production of Components for Libya's Uranium Enrichment Programme"; Dahlkamp *et al.*, "Network of Death on Trial," *Der Spiegel*, March 13, 2006; and Raymond Bonner and Craig Smith, "Pakistani Said to Have Given Libya Uranium," *New York Times*, February 21, 2004; and Corera, *Shopping for Bombs*, pp. 107–108.
71. Polis Diraja Malaysia, "Press Release by Inspector General of Police in Relation to Investigation on the Alleged Production of Components for Libya's Uranium Enrichment Programme"; Dahlkamp et al., "Network of Death on Trial"; Bonner and Smith, "Pakistani Said to Have Given Libya Uranium"; Judith Miller, "How Gadhafi Lost his Groove," *Wall Street Journal*, May 16, 2006.
72. Polis Diraja Malaysia, "Press Release by Inspector General of Police in Relation to Investigation on the Alleged Production of Components for Libya's Uranium Enrichment Programme"; Dahlkamp et al., "Network of Death on Trial"; and Bonner and Smith, "Pakistani Said to Have Given Libya Uranium"; Powell and McGirk, "The Man Who Sold the Bomb," *Time*, February 6, 2005; William Broad and David Sanger, "As Nuclear Secrets Emerge in Khan Inquiry, More Are Suspected," *New York Times*, December 26, 2004.
73. SMB Europe Ltd. corporate records, Companies House, London; Testimony of Abu Bakr Siddiqui, Southwark Crown Court, London, August 16, 2001; and Summary of Interviews, R v. Abu Bakr Siddiqui.
74. Testimony of Abu Bakr Siddiqui, Southwark Crown Court, London, August 16, 2001; Testimony of Abu Bakr Siddiqui, Southwark Crown Court, London, August 17, 2001; and Summary of Interviews, R v. Abu Bakr Siddiqui. In August 1996, Tahir placed an order with Orland for a gravity meter, a device that measures the pull of gravity at a particular location. Siddiqui ordered the meter from a Canadian manufacturer, telling them it was going to Bangladesh, according to records from Siddiqui's trial. When it arrived in London, Siddiqui sent it on to Tahir in Dubai. Later that year, Siddiqui received an inquiry from one of Khan's associates in Pakistan indicating that there was difficulty setting up the instrument. The inquiry indicated the meter was being used at latitude 33.2°. Siddiqui later told investigators he did not know what the instrument was for. Told by one of the investigating officers that it was "set up at a latitude consistent with Kahuta," Siddiqui said he did not know what the latitude reading meant. He denied knowing the machine was going to Pakistan. Asked at trial if the latitude reading had meant anything to him,

Siddiqui replied "no." Asked if he had noticed the lack of a longitude measurement, Siddiqui apparently did not offer an answer. Asked if he knew what longitude was, he replied "no." (Summary of Interviews, R v. Abu Bakr Siddiqui, Southwark Crown Court, London; and Testimony of Abu Bakr Siddiqui, Southwark Crown Court, London, 8/20/01.)

75. Testimony of Abu Bakr Siddiqui, Southwark Crown Court, London, August 16, 2001; Testimony of Abu Bakr Siddiqui, Southwark Crown Court, London, August 17, 2001; Summary of Interviews, R v. Abu Bakr Siddiqui; and Evidence of Alan Bicker, R v. Abu Siddiqui, Southwark Crown Court, London, August 13, 2001. Siddiqui testified that he was occasionally asked to send money to Dina Khan and her professor/tutor at the University of Kent in Canterbury. At least once, he testified, Dina Khan and her then-husband, Noman Shah, who is reported to have had business dealings with A.Q. Khan, were invited to the Siddiquis' for dinner. According to Siddiqui, shortly after Dina Khan and Shah were married, they stayed at the apartment used by Farook and Tahir and stopped by the Siddiquis' to pick up the keys. (Testimony of Abu Bakr Siddiqui, Southwark Crown Court, London, August 16, 2001; and "Khan Built Hotel in Timbuktu," *Times of India*, February 1, 2004.)
76. Testimony of Abu Bakr Siddiqui, Southwark Crown Court, London, August 16, 2001; Testimony of Abu Bakr Siddiqui, Southwark Crown Court, London, August 17, 2001; and Summary of Interviews, R v. Abu Bakr Siddiqui.
77. Victoria Burnett and Stephen Fidler, "Connections Across The World Were Involved . . . ," *Financial Times* (London), April 7, 2004; and Melvyn Howe, "Metal Bars Destined for Nuclear Programme, Court Told," Press Association (UK), August 7, 2001.
78. Testimony of Abu Bakr Siddiqui, Southwark Crown Court, London, August 17, 2001; and Summary of Interviews, R v. Abu Bakr Siddiqui.
79. Weiss, "Pakistan, It's Déjà Vu All Over Again"; Leonard Weiss, *Testimony on the A.Q. Khan Network*, before the House Committee on International Relations, Subcommittee on International Terrorism and Nonproliferation, May 25, 2006.
80. "Interview with Adbul Qadeeer Khan," *The News* (Islamabad), May 30, 1998.
81. Summary of Interviews, R v. Abu Bakr Siddiqui; and Testimony of Abu Bakr Siddiqui, Southwark Crown Court, London, August 20, 2001
82. With apologies to Preston Sturges.
83. Testimony of Abu Bakr Siddiqui, Southwark Crown Court, London, August 16, 2001; and Summary of Interviews, R v. Abu Bakr Siddiqui.
84. Abdul Mabood Siddiqui, *Timbuktu, City in the Middle of Nowhere*; Shyam Bhatia, "UK Accountant Writes of Travel to Timbuktu with Khan," *Deccan Herald*, February 19, 2004; Edward Harris and Ellen Knickmeyer, "Pakistani's Trips to Africa Raise Concerns," Associated Press, April 27, 2004; "Pakistan Releases Three Men Detained for Questioning in Nuclear Smuggling Investigations," Global Security Newswire, July 26, 2004; "Pakistan Nuke Scientist Khan, Colleagues Met Libyans," Dow Jones International, March 17, 2005; Burnett and Fidler, "Connections Across the World Were Involved . . ."; "Dutch Businessman Jailed for Nuclear Exports to Pakistan," Agence France Presse, December 16, 2005; "Businessman Accussed Of Illegal Shipments," *Brandon Sun* (Canada), November 6, 2005.
85. Siddiqui, *Timbuktu*; and "Pakistan Nuke Scientist Khan, Colleagues Met Libyans."
86. Siddiqui, *Timbuktu*; and Shyam Bhatia, "UK Accountant Writes of Travel to Timbuktu with Khan."
87. Siddiqui, *Timbuktu*; Raymond Bonner, "Salesman on Nuclear Circuit Casts Blurry Corporate Shadow," *New York Times*, February 18, 2004; and Nakashima and Sipress, "Insider Tells of Nuclear Deals, Cash." Regarding Tahir's relationship with Khan, see Dahlkamp et al., "Network of Death on Trial."
88. Siddiqui, *Timbuktu*; Summary of Interviews, R v. Abu Bakr Siddiqui; Testimony of Abu Bakr Siddiqui, Southwark Crown Court, London, August 16, 2001; and Corera, *Shopping for Bombs*, p. 106.

89. Summary of Interviews, R v. Abu Bakr Siddiqui; Witness Evidence of Atif Amin, R v. Abu Siddiqui, Southwark Crown Court, London, August 6, 2001. Regarding the Internet ad, see http://www.action2000.com/newpg/ORLAND.EUROPE.html.
90. Summary of Interviews, R v. Abu Bakr Siddiqui; Testimony of Abu Bakr Siddiqui, Southwark Crown Court, London, August 20, 2001.
91. Witness Evidence of Maxine Crook, R v. Abu Siddiqui, Southwark Crown Court, London, August 6, 2001; and Testimony of Abu Bakr Siddiqui, Southwark Crown Court, London, August 20, 2001.
92. Ibid.
93. Witness Evidence of Maxine Crook, R v. Abu Siddiqui, Southwark Crown Court, London, August 6, 2001.
94. Testimony of Abu Bakr Siddiqui, Southwark Crown Court, London, August 20, 2001; Witness Evidence of Atif Amin, R v. Abu Siddiqui, Southwark Crown Court, London, August 6, 2001; and Testimony of Abu Bakr Siddiqui, Southwark Crown Court, London, August 16, 2001.
95. Witness Evidence of Atif Amin, R v. Abu Siddiqui, Southwark Crown Court, London, August 6, 2001; Witness Evidence of Maxine Crook, R v. Abu Siddiqui, Southwark Crown Court, London, August 6, 2001; Summary of Interviews, R v. Abu Bakr Siddiqui; Testimony of Abu Bakr Siddiqui, Southwark Crown Court, London, August 17, 2001; Testimony of Abu Bakr Siddiqui, Southwark Crown Court, London, August 20, 2001; Shyam Bhatia, "Nuclear Links," *Deccan Herald*, September 13, 2001; Helen William, "Exporter Aided Pakistan N-Weapon Drive," Press Association (UK), August 1, 2001.
96. "Operation Akin" Witness List, R v. Abu Siddiqui, Southwark Crown Court, London; Witness Evidence of Atif Amin, R v. Abu Siddiqui, Southwark Crown Court, London, August 6, 2001; Tim Spicer, *An Unorthodox Soldier* (Edinburgh: Mainstream Publishing, 1999), pp. 208–209. One of the investigations Amin headed was a breech-of-sanctions case in which Western military equipment had been smuggled to Montenegrin paramilitary forces during the Bosnian conflict. The other was the so-called Arms for Africa affair, an investigation into whether a private military company had violated a UN embargo by supplying arms to the government-in-exile of Sierra Leone.
97. Witness Evidence of Atif Amin, R v. Abu Siddiqui, Southwark Crown Court, London, August 6, 2001; Witness Evidence of Atif Amin, R v. Abu Siddiqui, Southwark Crown Court, London, August 15, 2001; Summary of Interviews, R v. Abu Bakr Siddiqui.
98. Ibid.
99. Raymond Bonner, "Salesman on Nuclear Circuit Casts Blurry Corporate Shadow"; and Burnett and Fidler, "Connections Across the World Were Involved . . . ," *Financial Times* (UK), April 7, 2004.
100. Summary of Interviews, R v. Abu Bakr Siddiqui.
101. Green Crest Industries (M.E.) Ltd. listing in the Dubai Chamber of Commerce and Industry registry, www.dcci.ae; and Green Crest Industries (M.E.) Ltd. corporate records, Registrar General's Department, Commonwealth of the Bahamas, Nassau, Bahamas.
102. Green Crest Industries (M.E.) Ltd. corporate records.
103. Author's telephone interviews with S.M. Farooq, March 3 and March 21, 2006; and "S.M. Farooq's Death Condoled," *Daily Times* (Pakistan), April 4, 2007.
104. Weissman and Krosney, *The Islamic Bomb*, pp. 203–206. Farooq confirmed in an interview that he and his brothers controlled Asiatic. (Author's telephone interview with S.M. Farooq, March 21, 2006.)
105. Cable, Secretary of State (Shultz), Secret State 239574, "Japanese Assistance to the Indian and Pakistani Nuclear Programs," Ref: State 230826, R 052230Z August 85. The cable identifies the managing director of Arshad Amjad & Abid as "S.M. Farooq." (Also see Government of Pakistan, Board of Investment, "Arshad Amjad & Abid (Private) Limited," http://www.pakboi.gov.pk/pak/xdetail.asp?ComID=7817.)
106. UK Security Service (MI5), "Companies and Organizations of Proliferation Concern."
107. Malik, *Dr. A.Q. Khan and the Islamic Bomb*, pp. 85–86.

108. "Karachi: Resumption of Academic Activities at KU Ordered," *Dawn*, September 15, 2002; Board of Governors, Indus Institute, www.indus.edu.pk/board_govern.htm (as retrieved August 14, 2005; accessed March 10, 2006); "Pak Nuclear Capability Indebted to Afghan War," *Dawn/The News International*, May 27, 2001; Ardeshire Cowasjee, "The Depths of Degradation," *Dawn*, December 22, 2002; Syed Shoaib Hasan, "Behaviour Problems" *Herald* (Pakistan), February 2003; and E-mail, Dr. S. Haroon Ahmed to author, May 2, 2007.
109. Author's interviews with K. Hafeez Uddin, February 27, March 8, and March 28, 2006.
110. Author's interviews with K. Hafeez Uddin, March 28, 2006.
111. Author's interviews with K. Hafeez Uddin, February 27, March 8, and March 28, 2006.
112. Ibid.
113. Author's interview with K. Hafeez Uddin, June 13, 2007.
114. Bhatia, "Nuclear Links"; William, "Exporter Aided Pakistan N-Weapon Drive"; Crawford and Stecklow, "How the Pakistani Nuclear Ring Managed to Skirt Export Law," *Wall Street Journal*, March 23, 2004.
115. HM Customs and Excise News Release, "Guilty of Contravening Customs Export Controls," August 29, 2001; and "Briton's N-Bomb Scandal," *The Express* (UK), August 30, 2001.
116. Melvyn Howe, "Businessman Who Aided Nuclear Race Escapes Jail Term," Press Association (UK), October 8, 2001; Lucy Rock, "Nuke Test Smuggler Goes Free," *The Mirror* (UK), October 9, 2001.
117. Nakashima and Sipress, "Insider Tells of Nuclear Deals, Cash"; and Burnett and Fidler, "Connections Across The World Were Involved."
118. See, for example Douglas Jehl, "CIA Says Pakistan Gave Iran Nuclear Aid," *New York Times*, November 23, 2004; Remarks by CIA Deputy Director for Operations James L. Pavitt at Foreign Policy Council, June 21, 2004, www.cia.gov; and "Review of Intelligence on Weapons of Mass Destruction," (Butler Committee Report), London: The Stationery Office, July 14, 2004, p. 19.
119. "Transcript of the Candidates' First Debate in the Presidential Campaign," *New York Times*, October 1, 2004.

Chapter 10

1. Jim Hoagland, "Briefing to Yeltsin on Iran," *Washington Post*, May 17, 1995; "Review of Intelligence on Weapons of Mass Destruction," (Butler Committee Report) (London: The Stationery Office, July 14, 2004); Ron Suskind, *The One Percent Doctrine* (New York: Simon & Schuster, 2006), p. 68; "Iraq and Weapons of Mass Destruction," Remarks of CIA Director George Tenet at Georgetown University, February 5, 2004, available at www.cia.gov; Remarks of CIA Deputy Director for Operations James Pavitt at the Foreign Policy Association, June 21, 2004, available at www.cia.gov; Douglas Jehl, "CIA Says Pakistan Gave Iran Nuclear Aid," *New York Times*, November 23, 2004; and "Remarks by the President on Weapons of Mass Destruction Proliferation," National Defense University, Washington, DC, February 11, 2004, available at www.whitehouse.gov.
2. "Review of Intelligence on Weapons of Mass Destruction." Also see "The Commission on the Intelligence Capabilities of the United States Regarding Weapons of Mass Destruction," Report to the President of United States, March 31, 2005, available at www.whitehouse.gov/wmd.
3. "Review of Intelligence on Weapons of Mass Destruction."
4. Gordon Corera, *Shopping for Bombs* (New York: Oxford University Press, 2006), pp. 136–37; and "Review of Intelligence on Weapons of Mass Destruction."
5. John Lancaster and Kamran Khan, "Musharraf Named in Nuclear Probe," *Washington Post*, February 3, 2004; Dennis Kux, *The United States and Pakistan 1947–2000* (Baltimore: Johns Hopkins University Press, 2001), p. 207; Dan Stober and Daniel Sneider, "US Knew about Nuclear Link between N. Korea, Pakistan," *San Jose Mercury News*, October 24, 2002; David Sanger, "In North Korea and Pakistan, Deep Roots of Nuclear Barter," *Los Angeles Times*, November 24, 2002.

6. Stephen Fidler, "Suspicions Deepen over Islamabad Connection," *Financial Times* (London), November 1, 2002.
7. Farhan Bokhari, et al., "Pakistan's 'Rogue Nuclear Scientist," *Financial Times* (London), April 6, 2004; Fidler, "Suspicions Deepen over Islamabad Connection"; Lancaster and Khan, "Musharraf Named in Nuclear Probe." Also see David Armstrong, "Friend Like This," *The New Republic*, November 29/December 6, 2004.
8. Jehl, "CIA Says Pakistan Gave Iran Nuclear Aid."
9. Suskind, *The One Percent Doctrine*, p. 261.
10. Douglas Frantz, "Pakistan's Role in Scientists Nuclear Trafficking Debated," *Los Angeles Times*, May 16, 2005; "Excerpts of the Speech Given by General Pervez Musharraf . . . at the Farewell Dinner in Honour of Dr. A.Q. Khan," March 27, 2001, in KRL, "Dr. A.Q. Khan Research Laboratories, 1976–2001: 25 Years of Excellence," July 31, 2001.
11. William Broad and David Sanger, "As Nuclear Secrets Emerge in Khan Inquiry, More Are Suspected," *New York Times*, December 26, 2004.
12. "Colin Powell Transcript," *Christian Science Monitor*, December 23, 2004.
13. Stephen Fidler and Edward Luce, "US Fears North Korea Could Gain Nuclear Capability through Pakistan," *Financial Times* (London), June 1, 2001.
14. Michael Abramowitz, "Bush Seeks Increased Cooperation," *Washington Post*, September 23, 2006; and Pervez Musharraf, *In the Line of Fire* (New York: Free Press, 2006), p. 201.
15. Leonard Weiss, "Pakistan, It's Déjà Vu All Over Again," *Bulletin of the Atomic Scientists*, May/June 2004.
16. Con Coughlin, "Our Safety Is in the Hands of Pakistan," *Daily Telegraph* (London), July 17, 2005.
17. Suskind, *The One Percent Doctrine*, pp. 66–69, 122; and Albright and Higgins, "A Bomb for the Ummah," *Bulletin of the Atomic Scientists*, March/April 2003.
18. Suskind, *The One Percent Doctrine*, pp. 260–261.
19. E-mail, Tait to Armstrong, May 11, 2007
20. The authors sent the letter to Paul Griffin via his attorney in Dubai, Christopher Mills. Mr. Mills indicated he would forward the letter to the most recent e-mail address he had for the younger Mr. Griffin. (E-mail, Mills to Armstrong, May 13, 2007.)
21. "Review of Intelligence on Weapons of Mass Destruction."
22. George Tenet with Bill Harlow, *At the Center of the Storm: My Years at the CIA* (New York: HarperCollins, 2007), p. 296.
23. Owen Bowcott, et al., "Businessman under Scrutiny 25 Years Ago after Ordering Unusual Supplies," *The Guardian* (London), March 5, 2004; Steve Coll, "The Atomic Emporium," *New Yorker*, August 7 and 14, 2006.
24. Juergen Dahlkamp et al., "Network of Death on Trial," *Der Spiegel*, March 13, 2006.
25. William J. Broad, et al., "A Tale of Nuclear Proliferation: How Pakistani Built His Network," *New York Times*, February 12, 2004; Ian Traynor, et al., "Customs Examine British Link in Nuclear Parts Trade," *The Guardian* (London), March 5, 2004.
26. Coll, "The Atomic Emporium"; Shaun Greaney, "We Didn't Help in Libya Arms Design," *South Wales Evening Post*, March 2, 2004.
27. "Statement in Open Court: Peter Griffin and Guardian Newspapers Limited," Carter-Ruck, http://www.carter-ruck.com/recentwork/Griffin_SOC.html; "Guardian Libel Case Settled," *The Guardian* (London), May 4, 2005; Chris Peregrine, "City Man Says 'Yes' to £50,000 Libel Payout," *South Wales Evening Post*, May 5, 2005.
28. Carter-Ruck, "Recent Work," http://www.carter-ruck.com/recentwork/; "Apology broadcast on *File on Four*, BBC Radio 4 on 13 December 2005," Carter-Ruck, http://www.carter-ruck.com/recentwork/Griffin_BBC-Apology03Jan06.html.
29. Bowcott, et al., "Businessman under Scrutiny 25 Years Ago after Ordering Unusual Supplies"; Ian Traynor, et al., "Customs Examine British Link in Nuclear Parts Trade," *The Guardian* (London), March 5, 2004.

30. Regarding Amin's work at the Home Office, see Atif Amin, Explanatory Memorandum, "The Regulation of Investigatory Powers (Communications Data) Order 2003," UK Home Office, September 2003.
31. For an account of the raid on Griffin's home, also see Richard Norton-Taylor, Owen Bowcott, and Ian Traynor, "MI6 Seized Computers from British Suspect," *The Guardian* (London), February 13, 2004.
32. Polis Diraja Malaysia, "Press Release by Inspector General of Police in Relation to Investigation on the Alleged Production of Components for Libya's Uranium Enrichment Programme," February 20, 2004, available at www.webarchive.org by entering www.rmp.gov.my/rmp03/040220sconi_eng.htm.
33. "Statement in Open Court: Peter Griffin and Guardian Newspapers Limited."
34. Polis Diraja Malaysia, "Press Release by Inspector General of Police in Relation to Investigation on the Alleged Production of Components for Libya's Uranium Enrichment Programme."
35. Wolfgang Frey, "German Engineer Incriminated in Trial on Nuclear Deals with Iran, Libya," *Spiegel* Online, May 17, 2006.
36. E-mail, Al Abbar Medical and Scientific Supplies to Armstrong, May 30, 2007. The e-mail from Al Abbar Medical states:

    To start we would like to express our appreciation for contacting us and allowing us to make a comment and in turn have some input on the information that will be included in your book and which pertains to our company.

    We would like to clarify some matters concerning our company and issues that your book will cover. The company "Al Abbar Medical and Scientific Supplies" does not currently and did not at any point in the past have an Engineering Division. . . . Also [Gulf Technical Industries] is currently closed and has been for some time, in other words it doesn't exist anymore. We would also like to clarify that the person who was involved in the business transactions that your book will discuss was an employee at Al Abbar Medical and Scientific Supplies and later worked for Gulf Technical Industries and any actions that he might have taken with regards to the transactions in question were an abuse of the trust placed in him by both companies. He is no longer an employee with the company and has nothing to do with the company or its owners.

    We are aware that his actions have been investigated by the concerned authorities in UK and as far as we know there was nothing on him. Al Abbar Medical and Scientific Supplies is a family business with a respectable reputation in the UAE. We are a company that conducts all its business with full transparency and compliancy to all UAE laws; and does not under any circumstances conduct any questionable business dealings. Again I would like to stipulate that with regards to the transactions your e-mail mentioned Al Abbar Medical and Scientific Supplies was abused by the individual involved.

    We again thank you for allowing us to explain our position and contribute to your book. We hope that you will consider the information above while finalizing your book so as to preserve our companies market reputation and goodwill. . . . We wish you the best of luck in your book and your future ventures.

    Best Regards,
    Al Abbar Medical & Scientific Supplies
    www.alabbarmed.com

37. Rempel and Frantz, "Global Nuclear Inquiry Stalls," *Los Angeles Times*, December 5, 2004.
38. Telephone interview with Harbil Echaniz (former co-owner of Comercial Nork), May 23, 2007.
39. William Rempel and Douglas Frantz, "Global Nuclear Inquiry Stalls."
40. Gordon Corera, *Shopping for Bombs* (New York: Oxford University Press, 2006), p. 117. Corera reports that Griffin said the lathes were for a Libyan National Oil Company "repair

shop" in Dubai. The 2005 British Customs report states that Griffin said they were for an LNOC "machine shop" in Dubai.

41. Indictment, State v. Daniel Geiges and Gerhard Wisser, High Court of South Africa, Transvaal Provincial Division, pp. 28–29.
42. Telephone interview with Harbil Echaniz, May 23, 2007; eInforma (www.einforma.com) listing for Comercial Nork. Echaniz noted that one piece of equipment that Comercial Nork sold to GTI, a cylindrical grinder, did not require an export license at the time her company sent it to Dubai but that new managers of the company that made the machine have since said that it could be dual-use.
43. Polis Diraja Malaysia, "Press Release by Inspector General of Police in Relation to Investigation on the Alleged Production of Components for Libya's Uranium Enrichment Programme."
44. Telephone interview with Harbil Echaniz.
45. Broad, et al., "A Tale of Nuclear Proliferation: How Pakistani Built His Network"; Bowcott, et al., "Businessman Under Scrutiny 25 Years Ago after Ordering Unusual Supplies"; Nakashima and Sipress, "Insider Tells of Nuclear Deals, Cash."
46. Bonner, "Salesman From Nuclear Circuit Casts Blurry Corporate Shadow"; and Nakashima and Sipress, "Insider Tells of Nuclear Deals, Cash."
47. Broad, et al., "A Tale of Nuclear Proliferation: How Pakistani Built His Network."
48. Polis Diraja Malaysia, "Press Release by Inspector General of Police in Relation to Investigation on the Alleged Production of Components for Libya's Uranium Enrichment Programme."
49. Suskind, *The One Percent Doctrine*, pp. 259–260; Associated Press, "U.S. Assists Swiss Probe of Family Accused of Aiding Libyan Nuclear Program," *International Herald Tribune*, November 28, 2006. Also see Mark Hosenball and Christopher Dicky, "Pyongyang Isn't about to Give Up Its Weapons," *Newsweek*, October 30, 2006.
50. Broad, et al., "A Tale of Nuclear Proliferation: How Pakistani Built His Network"; Raymond Bonner, "Libyan Nuclear Components Are Traced to Malaysian Firm," *International Herald Tribune*, February 7, 2004; Raymond Bonner, "'Business as Usual' at Plant that Tenet Says Was Shut," *New York Times*, February 7, 2004; "SCOPE's Press Statement on Its Contract in Dubai," February 4, 2004, www.scomigroup.com/my/publish/04news008.shtml; Anwar Faruqi, "Dubai Businessman Accused of Nuclear Smuggling Lived a Flashy but Secretive Life," Associated Press, February 14, 2004; and Polis Diraja Malaysia, "Press Release by Inspector General of Police in Relation to Investigation on the Alleged Production of Components for Libya's Uranium Enrichment Programme."
51. Polis Diraja Malaysia, "Press Release by Inspector General of Police in Relation to Investigation on the Alleged Production of Components for Libya's Uranium Enrichment Programme."
52. Victoria Burnett and Stephen Fidler, "Connections Across the World Were Involved . . . ," *Financial Times* (UK), April 7, 2004. Also see Bonner, "Salesman From Nuclear Circuit Casts Blurry Corporate Shadow"; Desert Electrical Equipment Factory listing at ecEmirates.com (http://www.ecemirates.com/yp/details.asp?companyid=23803); and Fast Way Advertising listing at AME Info (http://www.ameinfo.com/db-147289.html). Desert Electrical did not respond to a letter seeking comment sent to the fax number by the authors (fax to Desert Electrical, May 20, 2007). Fast Way confirmed that the phone and fax numbers are now theirs (telephone interview, May 20, 2007).
53. Coll, "The Atomic Emporium"; Corera, *Shopping for Bombs*, pp. 113–14.
54. Broad, et al., "A Tale of Nuclear Proliferation: How Pakistani Built His Network"; Miller, "How Gadhafi Lost His Groove"; Suskind, *The One Percent Doctrine*, p. 268; Robin Wright, "Ship Incident May Have Swayed Libya," January 1, 2004; "Remarks by the President on Weapons of Mass Destruction Proliferation"; and Polis Diraja Malaysia, "Press Release by Inspector General of Police in Relation to Investigation on the Alleged Production of Components for Libya's Uranium Enrichment Programme."

55. Miller, "How Gadhafi Lost His Groove."
56. John Lancaster and Kamran Khan, "Pakistani Scientist is Pardoned," *Washington Post*, February 6, 2004.
57. David Rohde and Talat Hussain, "Delicate Dance For Musharraf in Nuclear Case," *New York Times*, February 8, 2004; "No Relief for Pak Scientist," *The Telegraph* (India), October 25, 2004; and Agence France-Presse, "Pakistan Says Nuclear Proliferation Chapter Is Closed," May 3, 2006. In 2004, three of Khan's KRL associates were said to be under house arrest after their release by Pakistani authorities. ("Pakistan Releases Three Men Detained for Questioning in Nuclear Smuggling Investigation," Global Security Newswire, July 26, 2004.)
58. Hardev Kaur, "Tahir Posed a Threat to Security," *New Straits Times*, May 30, 2004; "Rogues' Gallery," *Bulletin of the Atomic Scientists*, November/December 2006.
59. "Rogues' Gallery."
60. Dafna Linzer and Craig Timberg, "S. African's Arrest Seen as Key to Nuclear Black Market," September 4, 2004; Indictment, State v Daniel Geiges and Gerhard Wisser, High Court of South Africa, Transvaal Provincial Division; and Nuclear Threat Initiative, "South Africa Profile: Nuclear Chronology," entry for September 2, 2004, http://www.nti.org/e_research/profiles/SAfrica/Nuclear/2149_4624.html.
61. South African Press Association, "Bid to Bar Media from Nuclear Trial," iafrica.com, March 1, 2007, http://www.iafrica.com/news/sa/646467.htm.
62. Ian Traynor, "First Trial over Libya's Nuclear Bomb Plan Collapses," *Guardian* (London) July 26, 2006; Stephen Graham, "Trial Opens against German . . ." Associated Press, March 17, 2006; Coll, "The Atomic Emporium"; "Retrial Ordered for German in Libyan Arms Deal," www.expatica.com, July 26, 2006.
63. Coll, "The Atomic Emporium."
64. E-mail, Griffin to Armstrong, June 18, 2006.

Epilogue

1. "Remarks by the President on Weapons of Mass Destruction Proliferation," National Defense University, Washington, DC, February 11, 2004, available at www.whitehouse.gov; P.S. Suryanarayana, "Ties with India Toned Up: Rumsfeld," *The Hindu* (India), June 6, 2004; and *CNN Late Edition* with Wolf Blitzer, Interview with Condolezza Rice, October 3, 2004.
2. Kenley Butler et al., "Where is the Justice?" *Bulletin of the Atomic Scientists*, November/December 2006; and Ian Traynor and Ian Cobain, "Intelligence Report Claims Nuclear Market Thriving," *Guardian* (London), January 4, 2006.
3. David Sanger, "Inspectors Cite Big Gain by Iran on Nuclear Fuel," *New York Times*, May 15, 2007; and William Broad and David Sanger, "With Eye on Iran, Rivals Also Want Nuclear Power," *New York Times*, April 15, 2007.
4. George Tenet with Bill Harlow, *At the Center of the Storm: My Years at the CIA* (New York: HarperCollins, 2007), p. 287; Douglas Jehl, "CIA Says Pakistan Gave Iran Nuclear Aid," *New York Times*, November 23, 2004; Remarks by the CIA Deputy Director for Operations James L. Pavitt at Foreign Policy Council, June 21, 2004, available at www.cia.gov; Remarks as prepared for delivery by Director of Central Intelligence George Tenet at Georgetown University, February 5, 2004, available at www.cia.gov; John McLaughlin, "The CIA Is No 'Rogue' Agency," *Washington Post*, November 24, 2004.
5. Brian Ross, "U.S. Was Aware of Secret Nuke," ABC News, March 4, 2007.
6. Author's telephone interview with Richard Barlow, August 27, 2006.
7. John Lancaster and Kamran Khan, "Musharraf Named in Nuclear Probe," *Washington Post*, February 3, 2004; and "Khan Had Run His Own Nuclear Proliferation Network," outlookindia.com, July 8, 2006.
8. "AQ Was Made Scapegoat to save Bigwigs: Benazir," *The Times of India*, July 10, 2006.
9. "U.S. Believes Musharraf that A Q Khan Acted Alone," *Malaysia Sun*, March 2, 2007; and "Khan Network Dismantled, Says Crocker," *The Nation* (Pakistan), March 31, 2006.

10. David Rhode, "U.S. Will Celebrate Pakistan As a 'Major Non-NATO Ally,'" *New York Times*, March 19, 2004; and Testimony of Leonard Weiss, on the A.Q. Khan Network before the House Committee on Internal Relations, Subcommittee on International Terrorism and Nonproliferation, May 25, 2006.
11. Rhode, "U.S. Will Celebrate Pakistan As a 'Major Non-NATO Ally.'"
12. William Broad and David Sanger, "As Nuclear Secrets Emerge in Khan Inquiry, More Are Suspected," *New York Times*, December 26, 2004.
13. Marie Colvin, "How an Insider Lifted the Veil on Saudi Plot for an 'Islamic Bomb,'" *Sunday Times* (London), July 24, 1994; Thomas Woodrow, "The Sino-Saudi Connection," *China Brief*, The Jamestown Foundation, October 24, 2002; Ian Black and Richard Norton-Taylor, "Saudis 'Trying to Buy Nuclear Weapons,'" *The Guardian* (London), August 4, 1999; Umar Faruq, "Saudi Defense Minister Visits Pakistani Nuclear Sites," *Al-Sharq al-Awsat* (London), May 9, 1999; Yossef Bodansky and Gregory Copley, "Pakistan Agrees to Station Nuclear Weapons, Long-Range Missiles in Saudi Arabia," *Defense & Foreign Affairs Daily*, October 22, 2003; Testimony of Andrew Koch, "AQ Khan Network: Case Closed," before the House Committee on Internal Relations, Subcommittee on International Terrorism and Nonproliferation, May 25, 2006; Arnaud de Borchgrave, "Pakistan, Saudi Arabia in Secret Nuke Pact," *Washington Times*, October 22, 2003; Broad and Sanger, "With Eye on Iran, Rivals Also Want Nuclear Power."
14. Testimony of Andrew Koch.
15. Ian Traynor and Ian Cobain, "Intelligence Report Claims Nuclear Market Thriving," *The Guardian* (London), January 4, 2006.
16. Testimony of Leonard Weiss, May 25, 2006.
17. Traynor and Cobain, "Intelligence Report Claims Nuclear Market Thriving."
18. Josh Meyer, "Illegal Nuclear Deals Alleged," *Los Angeles Times*, March 26, 2005; and Josh Meyer, "Case Reveals Nuts and Bolts of Nuclear Network, Officials Say," *Los Angeles Times*, May 24, 2004.
19. Ibid.
20. Meyer, "Illegal Nuclear Deals Alleged"; Josh Meyer, "Indictment Points to Arms Danger," *Los Angeles Times*, April 9, 2005; Josh Meyer, "Man Gets 3 Years for Sales Linked to Nuclear Arms," *Los Angeles Times*, August 5, 2005; Testimony of Leonard Weiss; U.S. State Department, *Treaties in Force 2006*, http://www.state.gov/documents/organization/66286.pdf, p. 246.
21. Testimony of Leonard Weiss, May 25, 2006.
22. Meyer, "Illegal Nuclear Deals Alleged."
23. Ibid.
24. David Wood, "Pakistan's Bombs a Dilemma for U.S.," *Times-Picayune* (New Orleans), April 7, 2006.
25. Al Santoli, ed., "North Korea — al Qaeda Nuclear Threat; Nepotism Could Hinder Thailand's Counter-terror Campaign," Asia *Security Monitor*, no. 73 (April 2, 2004), http://www.afpc.org/asm/asm73.shtml.
26. Traynor and Cobain, "Intelligence Report Claims Nuclear Market Thriving"; Testimony of Andrew Koch.
27. Broad and Sanger, "With Eye on Iran, Rivals Also Want Nuclear Power," *New York Times*, April 15,2007; William Broad and David Sanger, "Restraints Fray and Risks Grow as Nuclear Club Gains Members," *New York Times*, October 15, 2006; and Associated Press, "Saudi Arabia Is Experimenting with Peaceful Nuclear Technology," *International Herald Tribune*, November 15, 2006.
28. Michael Jordan, "UN Nuclear Watchdog Ponders International 'Fuel Bank,'" *Christian Science Monitor*, September 18, 2006; Broad and Sanger, "With Eye on Iran, Rivals Also Want Nuclear Power"; ITAR-TASS, Interfax, "Russia Offers to Host International Nuclear Fuel-Cycle Center," Radio Free Europe, January 25, 2006; "Nuclear Threat Initiative Commits $50 Million to Create IAEA Nuclear Fuel Bank," press release, Nuclear Threat Initiative, September 19, 2006; and James Wensits, "Indiana U.S. Sens. Lugar, Bayh Push for Nuclear Fuel Bank," *South Bend* (Indiana) *Star Tribune*, April 19, 2007.

29. "A Report on the International Control of Atomic Energy," (The Acheson-Lilienthal Report), March 16, 1946, http://honors.umd.edu/HONR269J/archive/AchesonLilienthal.html.
30. "Top Officials Warn Iran over Nukes," MSNBC News Services, March 7, 2006.
31. Seymour Hersh, "The Iran Plans," *New Yorker*, April 17, 2006. In early 2007, the *Times* of London reported that the Israeli air force was practicing for air strikes on Iran's nuclear facilities that could include the use of "low-yield" tactical nuclear weapons. (Uzi Mahnaimi and Sarah Baxter, "Focus: Mission Iran," *Times* [London], January 7, 2007.) Also see "Is Washington Planning a Nuclear Strike?" *Der Spiegel*, December 30, 2005.
32. Ralph Vartaedian, "New Nuke Plans Are Up in the Air," *Los Angeles Times*, January 23, 2007.
33. James Risen, *State of War* (New York: Simon and Schuster, 2006), pp. 194–212. The German magazine *Focus* reported what may have been a US effort to target the enrichment programs of both Iran and Libya. In 2005, the magazine reported that IAEA investigators had discovered German-made vacuum pumps used for uranium-enrichment centrifuges at secret nuclear facilities in Iran and Libya that had originally been ordered by Los Alamos National Laboratories, America's best-known nuclear weapons lab and the birthplace of the atomic bomb. Los Alamos reportedly ordered seven of the pumps from the US subsidiary of the German firm Pfeiffer Vacuum in early 2000. The lab received the pumps in mid-April of that year. In 2003, the IAEA inspectors found one of those pumps at Iran's centrifuge enrichment facility in Natanz. In early 2004, the inspectors found the other six Los Alamos pumps at one of Libya's nuclear sites. *Focus* speculated that "US intelligence may have deliberately released these products . . . onto the nuclear black market for the purpose of infiltrating the Libyan's and Iranians' procurement sources." In support of that idea, the magazine noted that although the standard voltage frequency in the United States is sixty hertz, Los Alamos had ordered the pumps with motors designed to operate at fifty hertz, the same frequency that is used in Iran and Libya. (T. Scheuer and M. Wisniewski, "US Sting Said To Be Likely for German Pumps at WMD Sites in Libya, Iran," *Focus*, November 28, 2005.)
34. Sen. John F. Kennedy, Nixon–Kennedy Debates, "The Presidential Countdown: Mr. Lodge: A Profile," CBS Television, October 10, 1960.
35. A Report to the President by the Committee on Nuclear Proliferation (The Gilpatric Committee), *Foreign Relations of the United States, 1964–1968, Volume XI: Arms Control and Disarmament* (Washington, DC: Department of State), entry 64.
36. Interview with Zbigniew Brzezinski, *Le Nouvel Observateur* (France), January 15–21, 1998, p. 76. This interview was translated from French by author Bill Blum. A slightly different translation of part of the interview appears in Cooley, *UnHoly Wars*, p. 20.
37. Dr. R. Nicholas Palarino and Mr. Jarrel Price, Briefing Memorandum, "Combating Terrorism: Preventing Nuclear Terrorism," for the House Committee on Government Reform, Subcommittee on National Security, Veterans Affairs, and International Relations, September 19, 2002.

Acknowledgments

WE ARE DEEPLY INDEBTED to a great many people who have helped make this book possible. First among them are the funders of the Public Education Center and the National Security News Service who have underwritten our efforts over many years and allowed us to do the work that became the genesis this project. Christian Ettinger, our sponsor at the Educational Foundation of America, and his colleagues at EFA have supported the reporting that made this investigation possible. We also wish to thank Edie Allen of the Colombe Foundation, the Town Creek Foundation, Sandra Silverman of the Sherman Foundation, and Adelaide Gomer of the Park Foundation for supporting our work.

We wish to pay special tribute to the late Sally Lilienthal, founder of the Ploughshares Fund and a driving force behind the movement to prevent the use and spread of chemical, biological and nuclear weapons. Sally, who passed away in 2006, was not only a longtime supporter of our work but also a major inspiration to the authors. Her colleagues Naila Bolus and Paul Carroll share Sally's passion for peace.

We also benefited greatly from the assistance of series of News Service interns and researchers who both supplemented and backstopped our work. Two, in particular, stand out. Scoville Peace Fellow Farah Mahesri put in prodigious hours researching the many individuals and institutions that fill these pages. Her careful attention to detail and great organizational skills made our jobs much simpler. Kimberlyn David provided enormous help in tracking down some of the more elusive characters with whom we've dealt and in drawing together many diverse strands of information.

Several outstanding legal practitioners have also contributed to this effort. Stuart Pierson of Troutman Sanders patiently fielded questions and, as always, provided us with wise and valuable counsel. Attorneys Peter Szyszko, Harris Bor, and Richard Mattick of the London offices of Covington & Burling LLP provided excellent guidance and Mark Stephens, a partner at Finers Stephens Innocent LLP in London, graciously contributed his very helpful insights.

At Steerforth Press, the team headed by Chip Fleischer, including managing editor Kris Sperber, sales director Helga Schmidt, publicity director Christa Demment-González, and Abrah Griggs and Peter Holm of Sterling Hill Productions have been a pleasure to work with. We greatly appreciate it.

There are many other contributors to this book whose names we cannot mention. They know who they are and we know how valuable they have been. They have risked much to help see that this story was told, and told properly. We are grateful to them for their assistance, their courage, and their commitment to seeing that the public is informed about these important matters.

Finally, we would like to offer special thanks to the many journalists and investigators who have worked diligently over the decades to expose the global trade in nuclear weapons technology. It is a scandal of almost unspeakable proportions that despite all that they have made known, the policy makers in Washington, London and other Western capitals continue to place political objectives above the goal of halting nuclear traffic. It is our hope that this book, building on the efforts of those who have come before us, can help bring about a more enlightened and rational approach.

Photo Credits

Page 1, top: Photo of Bhutto at Friday prayers © Associated Press

Page 1, bottom: December 26, 1979, memoranda from Zbigniew Brzezinski to the President, "Reflections on the Soviet Invasion of Afghanistan," published by the Cold War International History Project (CWIHP) in volume one of the Document Reader prepared for the international conference "Toward an International History of the War in Afghanistan, 1979–1989" (April 2002), edited by Christian Ostermann and Mircea Munteanu.

Page 1, bottom right: The photo of Brzezinski and Zia appeared in the *New York Times* on February 3, 1980. Provided by General Research Division, The New York Public Library, Astor Lenox and Tilden Foundations.

Page 2, top: Photograph of Reagan and Zia courtesy Ronald Reagan Library

Page 2, bottom: Photograph of George H.W. Bush and General Zia courtesy George Bush Presidential Library.

Page 3, top: The Weargate and Source Reliance International records are on file at Companies House, London.

Page 3, bottom: These orders were depicted in "How Pakistan Fooled the World and Got the Bomb," *8 Days* (London), June 23, 1979.

Page 4, top: Published in Zahid Malik's *Dr. A.Q. Khan, From Kahuta to Hills of Chagai* (Islamabad: Pakistan Observer, 1998).

Page 4, bottom: The photo of Abdus Salam © Tampa Tribune. The BCCI judgment comes from records of the 11th Judicial Circuit Court of Florida, Miami-Dade County.

Page 5, top: Vaid photo courtesy Mark Obenhaus, printed in the *New York Times* on February 25, 1985. Provided by General Research Division, The New York Public Library, Astor Lenox and Tilden Foundations. The telexes were depicted in the *New York Times* on February 25, 1985.

Page 5, bottom: The cables were obtained by the National Security News Service through a Freedom of Information Act request. Vaid's more recent photo can be found at www.ehealth-services.com/files/Nazir%20Vaid.pdf.

Page 6, top: Richard Barlow photo is reprinted courtesy of Richard Barlow; the Peter Griffin photo appeared in Pakistan's *Weekly Hurmat*, an Urdu-language publication.

Page 6, bottom: Published in Zahid Malik's *Dr. A.Q. Khan, From Kahuta to Hills of Chagai* (Islamabad: Pakistan Observer, 1998).

Page 7, top: This photo was published in Abdul Mabood Siddiqui's *Timbuktu: City in the Middle of Nowhere* (Islamabad: Hurmat Publications, 2000).

Page 7, bottom: These documents were included in the Malaysian police report "Press Release by Inspector General of Police in Relation to Investigation on the Alleged Production of Components for Libya's Uranium Enrichment Programme" (February 20, 2004).

Page 8: The photo of Khan with Aslam Beg was published in Zahid Malik's *Dr. A.Q. Khan, From Kahuta to Hills of Chagai* (Islamabad: Pakistan Observer, 1998); other photos on the page have been published in Khan Research Laboratory promotional materials.

Index